P9-BYC-608

# HANDBOOK
# OF
# ENERGY
# ENGINEERING

**Fourth Edition**

# HANDBOOK
# OF
# ENERGY
# ENGINEERING
## Fourth Edition

**Albert Thumann, P.E., C.E.M.**
**D. Paul Mehta, Ph.D.**

Published by
**THE FAIRMONT PRESS, INC.**
700 Indian Trail
Lilburn, GA 30247

**Library of Congress Cataloging-in-Publication Data**

Thumann, Albert.
  Handbook of energy engineering / Albert Thumann, D. Paul Mehta. --
4th ed.
  p. cm.
  Includes bibliographical references and index.
  ISBN 0-88173-260-5
  1. Power resources--Handbooks, manuals, etc. 2. Power (Mechanics)--
Handbooks, manuals, etc. I. Mehta, D. Paul, 1940-  . II. Title.
TJ163.2.T49      1997      621.042--dc21      96-48535
                                              CIP

*Handbook of Energy Engineering, Fourth Edition.*
©1997 by The Fairmont Press, Inc. All rights reserved. No part of
this publication may be reproduced or transmitted in any form or by
any means, electronic or mechanical, including photocopy, recording,
or any information storage and retrieval system, without permission in
writing from the publisher.

Published by The Fairmont Press, Inc.
700 Indian Trail
Lilburn, GA 30247

Printed in the United States of America

10  9  8  7  6  5  4  3  2


ISBN 0-88173-260-5 FP

ISBN 0-13-858622-5 PH

While every effort is made to provide dependable information, the publisher, authors, and
editors cannot be held responsible for any errors or omissions.

Distributed by Prentice Hall PTR
Prentice-Hall, Inc.
A Simon & Schuster Company
Upper Saddle River, NJ 07458

Prentice-Hall International (UK) Limited, London
Prentice-Hall of Australia Pty. Limited, Sydney
Prentice-Hall Canada Inc., Toronto
Prentice-Hall Hispanoamericana, S.A., Mexico
Prentice-Hall of India Private Limited, New Delhi
Prentice-Hall of Japan, Inc., Tokyo
Simon & Schuster Asia Pte. Ltd., Singapore
Editora Prentice-Hall do Brasil, Ltda., Rio de Janeiro

# Contents

# 1
# Fundamentals of Energy Engineering

## INTRODUCTION

Energy engineering is a profession which applies scientific knowledge for the improvement of the overall use of energy. It combines the skills of engineering with the knowledge of energy problems. The energy engineer must be able to identify problems in the use of energy, creatively design solutions, and implement the process.

In order to develop economic and socially acceptable ways to use available resources for the benefit of mankind, the energy engineer searches for a better way to combine a broad base of knowledge with experience.

Energy engineering requires a "system approach" and is multi-disciplinary in nature. Thus an energy engineer must have both engineering skills in, for example, electrical, mechanical and process engineering, as well as good management knowledge.

Problems such as the high cost of energy, the depletion of resources, and the degradation of the environment are not transitory but will plague mankind for decades to come. The energy engineering profession is addressing these and other problems in a systematic and cost-effective manner.

## ENERGY CONVERSION FACTORS

In order to communicate energy engineering goals and to analyze the literature in the field, it is important to understand the conversion factors used in energy engineering and how they are applied.

Each fuel has a heating value, expressed in terms of the British thermal unit, Btu. The Btu is the heat required to raise the temperature of one pound of water 1°F. Table 1-1 illustrates the heating values of various fuels. To compare efficiencies of various fuels, it is best to convert fuel usage in terms of Btu's. Table 1-2 illustrates conversions used in energy engineering calculations.

When comparing costs of fuels, the term "cents or dollars per therm" (100,000 Btu) is commonly used.

The chemical energy stored in fuel is sometimes expressed as Higher Heating Value (HHV) or Lower Heating Value (LHV).

### Table 1-1. Heating Values for Various Fuels

| Fuel | Average Heating Value |
|------|----------------------|
| **Fuel Oil** | |
| Kerosene | 134,000 Btu/gal. |
| No. 2 Burner Fuel Oil | 140,000 Btu/gal. |
| No. 4 Heavy Fuel Oil | 144,000 Btu/gal. |
| No. 5 Heavy Fuel Oil | 150,000 Btu/gal. |
| No. 6 Heavy Fuel Oil 2.7% sulfur | 152,000 Btu/gal. |
| No. 6 Heavy Fuel Oil 0.3% sulfur | 143,800 Btu/gal. |
| **Coal** | |
| Anthracite | 13,900 Btu/lb |
| Bituminous | 14,000 Btu/lb |
| Sub-bituminous | 12,600 Btu/lb |
| Lignite | 11,000 Btu/lb |
| **Gas** | |
| Natural | 1,000 Btu/cu ft |
| Liquefied butane | 103,300 Btu/gal. |
| Liquefied propane | 91,600 Btu/gal. |

*Source: Brick & Clay Record, October 1972.*

## Table 1-2. List of Conversion Factors

| | | |
|---|---|---|
| 1 U.S. barrel | = | 42 U.S. gallons |
| 1 atmosphere | = | 14.7 pounds per square inch absolute (psia) |
| 1 atmosphere | = | 760 mm (29.92 in.) mercury with density of 13.6 grams per cubic centimeter |
| 1 pound per square inch | = | 2.04 inches head of mercury |
| | = | 2.31 feet head of water |
| 1 inch head of water | = | 5.20 pounds per square foot |
| 1 foot head of water | = | 0.433 pound per square inch |
| 1 British thermal unit (Btu) | = | heat required to raise the temperature of 1 pound of water by 1°F |
| 1 therm | = | 100,000 Btu |
| 1 kilowatt (kW) | = | 1.341 horsepower (hp) |
| 1 kilowatt-hour (kWh) | = | 1.34 horsepower-hour |
| 1 horsepower (hp) | = | 0.746 kilowatt (kW) |
| 1 horsepower-hour | = | 0.746 kilowatt hour (kWh) |
| 1 horsepower-hour | = | 2545 Btu |
| 1 kilowatt-hour (kWh) | = | 3412 Btu |

To generate 1 kilowatt-hour (kWh) requires 10,000 Btu of fuel burned by average utility

| | | |
|---|---|---|
| 1 ton of refrigeration | = | 12,000 Btu per hr |

1 ton of refrigeration requires about 1 kW (or 1.341 hp) in commercial air conditioning

1 standard cubic foot is at standard conditions of 60°F and 14.7 psia

| | | |
|---|---|---|
| 1 degree day | = | 65°F minus mean temperature of the day, °F |
| 1 year | = | 8760 hours |
| 1 year | = | 365 days |
| 1 MMBtu | = | 1 million Btu |
| 1 kW | = | 1000 watts |
| 1 trillion barrels | = | 1 X $10^{12}$ barrels |
| 1 KSCF | = | 1000 standard cubic feet |
| 1 Quad | = | one quadrillion ($10^{15}$) Btu's |

*Note:* In these conversions, inches and feet of water are measured at 62°F (16.7°C), and inches and millimeters of mercury at 32°F (0°C).

The higher heating value is obtained by burning a small sample of fuel in an oxygen environment and recording the heat transferred to the water sample surrounding it. This test includes the latent heat of vaporization of the condensed vapor.

The lower heating value subtracts the latent heat of vaporization since this energy is usually unavailable in practice.

European countries usually use lower heating values for fuels while in the United States higher heating values are used. Thus a heating device tested in Europe will be approximately 10% more efficient than if tested in the United States due to the standard in heating values assumed.

### Example Problem 1-1

The supply of fuel oil is projected at 11.5 million barrels per day. What is the supply in Btu per year, assuming an average fuel oil value of 140,000 Btu/gallon?

### Answer

Fuel supply = 140,000 Btu/gallon X 11.5 X $10^6$ barrels/day
X 42 gallons/barrel X 365 days/year = 24.7 X
$10^{15}$ Btu/year.

### Example Problem 1-2

The warehouse of a plant is required to be minimally heated at night. Two methods of heating are being considered. One method is electric heaters. The second is to operate the oil-fired boiler (using No. 2 fuel oil) which keeps the radiators warm all night. Electricity costs, excluding demand charges, is 10 cents per kilowatt hour; No. 2 fuel oil costs $1.00 per gallon. Assume the boiler is 70% efficient and exclude the costs of running pumps and fans to distribute heat. What is the relative cost to heat the building using electricity or the oil-fired boiler?

### Answer

To compare fuel costs, a common base of $ per therm will be used.
*Electricity Cost*
$$1 \text{ kWh} = 3412 \text{ Btu}$$

$$\text{Cost} = \frac{100,000}{3412} \text{ X } .10 = \$2.93/\text{therm}$$

*No. 2 Fuel Oil Cost*
$$\text{No. 2 fuel oil} = 140,000 \text{ Btu/gallon}$$

$$\text{Cost} = \frac{100,000}{140,000} \text{ X } \$1.00 = 71 \text{ cents/therm}$$

Taking into account an efficiency of combustion of 0.7, the relative cost becomes

$$\text{Cost} = \frac{71}{0.7} = \$1.02/\text{therm}$$

## CODES, STANDARDS & LEGISLATION

This section presents a historical perspective on key codes, standards and regulations which have impacted energy policy and are still playing a major role in shaping energy usage. The Energy Policy Act of 1992 is far-reaching and its implementation is impacting electric power deregulation, building codes and new energy-efficient products. Sometimes policy makers do not see the far-reaching impact of their legislation. The Energy Policy Act, for example, has created an environment for retail competition. Electric utilities will drastically change the way they operate in order to provide power and lowest cost. This in turn will drastically reduce utility-sponsored incentive and rebate programs which have influenced energy conservation adoption.

## THE ENERGY POLICY ACT OF 1992

This comprehensive legislation is far-reaching and impacts energy conservation, power generation and alternative-fuel vehicles as well as energy production. The federal as well as private sectors are impacted by this comprehensive energy act. Highlights are described below:

### Energy Efficiency Provisions
Buildings

• Requires states to establish minimum commercial building energy codes and to consider minimum residential codes based on current voluntary codes.

Utilities

• Requires states to consider new regulatory standards that would require utilities to undertake integrated resource planning, allow efficiency programs to be at least as profitable as new supply options and encourage improvements in supply system efficiency.

## Equipment Standards
- Establishes efficiency standards for commercial heating and air-conditioning equipment, electric motors, and lamps.

- Gives private sector an opportunity to establish voluntary efficiency information/labeling programs for windows, office equipment and luminaires, or the Department of Energy will establish such programs.

## Renewable Energy
- Establishes a program for providing federal support on a competitive basis for renewable energy technologies. Expands program to promote export of these renewable energy technologies to emerging markets in developing countries.

## Alternative Fuels
- Gives Department of Energy authority to require a private and municipal alternative fuel fleet program starting in 1998. Provides a federal alternative fuel fleet program with phased-in acquisition schedule; also provides state fleet program for large fleets in large cities.

## Electric Vehicles
- Establishes comprehensive program for the research and development, infrastructure promotion and vehicle demonstration for electric motor vehicles.

## Electricity
- Removes obstacles to wholesale power competition in the Public Utilities Holding Company Act by allowing both utilities and non-utilities to form exempt wholesale generators without triggering the PUHCA restrictions.

## Global Climate Change
- Directs the Energy Information Administration to establish a baseline inventory of greenhouse gas emissions and establishes a program for the voluntary reporting of those emissions. Directs the Department of Energy to prepare a report analyzing the strategies for mitigating global climate change and to develop a least-cost energy strategy for reducing the generation of greenhouse gases.

## Research and Development
- Directs the Department of Energy to undertake research and development on a wide range of energy technologies, including energy

efficiency technologies, natural gas end-use products, renewable energy resources, heating and cooling products, and electric vehicles.

## STATE CODES

More than three quarters of the states have adopted ASHRAE Standard 90-80 as a basis for their energy efficiency standard for new building design. The ASHRAE Standard 90-80 is essentially "prescriptive" in nature. For example, the energy engineer using this standard would compute the average conductive value for the building walls and compare it against the value in the standard. If the computed value is above the recommendation, the amount of glass or building construction materials would need to be changed to meet the standard.

Most states have initiated "Model Energy Codes" for efficiency standards in lighting and HVAC. Probably one of the most comprehensive building efficiency standards is California Title 24. Title 24 established lighting and HVAC efficiency standards for new construction, alterations and additions of commercial and noncommercial buildings.

ASHRAE Standard 90-80 has been updated into two new standards:

ASHRAE 90.1-1989 Energy-Efficient Design of New Buildings Except New, Low-Rise Residential Buildings

ASHRAE 90.2 Energy-Efficient Design of New, Low-Rise Residential Buildings

The purposes of ASHRAE Standard 90.1-1989 are:

(a)    set minimum requirements for the energy-efficient design of new buildings so that they may be constructed, operated and maintained in a manner that minimizes the use of energy without constraining the building function or the comfort or productivity of the occupants;

(b)    provide criteria for energy-efficient design and methods for determining compliance with these criteria;

(c)    provide sound guidance for energy-efficient design.

In addition to recognizing advances in the performance of various components and equipment, the Standard encourages innovative energy-conserving designs. This has been accomplished by allowing the building designer to take into consideration the dynamics that exist between the many components of a building through use of the System Performance Method or the Building Energy Cost Budget Method compliance paths. The standard, which is cosponsored by the Illuminating Engineering Society of North America, includes an extensive section on lighting efficiency, utilizing the Unit Power Allowance Method.

## MODEL ENERGY CODE

The American Building Officials (ABO) have upgraded their Model Energy Code (MEC). The MEC is a widely accepted model energy standard that is developed in an open process involving code officials, builders, manufacturers, government agencies and researchers. In a July 1991 report by the Alliance to Save Energy it found 34 states currently using less-stringent codes and standards. The report recommended:

- States should review their existing building code standards and use the MEC as a minimum baseline for upgrading their energy standards for new housing.

- The federal government should help states upgrade their energy standards. The Department of Energy should increase its efforts to provide technical assistance to states, and it should do further studies of the benefit of various model-building energy standards and codes.

- Model Energy Code standards should be strengthened in the southern region. The studies found that in many southern states the MEC's thermal requirements were too lax, especially considering the heavy use of electricity for air conditioning in new homes.

- The MEC's standards for multifamily homes should be reassessed. The study found that MEC standards for multifamily housing, while beneficial overall, showed less cost-effectiveness and affordability benefits than single-family housing. This suggests that the MEC's multifamily standards should be refined and strengthened where necessary.

## NATIONAL ENERGY CONSERVATION
## POLICY ACT (NECPA)—1978.

This act is probably the most important legislation passed. Programs initiated by NECPA are still in progress today. One of the first successful aspects of NECPA is the: "Matching Grants Program for Schools & Hospitals," is a voluntary program to assist these nonprofit institutions in saving energy.

Authorized by Part 1, Title III of the NECPA and administered by the Department of Energy's Institutional Conservation Programs (ICP) through state energy offices, the ICP provides grants for detailed energy analyses and for installation of energy-saving capital improvements to eligible institutions. Participation in the grant program requires a 50% match of funds from recipient institutions, except in hardship cases.

### Program Activities

NECPA divided the program into two main phases: Phase I, Preliminary Energy Audits and Energy Audits, and Phase II, Technical Assistance and Energy Conservation Measures. Funding for Phase I activities has expired; however, the program still administers the Technical Assistance and Energy Conservation Measures phase.

The Preliminary Energy Audit (PEA) was a data-gathering activity to determine basic information about the number, size, type and rate of energy use in eligible buildings and the actions taken to conserve energy. The Energy Audit (EA) consisted of an on-site visit which confirmed and expanded upon information gathered in the PEA and provided further information to establish a "building profile."

The Technical Assistance (TA) Audit consists of detailed engineering analyses and reports specific costs, payback periods and projected energy savings possible from the purchase and installation of devices or systems, or from modification to building operations and the building shell. Examples of this type of activity are the installation of storm windows, insulation, solar hot water or space heating systems or automatic setback devices.

Energy Conservation Measure (ECM) grants provide for the actual purchase and installation of energy measures recommended as a result of the TA Audit.

## INDOOR AIR QUALITY (IAQ) STANDARDS[1]

Indoor Air Quality (IAQ) is an emerging issue of concern to building managers, operators and designers. Recent research has shown that indoor air is often less clean than outdoor air and federal legislation has been proposed to establish programs to deal with this issue on a national level. This, like the asbestos issue, will have an impact on building design and operations. Americans today spend long hours inside buildings, and building operators, managers and designers must be aware of potential IAQ problems and how they can be avoided.

IAQ problems, sometimes termed "Sick Building Syndrome," have become an acknowledged health and comfort problem. Buildings are characterized as sick when occupants complain of acute symptoms such as headache, eye, nose and throat irritation, dizziness, nausea, sensitivity to odors and difficulty in concentrating. The complaints may become more clinically defined so that an occupant may develop an actual building-related illness that is believed to be related to IAQ problems.

The most effective means to deal with an IAQ problem is to remove or minimize the pollutant source, when feasible. If not, dilution and filtration may be effective.

Dilution (increased ventilation) is to admit more outside air to the building. ASHRAE's 1981 standard recommended 5 CFM/person outside air in an office environment. The new ASHRAE ventilation standard, 62-1989, now requires 20 CFM/person for offices if the prescriptive approach is used.

Increased ventilation will have an impact on building energy consumption. However, this cost need not be severe. If an airside economizer cycle is employed and the HVAC system is controlled to respond to IAQ loads as well as thermal loads, 20 CFM/person need not be adhered to and the economizer hours will help attain air-quality goals with energy savings at the same time. Incidentally, it was the energy cost of treating outside air that led to the 1981 standard. The superseded 1973 standard recommended 15-25 CFM/person.

Energy savings can be realized by the use of improved filtration in lieu of the prescriptive 20 CFM/person approach. Improved filtration can occur at the air handler, in the supply and return ductwork or in the spaces via self-contained units. Improved filtration can include enhance-

---

[1]Source: Indoor Air Quality: Problems & Cures, M. Black & W. Robertson, Presented at 13th World Energy Engineering Congress.

ments such as ionization devices to neutralize airborne biological matter and to electrically charge fine particles, causing them to agglomerate and be more easily filtered.

The Occupational Safety and Health Administration (OSHA) announced a proposed rule on March 25, 1994 that would regulate indoor air quality (IAQ) in workplaces across the nation. The proposed rule addresses all indoor contaminants but a significant step would ban all smoking in the workplace or restrict it to specially designed lounges exhausted directly to the outside. The smoking rule would apply to all workplaces while the IAQ provisions would impact "non-industrial" indoor facilities.

There is growing consensus that the most promising way to achieve good indoor air quality is through contaminant source control. Source control is more cost-effective than trying to remove a contaminant once it has disseminated into the environment. Source control options include chemical substitution or product reformulating, product substitution, product encapsulation, banning some substances or implementing material emission standards. Source control methods except emission standards are incorporated in the proposed rule.

## REGULATIONS & STANDARDS IMPACTING CFCs

For years, chlorofluorocarbons (CFCs) have been used in air-conditioning and refrigeration systems designed for long-term use. However, because CFCs are implicated in the depletion of the earth's ozone layer, regulations will require the complete phaseout of the production of new CFCs by the turn of the century. Many companies, like DuPont, are developing alternative refrigerants to replace CFCs. The need for alternatives will become even greater as regulatory cutbacks cause continuing CFC shortages.

Air-conditioning and refrigeration systems designed to operate with CFCs will need to be retrofitted (where possible) to operate with alternative refrigerants so that these systems can remain in use for their intended service life.

DuPont and other companies are commercializing their series of alternatives—hydrochlorofluorocarbon (HCFC) and hydrofluorocarbon (HFC) compounds. See Table 1-3.

The Montreal Protocol which is being implemented by the United Nations Environment Program (UNEP) is a worldwide approach to the

**Table 1-3.**
**Candidate Alternatives for CFCs in Existing Cooling Systems**

| CFC | Alternative | Potential Retrofit Applications |
|-----|-------------|--------------------------------|
| CFC-11 | HCFC-123 | Water and brine chillers; process cooling |
| CFC-12 | HFC-134a or Ternary Blends | Auto air conditioning; medium temperature commercial food display and transportation equipment; refrigerators/freezers; dehumidifiers; ice makers; water fountains |
| CFC-114 | HCFC-124 | Water and brine chillers |
| R-502 | HFC-125 | Low-temperature commercial food equipment |

phaseout of CFCs. A major revision to the Montreal Protocol was implemented at the 1992 meeting in Copenhagen which accelerated the phaseout schedule.

The reader is advised to carefully consider both the "alternate" refrigerants entering the marketplace *and* the alternate technologies available. Alternate refrigerants come in the form of HCFCs and HFCs. HFCs have the attractive attribute of having no impact on the ozone layer (and accordingly are not named in the Clean Air Act). Alternative technologies include absorption and ammonia refrigeration (established technologies since the early 1900s), as well as desiccant cooling.

Taxes on CFCs originally took effect January 1, 1990. The Energy Policy Act of 1992 revised and further increased the excise tax effective January 1, 1993.

Another factor to consider in ASHRAE Guidelines 3-1990—Reducing Emission of Fully Halogenated Chlorofluorocarbon (CFC) Refrigerants in Refrigeration and Air-Conditioning Equipment and Applications:

> The purpose of this guideline is to recommend practices and procedures that will reduce inadvertent release of fully halogenated chlorofluorocarbon (CFC) refrigerants during manufacture, installation, testing, operation, maintenance and disposal of refrigeration and air-conditioning equipment and systems.

> The guideline is divided into 13 sections. Highlights are as follows: The Design Section deals with air-conditioning and refrigeration

systems and components and identifies possible sources of loss of refrigerants to the atmosphere. Another section outlines refrigerant recovery reuse and disposal. The Alternative Refrigerant section discusses replacing R11, R12, R113, R114, R115 and azeotropic mixtures R500 and R502 with HCFCs such as R22.

## CLEAN AIR ACT AMENDMENT

On November 15, 1990, the new Clean Air Act (CAA) was signed by President Bush. The legislation includes a section entitled Stratospheric Ozone Protection (Title VI). This section contains extraordinarily comprehensive regulations for the production and use of CFCs, halons, carbon tetrachloride, methyl chloroform, and HCFC and HFC substitutes. These regulations will be phased in over the next 40 years, and they will impact every industry that currently uses CFCs.

The seriousness of the ozone depletion is such that as new findings are obtained, there is tremendous political and scientific pressure placed on CFC end-users to phase out use of CFCs. This has resulted in the U.S., under the signature of President Bush in February 1992, to have accelerated the phaseout of CFCs.

## REGULATORY AND LEGISLATIVE ISSUES IMPACTING COGENERATION & INDEPENDENT POWER PRODUCTION[2]

Federal, state and local regulations must be addressed when considering any cogeneration project. This section will provide an overview of the federal regulations that have the most significant impact on cogeneration facilities.

### Federal Power Act
The Federal Power Act asserts the federal government's policy toward competition and anti-competitive activities in the electric power industry. It identifies the Federal Energy Regulatory Commission (FERC) as the agency with primary jurisdiction to prevent undesirable anti-competitive behavior with respect to electric power generation. Also, it provides cogenerators and small power producers with a judicial means to overcome obstacles put in place by electric utilities.

---

[2]Source: *Georgia Cogeneration Handbook*, published by the Governor's Office of Energy Resources.

**Public Utility Regulatory Policies Act (PURPA)**

This legislation was part of the 1978 National Energy Act and has had perhaps the most significant effect on the development of cogeneration and other forms of alternative energy production in the past decade. Certain provisions of PURPA also apply to the exchange of electric power between utilities and cogenerators.

PURPA provides a number of benefits to those cogenerators who can become Qualifying Facilities (QFs) under the act. Specifically, PURPA:

• Requires utilities to purchase the power made available by cogenerators at reasonable buy-back rates. These rates are typically based on the utilities' cost.

• Guarantees the cogenerator or small power producer interconnection with the electric grid and the availability of backup service from the utility.

• Dictates that supplemental power requirements of the cogenerator must be provided at a reasonable cost.

• Exempts cogenerators and small power producers from federal and state utility regulations and associated reporting requirements of these bodies.

In order to assure a facility the benefits of PURPA, a cogenerator must become a Qualifying Facility. To achieve Qualifying Status, a cogenerator must generate electricity and useful thermal energy from a single fuel source. In addition, a cogeneration facility must be less than 50% owned by an electric utility or an electric utility holding company. Finally, the plant must meet the minimum annual operating efficiency standard established by FERC when using oil or natural gas as the principal fuel source. The standard is that the useful electric power output plus one half of the useful thermal output of the facility must be no less than 42.5% of the total oil or natural gas energy input. The minimum efficiency standard increases to 45% if the useful thermal energy is less than 15% of the total energy output of the plant.

**Natural Gas Policy Act (NGPA)**

The major objective of this legislation was to create a deregulated national market for natural gas. It provides for incremental pricing of higher-cost natural gas supplies to industrial customers who use gas, and it allows the cost of natural gas to fluctuate with the cost of fuel oil. Cogenerators classified as Qualifying Facilities under PURPA are exempt from the incremental pricing schedule established for industrial customers.

### Resource Conservation and Recovery Act of 1976 (RCRA)

This act requires that disposal of non-hazardous solid waste be handled in a sanitary landfill instead of an open dump. It affects only cogenerators with biomass and coal-fired plants. This legislation has had little, if any, impact on oil and natural gas cogeneration projects.

### Public Utility Holding Company Act of 1935

The Public Utility Holding Company Act of 1935 (the 35 Act) authorizes the Securities and Exchange Commission (SEC) to regulate certain utility "holding companies" and their subsidiaries in a wide range of corporate transactions.

The Energy Policy Act of 1992 creates a new class of wholesale-only electric generators—"exempt wholesale generators" (EWGs)—which are exempt from the Public Utility Holding Company Act (PUHCA). The Act dramatically enhances competition in U.S. wholesale electric generation markets, including broader participation by subsidiaries of electric utilities and holding companies. It also opens up foreign markets by exempting companies from PUHCA with respect to retail sales as well as wholesale sales.

### Moving towards a deregulated electric power marketplace

The Energy Policy Act set into motion a widespread movement for utilities to become more competitive. Retail wheeling proposals were set into motion in states such as California, Wisconsin, Michigan, New Mexico, Illinois and New Jersey. There are many issues involved in a deregulated power marketplace and public service commission rulings and litigation will certainly play a major role in the power marketplace of the future. Deregulation has already brought about several important developments:

- Utilities will need to become more competitive. Downsizing and minimization of costs including elimination of rebates are the current trend. This translates into lower costs for consumers. For example, Southern California Edison announced that the system average price will be reduced from 10.7 cents/kWh to lower than 10 cents by the year 2000. This would be a 25% reduction after adjusting for inflation.

- Utilities will merge to gain a bigger market share. Wisconsin Electric Power Company recently announced a merger with Northern States Power; this is the largest merger of two utilities of its kind in the nation, resulting in a savings of $2 billion over 10 years.

- Utilities are forming new companies to broaden their services. Energy service companies, financial loan programs and purchasing of related companies are all part of the new utility strategy.

- In 1995 one hundred power marketing companies submitted applications to FERC. Power marketing companies will play a key role in brokering power between end users and utilities in different states and in purchasing of new power generation facilities.

- Utilities will need to restructure to take advantage of deregulation. Generation companies may be split away from other operating divisions such as transmission and distribution. Vertical disintegration will be part of the new utility structure.

- Utilities will weigh the cost of repowering and upgrading existing plants against purchasing power from a third party.

## OPPORTUNITIES IN THE SPOT MARKET[3]

### Basics of the Spot Market

A whole new method of contracting has emerged in the natural gas industry through the spot market. The market has developed because the Natural Gas Policy Act of 1978 (NGPA) guaranteed some rights for end users and marketers in the purchasing and transporting of natural gas. It also put natural gas supplies into a more competitive position with deregulation of several categories.

The Federal Energy Regulatory Commission (FERC) provided additional rulings that facilitated the growth of the spot market. These rulings included provisions for the Special Marketing Programs in 1983

---

[3]Source: *New Opportunities for Purchasing Natural Gas*, Fairmont Press, Lilburn, GA

(Order 2346) and Order No. 436 in 1985, which encouraged the natural gas pipelines to transport gas for end-users through blanket certificates.

The change in the structure of markets in the natural gas industry has been immense in terms of both volumes and the participants in the market. By year-end 1986, almost 40% of the interstate gas supply was being transported on a carriage basis. Not only were endusers participating in contract carriage, but local distribution companies (LDCs) were accounting for about one half of the spot volumes on interstate pipelines.

The "spot market" or "direct purchase" market refers to the purchase of gas supplies directly from the producer by a marketer, enduser or LDC. (The term "spot gas" is often used synonymously with "best efforts gas," "interruptible gas," "direct purchase gas" and "self-help gas.") This type of arrangement cannot be called new because the pipelines have always sold some supplies directly to endusers.

The new market differs from the past arrangements in terms of the frequency in contracting and the volumes involved in such contracts. Another characteristic of the spot market is that contracts are short-term, usually only 30 days, and on an interruptible basis. The interruptible nature of spot market supplies is an important key to understanding the operation of the spot market and the costs of dealing in it. On both the production and transportation sides, all activities in transportation or purchasing supplies are on a "best efforts" basis. This means that when a cold snap comes the direct purchaser may not get delivery on his contracts because of producer shutdowns, pipeline capacity and operational problems or a combination of these problems. The "best efforts" approach to dealing can also lead to problems in transporting supplies when demand is high and capacity limited.

### FERC's Order No. 436

The impetus for interstate pipeline carriage came with FERC's Order No. 436, which provided more flexibility in pricing and transporting natural gas. In passing the 1986 ruling, FERC was attempting to get out of the day-to-day operations of the market and into more generic rule making. More significantly, FERC was trying to get interstate pipelines out of the merchant business and into the transportation business—a step requiring a major restructuring of contracting in the gas industry.

FERC has expressed an intent to create a more competitive market so that prices would signal adjustments in the markets. The belief is that direct sales ties between producers and end users will facilitate market adjustments without regulatory requirements clouding the market. As more gas is deregulated, FERC reasoned that natural gas prices will respond to the demand: Lower prices would assist in clearing excess supplies; then as markets tightened, prices would rise, drawing further investment into supply development.

## FERC Order No. 636

Order 636 required significant "Restructuring" in interstate pipeline services, starting in the fall of 1993. The original Order 636:
- Separates (unbundles) pipeline gas sales from transportation
- Provides open access to pipeline storage
- Allows for "no notice" transportation service
- Requires access to upstream pipeline capacity
- Uses bulletin boards to disseminate information
- Provides for a "capacity release" program to temporarily sell firm transportation capacity
- Pregrants a pipeline the right to abandon gas sales
- Bases rates on straight fixed variable (SFV) design
- Passes through 100% of transition costs in fixed, monthly charges to firm transport customers

## FERC Order No. 636A

Order 636A makes several relatively minor changes in the original order and provides a great deal of written defense of the original order's terms. The key changes are:

- Concessions on transport and sales rates for a pipeline's traditional "small sales" customers (like municipalities).
- The option to "release" (sell) firm capacity for less than one month—*without posting it on a bulletin board system or bidding.*
- Greater flexibility in designing special transportation rates (i.e., off-peak service) while still requiring overall adherence to the straight fixed variable rate design.
- Recovery of 10% of the transition costs from the interruptible transportation customers (Part 284).

Court action is still likely on the Order. Further, each pipeline will submit its own unique tariff to comply with the Order. As a result, additional changes and variations are likely to occur.

## THE CLIMATIC CHANGE ACTION PLAN

The Climatic Change Action Plan was established April 21, 1993 and includes the following:

- Returns U.S. greenhouse gas emissions to 1990 levels by the year 2000 with cost-effective domestic actions.

- Includes measures to reduce all significant greenhouse gases, carbon dioxide, methane, nitrous oxide, hydrofluorocarbons and other gases.

## NORTH AMERICAN ENERGY MEASUREMENT AND VERIFICATION PROTOCOL

### Purpose

The long-term success of energy management projects has been hampered by the ability of project partners to agree on an accurate, successful Measurement & Verification (M&V) plan. This M&V protocol discusses procedures that, when implemented, allow buyers, sellers and financiers of energy projects to quantify energy conservation measure (ECM) performance and energy savings. By using one of the different M&V options discussed in this document, readers can allocate various risks associated with achieving energy cost savings to either the buyer or seller of the project. Additionally, this protocol:

- is designed to be consistent with EPA Acid Rain Program verification protocols for conservation and renewable energy reserve allowances,

- gives buyers, sellers and financiers a basis to discuss key M&V project-related issues,

- helps ensure the accurate verification of projects, with respect to anticipated versus achieved savings, and

- uses procedures which i) are consistently applicable to similar projects throughout all geographic regions, and ii) are nationally accepted, impartial and reliable.

Two basic aspects of ECM performance verification are addressed in this document:

1.    Verification of: i) the accuracy of baseline conditions as specified in the contract between buyer and seller, and ii) the complete installation and proper operation of new equipment/systems specified in the contract.
2.    Verification of the quantity of energy savings and/or energy cost savings that occur during the term of the contract.

This M&V protocol is not intended to prescribe contractual terms between buyers and sellers. Once other contractual issues are decided, this document may be used to select the verification plan that best matches: i) project costs and savings magnitude, ii) technology-specific requirements and iii) risk allocation between buyer and seller, i.e., which party is responsible for installed equipment performance and which party is responsible for achieving long-term energy savings.

The protocol provides an overview of current techniques available for verifying aspects of third-party, financed energy projects. It may also be used by building operators to assess and improve facility performance.

Coordinated with the efforts of the ASHRAE GPC 14P Committee, this document will be maintained and revised under the sponsorship of the Department of Energy on a biannual basis, based on recommendations developed by a consensus group of facility owners/operators, financiers, contractors or energy service companies (ESCOs) and other stakeholders. As the document is updated, it will incorporate new measurement techniques as they become available. Changes, modifications and additions will be reviewed on a regular basis and, if approved, will be incorporated directly into this document, so that buyers, sellers and financiers of energy projects are kept abreast of the latest options.

## SUMMARY

The dynamic process of revisions to existing codes plus the introduction of new legislation will impact the energy industry and bring a dramatic change. Energy conservation and creating new power generation supply options will be required to meet the energy demands of the 21st century.

# 2

# Energy Economic Analysis

To justify the energy investment cost, a knowledge of life-cycle costing is required.

The life-cycle cost analysis evaluates the total owning and operating cost. It takes into account the "time value" of money and can incorporate fuel cost escalation into the economic model. This approach is also used to evaluate competitive projects. In other words, the life-cycle cost analysis considers the cost over the life of the system rather than just the first cost.

## THE TIME VALUE OF MONEY CONCEPT

To compare energy utilization alternatives, it is necessary to convert all cash flow for each measure to an equivalent base. The life-cycle cost analysis takes into account the "time value" of money; thus a dollar in hand today is more valuable than one received at some time in the future. This is why a time value must be placed on all cash flows into and out of the company.

## DEVELOPING CASH FLOW MODELS

The cash flow model assumes that cash flows occur at discrete points in time as lump sums and that interest is computed and payable at discrete points in time.

To develop a cash flow model which illustrates the effect of "compounding" of interest payments, the cash flow model is developed as follows:

End of  Year 1:  $P + i(P) = (1 + i)P$
        Year 2:  $(1 + i) P + (1 + i) Pi = (1 + i)P [(1 + i)]$
                $= (1 + i)^2 P$
        Year 3:  $(1 + i)^3 P$
        Year $n$   $(1 + i)^n P$ or $F = (1 + i)^n P$
        Where  $P$ = present sum
                $i$ = interest rate earned at the end of each interest period
                $n$ = number of interest periods
                $F$ = future value

$(1 + i)^n$ is referred to as the "Single Payment Compound Amount" factor $(F/P)$ and is tabulated for various values of $i$ and $n$ in Tables 15-1 through 15-8.

The cash flow model can also be used to find the present value of a future sum $F$.

$$P = \left(\frac{1}{(1 + i)^n}\right) \times F$$

Cash flow models can be developed for a variety of other types of cash now as illustrated in Figure 2-1.

To develop the Cash Flow Model for the "Uniform Series Compound Amount" factor, the following cash flow diagram is drawn.

Where $A$ is a Uniform Series of year-end payments and $F$ is the future sum of $A$ payments for $n$ interest periods.

The $A$ dollars deposited at the end of the $n$th period earn no interest and, therefore, contribute $A$ dollars to the fund. The $A$ dollars deposited at the end of the $(n - 1)$ period earn interest for 1 year and will, therefore, contribute $A (1 + i)$ dollars to the fund. The $A$ dollars deposited at the end of the $(n - 2)$ period earn interest for 2 years and will, therefore, contribute

$A(1 + i)^2$. These years of earned interest in the contributions will continue to increase in this manner, and the $A$ deposited at the end of the first period will have earned interest for $(n-1)$ periods. The total in the fund $F$ is, thus, equal to $A + A(1+i)+A(1+i)^2 + A(1+i)3 + A(1+i)4 +...+ A(1+i)^{n-2} + A(1+i)^{n-1}$. Factoring out $A$,

(1)   $F = A [1+ (1 + i) + (1 + i)2+ ... +(1 + i)n.^{-2} + (1 + i)n^{-1}]$
      Multiplying both sides of this equation by $(1 = i)$;

(2)   $(1 + i)F = A [(1 + i)+(1+i)2 + (1+i)3 + ... +( 1 + i)n^{-1} + (1 + i)n]$
      Subtracting equation (1) from (2):

$$(1 + i)F - F = A [(1 + i) + (1 + i)2 + (1 + i)3$$
$$+(1 + i)n^{-}1+(1 + i)n]-A [1 + (1 + i)$$
$$+ (1 + i)2 +,, . + (1 + i)n^{-}2 + (1 + i)n^{-1}]$$
$$iF = A [(1 + i)n^{-1}]$$

$$F = A \left| \frac{(1 + i)^n - 1}{i} \right|$$

Interest factors are seldom calculated. They can be determined from computer programs, and interest tables included in Chapter 15, Appendix. Each factor is defined when the number of periods $(n)$ and interest rate $(i)$ are specified. In the case of the Gradient Present Worth factor the escalation rate must also be stated.

The three most commonly used methods in life-cycle costing are the annual cost, present worth and rate-of-return analysis.

In the present worth method a minimum rate of return $(i)$ is stipulated. All future expenditures are converted to present values using the interest factors. The alternative with lowest effective first cost is the most desirable.

A similar procedure is implemented in the annual cost method. The difference is that the first cost is converted to an annual expenditure. The alternative with lowest effective annual cost is the most desirable.

In the rate-of-return method, a trial-and-error procedure is usually required. Interpolation from the interest tables can determine what rate of return $(i)$ will give an interest factor which will make the overall cash flow

**Single Payment Compound Amount - *F/P***

The *F/P* factor is the future value of one dollar in "*n*" periods at interest of "*i*" percent.

$$F/P = (1+i)^n \qquad (2\text{-}1)$$

**Single Payment Present Worth - *P/F***

The *P/F* factor is the present worth of one dollar, in "*n*" periods from now at interest of "*i*" percent.

$$P/F = \frac{1}{(1+i)^n} \qquad (2\text{-}2)$$

**Uniform Series Compound Amount - *F/A***

The *F/A* factor is the future value of a uniform series of one dollar deposits.

$$F/A = \frac{(1+i)^n - 1}{i} \qquad (2\text{-}3)$$

**Sinking Fund Payment - *A/F***

The *A/F* factor is the uniform series of deposits whose future value is one dollar.

$$A/F = \frac{i}{(1+i)^n - 1} \qquad (2\text{-}4)$$

**Uniform Series Present Worth - *P/A***

The *P/A* factor is the present value of uniform series of one dollar deposits.

**Capital Recovery - A/P**

The A/P factor is the uniform series of deposits whose present value is one dollar.

$$P/A = \frac{(1+i)^n - 1}{i\,(1+i)^n} \qquad (2\text{-}5)$$

$$A/P = \frac{i\,(1+i)^n}{(1+i)^n - 1} \qquad (2\text{-}6)$$

**Gradient Present Worth - GPW**

The GPW factor is the present value of a gradient series.

$$GPW = P/A = \frac{\frac{1+e}{1+i}\left[1 - \left(\frac{1+e}{1+i}\right)^n\right]}{1 - \frac{1+e}{1+i}} \qquad (2\text{-}7)$$

**NOTES**

where

P   is the present worth (occurs at the beginning of the interest period).
F   is the future worth (occurs at the end).
n   is the number of periods that the interest is compounded.
i   is the interest rate or desired rate of return.
A   is the uniform series of deposits (occurs at the end of the interest period).
e   is the escalation rate

**Figure 2-1. Interest Factors**

balance. The rate-of-return analysis gives a good indication of the overall ranking of independent alternates.

The effect of escalation in fuel costs can influence greatly the final decision. When an annual cost grows at a steady rate it may be treated as a gradient and the Gradient Present Worth factor can be used.

Special appreciation is given to Rudolph R. Yaneck and Dr. Robert Brown for use of their specially designed interest and escalation tables used in this text.

When life-cycle costing is used to compare several alternatives, the differences between costs are important. For example, if one alternate forces additional maintenance or an operating expense to occur, then these factors as well as energy costs need to be included. Remember, what was previously spent for the item to be replaced is irrelevant. The only factor to be considered is whether the new cost can be justified based on projected savings over its useful life.

## PAYBACK ANALYSIS

The simple payback analysis is sometimes used instead of the methods previously outlined. The simple payback is defined as initial investment divided by annual savings after taxes. The simple payback method does not take into account the effect of interest or escalation rate.

Since the payback period is relatively simple to calculate, and due to the fact managers wish to recover their investment as rapidly as possible, the payback method is frequently used.

It should be used in conjunction with other decision-making tools. When used by itself as the principal criterion, it may result in choosing less profitable investments which yield high initial returns for short periods as compared with more profitable investments which provide profits over longer periods of time.

### Example Problem 2-1

An electrical energy audit indicates electrical motor consumption is 4 x $10^6$ kWh per year. By upgrading the motor spares with high efficiency motors, a 10% savings can be realized. The additional cost for these motors is estimated at \$80,000. Assuming an 8¢ per kWh energy charge and 20-year life, is the expenditure justified based on a minimum rate of return of 20% before taxes? Solve the problem using the present worth, annual cost, and rate-of-return methods.

## Analysis

### Present Worth Method

|  | Alternate 1 Present Method | Alternate 2 Use High Efficiency Motor Spares |
|---|---|---|
| (1) First Cost (*P*) | – | $80,000 |
| (2) Annual Cost (*A*) | $4 \times 106 \times .08$ = $320,000 | $.9 \times \$320,000$ = $288,000 |
| USPW (Table 15-4) | 4.87 | 4.87 |
| (3) *A* × 4.87 = | $1,558,400 | $1,402,560 |
| Present Worth (1) + (3) | $1,558,400 | $1,482,560 |

Choose Alternate with Lowest Present Worth Cost

### Annual Cost Method

|  | Alternate I | Alternate 2 |
|---|---|---|
| (1) First Cost (*P*) | – | $80,000 |
| (2) Annual Cost (*A*) | $320,000 | $288,000 |
| CR (Table 15-4) | .2 | .2 |
| (3) *P* × .2 | – | $16,000 |
| Annual Cost (2) + (3) | $320,000 | $304,000 |

Choose Alternate with Lowest Annual Cost

### Rate of Return Method

$$P = (\$320,000 - \$288,000)$$

$$P/A = \frac{80,000}{32,000} = 2.5$$

What value of *i* will make $P/A$ = 2.5? *i* = 40% (Table 15-7).

## Example Problem 2-2

Show the effect of 10% escalation on the rate-of-return analysis given the

| Energy equipment investment | = | $20,000 |
|---|---|---|
| After tax savings | = | $ 2,600 |
| Equipment life (*n*) | = | 15 years |

**Analysis**
   Without escalation

$$\text{CR} = \frac{A}{P} = \frac{2,600}{20,000} = .13$$

From Table 15-1, the rate of return in 10%.
   With 10% escalation assumed:

$$\text{GPW} = \frac{P}{A} = \frac{20,000}{2,600} = 7.69$$

From Table 15-11, the rate of return is 21%.

   Thus we see that taking into account a modest escalation rate can dramatically affect the justification of the project.


**TAX CONSIDERATIONS**

**Depreciation**
   Depreciation affects the "accounting procedure" for determining profits and losses and the income tax of a company. In other words, for tax purposes the expenditure for an asset such as a pump or motor cannot be fully expensed in its first year. The original investment must be charged off for tax purposes over the useful life of the asset. A company wishes to expense an item as quickly as possible.
   The Internal Revenue Service allows several methods for determining the annual depreciation rate.
   *Straight-Line Depreciation:* The simplest method is referred to as a straight-line depreciation and is defined as

$$D = \frac{P-L}{n} \qquad\qquad \textit{Formula (2-8)}$$

Where
   $D$   is the annual depreciation rate
   $L$   is the value of equipment at the end of its useful life, commonly
         referred to as salvage value

$n$   is the life of the equipment which is determined by Internal Revenue Service Guidelines

$P$   is the initial expenditure.

*Sum-of-Years Digits:* Another method is referred to as the sum-of years digits. In this method the depreciation rate is determined by finding the sum of digits using the following formula:

$$N = n \ \frac{n+1}{2}$$

Formula (2-9)

Where $n$ is the life of equipment.

Each year's depreciation rate is determined as follows:

First year

$$D = \frac{n}{N} (P - L)$$

Formula (2-10)

Second year

$$D = \frac{n-1}{N} (P - L)$$

Formula (2-11)

$n$ year

$$D = \frac{1}{N} (P - L)$$

Formula (2-12)

*Declining-Balance Depreciation:* The declining-balance method allows for larger depreciation charges in the early years, which is sometimes referred to as fast write-off.

The rate is calculated by taking a constant percentage of the declining undepreciated balance. The most common method used to calculate the declining balance is to predetermine the depreciation rate. Under certain circumstances a rate equal to 200% of the straight-line depreciation rate may be used. Under other circumstances the rate is limited to 1-1/2 or 1-1/4 times as great as straight-line depreciation. In this method the salvage value or undepreciated book value is established once the depreciation rate is pre-established.

To calculate the undepreciated book value, Formula 2-13 is used:

$$D = 1 - \left(\frac{L}{P}\right)^{1/N}$$

Formula (2-13)

Where
$D$    is the annual depreciation rate
$L$    is the salvage value
$P$    is the first cost

## Example Problem 2-3

Calculate the depreciation rate using the straight-line, sum-of-years digit, and declining-balance methods.
Salvage value is 0.
$n = 5$ years
$P = 150,000$
For declining balance use a 200% rate.

*Straight-Line Method.*

$$D = \frac{P - L}{n} = \frac{150,000}{5} = \$30,000 \text{ per year}$$

*Sum-of-Years Digits*

$$N = \frac{n(n + 1)}{2} = \frac{5(6)}{2} = 15$$

$$D_1 = \frac{n}{N}(P) = \frac{5}{15}(150,000) = 50,000$$

| $N$ | $P$ |
|---|---|
| 1 = | $54,000 |
| 2 = | 40,000 |
| 3 = | 30,000 |
| 4 = | 20,000 |
| 5 = | 10,000 |

$$D = 2 \times 20\% = 40\%$$

Declining-Balance Method
(Straight-Line Depreciation Rate = 20%)

| Year | Undepreciated Balance At Beginning of Year | Depreciation Charge |
|------|------|------|
| 1 | 150,000 | 60,000 |
| 2 | 90,000 | 36,000 |
| 3 | 54,000 | 21,600 |
| 4 | 32,400 | 12,960 |
| 5 | 19,440 | 7,776 |
| | TOTAL | 138,336 |

Undepreciated Book Value (150,000 – 138,336) = $11,664

## Cogeneration Equipment Depreciation

Most cogeneration equipment is depreciated over a 15- or 20-year period, depending on the particular type of equipment involved, using the 150% declining balance method switching to straight-line to maximize deductions. Gas and combustion turbine equipment used to produce electricity for sale is depreciated over a 15-year period. Equipment used in the steam power production of electricity for sale (including combustion turbines operated in combined cycle with steam units), as well as assets used to produce steam for sale, are normally depreciated over a 20-year period.

However, most electric and steam generation equipment owned by a taxpayer and producing electric or thermal energy for use by the taxpayer in its industrial process and plant activity, and not ordinarily for sale to others, is depreciated over a 15-year period. Electrical and steam transmission and distribution equipment will be depreciated over a 20-year period at the same 150 percent declining balance rate.

## Energy Efficiency Equipment and Real Property Depreciation

Energy conservation equipment, still classified as real property, is depreciated on a straight line basis over a recovery period. Equipment installed in connection with residential real property qualifies for a 27-1/2-year period, while equipment placed in nonresidential facilities is subject to a 31-1/2-year period. Other real property assets are depreciated over the above period, depending on their residential or nonresidential character.

**After-Tax Analysis**

Tax-deductible expenses such as maintenance, energy, operating costs, insurance and property taxes reduce the income subject to taxes.

For the after-tax life-cycle cost analysis and payback analysis, the actual incurred annual savings is given as follows:

$$AS = (1 - I) E + ID \qquad\qquad \textit{Formula (2-14)}$$

*Where*

| | | |
|---|---|---|
| $AS$ | = | yearly annual after-tax savings (excluding effect of tax credit) |
| $E$ | = | yearly annual energy savings (difference between original expenses and expenses after modification) |
| $D$ | = | annual depreciation rate |
| $I$ | = | income tax bracket |

Formula 2-14 takes into account that the yearly annual energy savings is partially offset by additional taxes which must be paid due to reduced operating expenses. On the other hand, the depreciation allowance reduces taxes directly.

To compute a rate of return which accounts for taxes, depreciation, escalation, and tax credits, a cash-flow analysis is usually required. This method analyzes all transactions including first and operating costs. To determine the after-tax rate of return, a trial and error or computer analysis is required.

The Present Worth factors tables in Chapter 15, Appendix, can be used for this analysis. All money is converted to the present assuming an interest rate. The summation of all present dollars should equal zero when the correct interest rate is selected, as illustrated in Figure 2-2.

This analysis can be made assuming a fuel escalation rate by using the Gradient Present Worth interest of the Present Worth Factor.

**Example Problem 2-4**

Comment on the after-tax rate of return for the installation of a heat-recovery system given the following:

- First Cost        $100,000
- Year Savings       36,363
- Straight-line depreciation life and equipment life of 5 years

- Income tax bracket 34%

| Year | 1 Investment | 2 Tax Credit | 3 After Tax Savings (AS) | 4 Single Payment Present Worth Factor | (2 + 3) x 4 Present Worth |
|------|------------|------------|------------|------------|------------|
| 0 | –P | | | | –P |
| 1 | | +TC | $AS_1$ | $SPPW_1$ | $+P_1$ |
| 2 | | | $AS_2$ | $SPPW_2$ | $P_2$ |
| 3 | | | $AS_3$ | $SPPW_3$ | $P_3$ |
| 4 | | | $AS_4$ | $SPPW_4$ | $P_4$ |
| Total | | | | | $\Sigma P$ |

$$AS = (1 - I) E + ID$$
Trial & Error Solution:
Correct $i$ when $\Sigma P = 0$

**Figure 2-2. Cash Flow Rate of Return Analysis**

## Analysis

$$D = 100,000/5 = 20,000$$
$$AS = (1 - I) E + ID = .66(36,363) + .34(20,000) = \$30,800$$

First Trial $i = 20\%$

| Investment | After Tax Savings | SPPW 20% | PW |
|------------|------------|------------|------------|
| 0–100,000 | | | –100,000 |
| 1 | 30,800 | .833 | 25,656 |
| 2 | 30,800 | .694 | 21,375 |
| 3 | 30,800 | .578 | 17,802 |
| 4 | 30,800 | .482 | 14,845 |
| 5 | 30,800 | .401 | 12,350 |
| | | | $\Sigma - 7,972$ |

Since summation is negative a higher present worth factor is required. Next try is 15%.

|             | *After Tax* | *SPPW* |          |
|             | *Savings*   | *15%*  | *PW*     |
| *Investment* |            |        |          |
| 0–100,000   |             |        | –100,000 |
| 1           | 30,800      | .869   | + 26,765 |
| 2           | 30,800      | .756   | + 23,284 |
| 3           | 30,800      | .657   | + 20,235 |
| 4           | 30,800      | .571   | + 17,586 |
| 5           | 30,800      | .497   | + 15,307 |
|             |             |        | +  3,177 |

Since rate of return is bracketed, linear interpolation will be used.

$$\frac{3177 + 7972}{-5} = \frac{3177 - 0}{15 - i\%}$$

$$i = \frac{3177}{2229.6} + 15 = 16.4\%$$

## Impact of Fuel Inflation on Life-Cycle Costing

As illustrated by Problem 2-2, a modest estimate of fuel inflation has a major impact on improving the rate of return on investment of the project. The problem facing the energy engineer is how to forecast what the future of energy costs will be. All too often no fuel inflation is considered because of the difficulty of projecting the future. In making projections the following guidelines may be helpful:

- Is there a rate increase that can be forecast based on new nuclear generating capacity? In locations such as Georgia, California, and Arizona electric rates will rise at a faster rate due to commissioning of new nuclear plants and rate increases approved by the Public Service Commission of that state.
- What has been the historical rate increase for the facility? Even with fluctuations there are likely to be trends to follow.
- What events on a national or international level would impact on your costs? New state taxes, new production quotas by OPEC and other factors will affect your fuel prices.
- What do the experts say? Energy economists, forecasting services, and your local utility projections all should be taken into account.

The rate of return on investment becomes more attractive when life-cycle costs are taken into account. Tables 15-9 through 15-12 can be used to show the impact of fuel inflation on the decision-making process.

## Example Problem 2-5

Develop a set of curves that indicate the capital that can be invested to give a rate of return of 15% after taxes for each $1000 saved for the following conditions:

1. The effect of escalation is not considered.
2. A 5% fuel escalation is considered.
3. A 10% fuel escalation is considered.
4. A 14% fuel escalation is considered.
5. A 20% fuel escalation is considered.

Calculate for 5-, 10-, 15-, 20-year life.

Assume straight-line depreciation over useful life, 34% income tax bracket, and no tax credit.

## Answer

$$AS = (1 - I)\, E + ID$$
$$I = 0.34,\ E = \$1000$$
$$AS = 660 + \frac{0.34P}{N}$$

Thus, the after-tax savings (*AS*) are composed of two components. The first component is a uniform series of $660 escalating at *e* percent/year. The second component is a uniform series of 0.34*P*/*N*.

Each component is treated individually and converted to present day values using the GPW factor and the USPW factor, respectively. The sum of these two present worth factors must equal *P*. In the case of no escalation, the formula is

$$P = 660\, P\,/A + \frac{0.34P}{N}\, P\,/A$$

In the case of escalation

$$P = 660\, GPW + \frac{0.34P}{N}\, P\,/A$$

Since there is only one unknown, the formulas can be readily solved. The results are indicated below.

|         | $N = 5$ $P$ | $N = 10$ $P$ | $N = 15$ $P$ | $N = 20$ $P$ |
|---------|-------------|--------------|--------------|--------------|
| $e = 0$     | 2869        | 4000         | 4459         | 4648         |
| $e = 10\%$  | 3753        | 6292         | 8165         | 9618         |
| $e = 14\%$  | 4170        | 7598         | 10,676       | 13,567       |
| $e = 20\%$  | 4871        | 10,146       | 16,353       | 23,918       |

Figure 2-3 illustrates the effects of escalation. This figure can be used as a quick way to determine after-tax economics of energy utilization expenditures.

### Example Problem 2-6

It is desired to have an after-tax savings of 15%. Comment on the investment that can be justified if it is assumed that the fuel rate escalation should not be considered and the annual energy savings is $2000 with an equipment economic life of 10 years.

Comment on the above, assuming a 10% fuel escalation.

### Answer

From Figure 2-3, for each $1000 energy savings, an investment of $3600 is justified or $8000 for a $2000 savings for which no fuel increase is accounted.

With a 10% fuel escalation rate on investment of $6300 justified for each $1000 energy savings, $12,600 can be justified for $2000 savings. Thus, a 57% higher expenditure is economically justifiable and will yield the same after tax rate of return of 15% when a fuel escalation of 10% is considered.

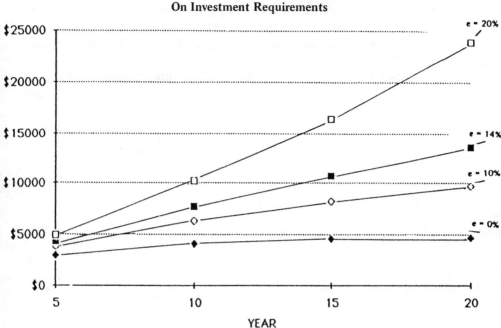

**Figure 2-3. Effects of Escalation**
**On Investment Requirements**

Note: Maximum investment in order to attain a 15% after-tax rate of return on investment for annual savings of $1000.

## COMPUTER ANALYSIS

The Alliance to Save Energy, 1925 K Street, NW, Suite 206, Washington, DC 20006, has introduced an investment analysis software package, ENVEST, which costs only $75 for 5-1/4-inch disc and $85 for 3-1/2-inch hard drive disc and includes a 170-page user manual and 30 days of telephone support. The program can be run on an IBM PC, PCXT, PCAT with 256K ram. The program enables the user to

- Generate spreadsheets and graphs showing the yearly cash flows from any energy-related investment.

- Compute payback, internal rate of return, and other important investment measures.
- Experiment with differing energy price projections.
- Perform sensitivity analysis on key assumptions.
- Compare alternative financing options, including loans, leases, and shared savings.
- Store data on over 100 energy efficiency investments.

# 3

# Energy Auditing and Accounting

## TYPES OF ENERGY AUDITS

The simplest definition for an energy audit is as follows: An energy audit serves the purpose of identifying where a building or plant facility uses energy and identifies energy conservation opportunities.

There is a direct relationship to the cost of the audit (amount of data collected and analyzed) and the number of energy conservation opportunities to be found. Thus, a first distinction is the cost of the audit which determines the type of audit to be performed.

The second distinction is the type of facility. For example, a building audit may emphasize the building envelope, lighting, heating, and ventilation requirements. On the other hand, an audit of an industrial plant emphasizes the process requirements.

Most energy audits fall into three categories or types, namely, Walk-Through, Mini-Audit, or Maxi-Audit.

*Walk-Through*—This type of audit is the least costly and identifies preliminary energy savings. A visual inspection of the facility is made to determine maintenance and operation energy saving opportunities plus collection of information to determine the need for a more detailed analysis.

*Mini-Audit*—This type of audit requires tests and measurements to quantify energy uses and losses and determine the economics for changes.

*Maxi-Audit*—This type of audit goes one step further than the mini-audit. It contains an evaluation of how much energy is used for each function such as lighting, process, etc. It also requires a model analysis, such as a computer simulation, to determine energy use patterns and predictions on a year-round basis, taking into account such variables as weather data.

As noted in the audit definition, there are two essential parts, namely, data acquisition and data analysis.

## Data Acquisition

This phase requires the accumulation of utility bills, establishing a baseline to provide historical documentation and a survey of the facility.

All energy flows should be accounted for; thus all "energy in" should equal "energy out." This is referred to as an energy balance.

All energy costs should be determined for each fuel type. The energy survey is essential. Instrumentation commonly used in conducting a survey is discussed at the conclusion of the chapter.

## Data Analysis

As a result of knowing how energy is used, a complete list of "Energy Conservation Opportunities" (ECOs) will be generated. The life-cycle costing techniques presented in Chapter 2 will be used to determine which alternative should be given priority.

A very important phase of the overall program is to continuously monitor the facility even after the ECOs have been implemented. Documentation of the cost avoidance or savings is essential to the audit.

Remember, in order to have a continuous ongoing program, *individuals must be made accountable* for energy use. As part of the audit, recommendations should be made as to where to add "root" or submetering.

## ENERGY USE PROFILES

The energy audit process for a building emphasizes building envelope, heating and ventilation, air conditioning, plus lighting functions. For an industrial facility the energy audit approach includes process consideration. Figures 3-1 through 3-3 illustrate how energy is used for a typical industrial plant. It is important to account for total consumption, cost, and how energy is used for each commodity such as steam, water, air and natural gas. This procedure is required to develop the appropriate energy conservation strategy.

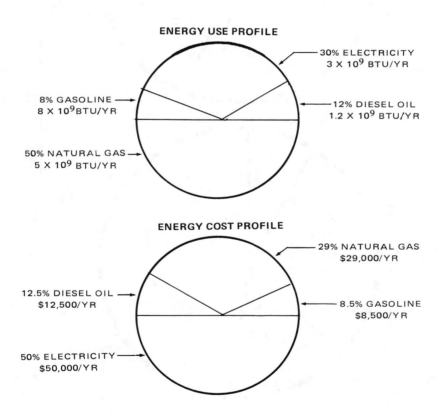

Figure 3-1. Energy Use and Cost Profile

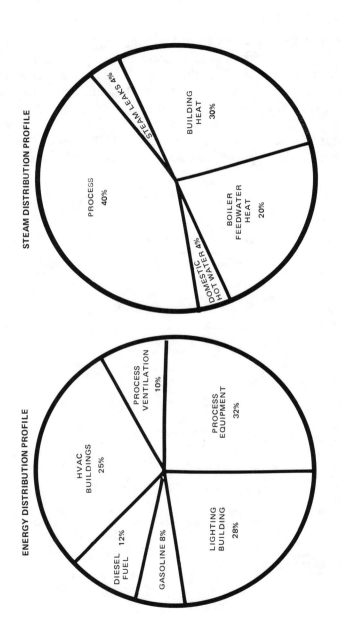

**Figure 3-2. Energy Profile by Function**

The top portion of Figure 3-1 illustrates how much energy is used by fuel type and its relative percentage. The pie chart below shows how much is spent for each fuel type. Using a pie-chart presentation or nodal flow diagram can be very helpful in visualizing how energy is being used.

Figure 3-2, on the other hand, shows how much of the energy is used for each function such as lighting, process, and building heating and ventilation. Pie charts similar to the right-hand side of the figure should be made for each category such as air, steam, electricity, water and natural gas.

Figure 3-3 illustrates an alternate representation for the steam distribution profile.

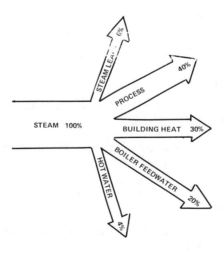

**Figure 3-3. Steam Distribution Nodal Diagram**

Several audits are required to construct the energy use profiles, such as:

*Envelope Audit*—This audit surveys the building envelope for losses or gains due to leaks, building construction, doors, glass, lack of insulation, etc.

*Functional Audit*—This audit determines the amount of energy required for a particular function and identifies energy conservation opportunities. Functional audits include:

- Heating, ventilation and air conditioning
- Building
- Lighting
- Domestic hot water
- Air distribution

*Process Audit*—This audit determines the amount of energy required for each process function and identifies energy conservation opportunities. Process functional audits include:

- Process machinery
- Heating, ventilation and air-conditioning process
- Heat treatment
- Furnaces

*Transportation Audit*—This audit determines the amount of energy required for forklift trucks, cars, vehicles, trucks, etc.

*Utility Audit*—This audit analyzes the monthly, daily or yearly energy usage for each utility.

## ENERGY USERS

Energy use profiles for several end-users are summarized in Tables 3-1 through 3-11.

Table 3-1. Energy Use in Apartment Buildings

|  | Range (%) | Norms (%) |
|---|---|---|
| Environmental Control | 50 to 80 | 70 |
| Lighting and Wall Receptacles | 10 to 20 | 15 |
| Hot Water | 2 to 5 | 3 |
| Special Functions |  |  |
| Laundry, Swimming Pool, Restaurants, |  |  |
| Parking, Elevators, Security Lighting | 5 to 20 | 10 |

Table 3-2. Energy Use in Bakeries

| Housekeeping Energy | Percent |
|---|---|
| Space Heating | 21.5 |
| Air Conditioning | 1.6 |
| Lighting | 1.4 |
| Domestic Hot Water | 1.8 |
| TOTAL | 26.3 |
| Process Energy | Percent |
| Baking Ovens | 49.0 |
| Pan Washing | 10.6 |
| Mixers | 4.1 |
| Freezers | 3.3 |
| Cooking | 2.0 |
| Fryers | 1.8 |
| Proof Boxes | 1.8 |
| Other Processes | 1.1 |
| TOTAL | 73.7 |

Data are for a 27,000-square-foot bakery in Washington, D.C.

Table 3-3. Energy Use in Die Casting Plants

| Housekeeping Energy | Percent |
|---|---|
| Space Heating | 24 |
| Air Conditioning | 2 |
| Lighting | 2 |
| Domestic Hot Water | 2 |
| TOTAL | 30 |
| Process Energy | Percent |
| Melting Hearth | 30 |
| Quiet Pool | 20 |
| Molding Machines | 10 |
| Air Compressors | 5 |
| Other Processes | 5 |
| TOTAL | 70 |

Table 3-4. Energy Use in Hospital Buildings

| | Range (%) | Norms (%) |
|---|---|---|
| Environmental Control | 40 to 65 | 58 |
| Lighting and Wall Receptacles | 10 to 20 | 15 |
| Laundry | 8 to 15 | 12 |
| Food Service, Kitchen Operations | 5 to 10 | 7 |
| Medical Equipment, Sterilization, Incinerator, Parking, Elevators, Security Lighting | 5 to 15 | 8 |

Table 3-5. Energy Use in Hotels and Motels

|  | Range (%) | Norms (%) |
|---|---|---|
| Space Heating | 45 to 70 | 60 |
| Lighting | 5 to 15 | 11 |
| Air Conditioning | 3 to 15 | 10 |
| Refrigeration | 0 to 10 | 4 |
| Special Functions | 5 to 20 | 15 |
|     Laundry, Kitchen, Restaurant, | | |
|     Swimming Pool, Garage, | | |
|     Security Lighting, Hot Water | | |

Table 3-6. Energy Use in Retail Stores

|  | Range (%) | Norms (%) |
|---|---|---|
| HVAC | 20 to 50 | 30 |
| Lighting | 40 to 75 | 60 |
| Special Functions | 5 to 20 | 10 |
|     Elevators, General Power, Parking | | |
|     Security Lighting, Hot Water | | |

Table 3-7. Energy Use in Restaurants

|  | Table Restaurant Norms (%) | Fast Food Restaurant Norms (%) |
|---|---|---|
| HVAC | 32 | 36 |
| Lighting | 8 | 26 |
| Special Functions |  |  |
| Food Preparation | 45 | 27 |
| Food Storage | 2 | 6 |
| Sanitation | 12 | 1 |
| Other | 1 | 4 |

Table 3-8. Energy Use in Schools

|  | Range (%) | Norms (%) |
|---|---|---|
| Environmental Control | 45 to 80 | 65 |
| Lighting and Wall Receptacles | 10 to 20 | 15 |
| Food Service | 5 to 10 | 7 |
| Hot Water | 2 to 5 | 3 |
| Special Functions | 0 to 20 | 10 |

Table 3-9. Energy Use in Transportation Terminals

|  | Range (%) | Norms (%) |
|---|---|---|
| Space Heating | 50 to 75 | 60 |
| Lighting | 5 to 25 | 15 |
| Air Conditioning | 5 to 25 | 15 |
| Special Functions | 3 to 20 | 10 |
| Elevators, General Power, Parking, Security Lighting, Hot Water |  |  |

Table 3-10. Energy Use in Warehouses and Storage Facilities

| (Vehicles Not Included) | Range (%) | Norms (%)* |
|---|---|---|
| Space Heating | 45 to 80 | 67 |
| Air Conditioning | 3 to 10 | 6 |
| Lighting | 4 to 12 | 7 |
| Refrigeration | 0 to 40 | 12 |
| Special Functions | 5 to 15 | 8 |
| Elevators, General Power, Parking, Security Lighting, Hot Water |  |  |

* Norms for a warehouse or storage facility are strongly dependent on the products and their specific requirements for temperature and humidity control.

Table 3-11. Comparative Energy Use by System

|  |  | Heating & Ventilation | Cooling & Ventilation | Lighting | Power & Process | Domestic Hot Water |
|---|---|---|---|---|---|---|
| Schools | A | 4 | 3 | 1 | 5 | — |
|  | B | 1 | 4 | 2 | 5 | 3 |
|  | C | 1 | 4 | 2 | 5 | 3 |
| Colleges | A | 5 | 2 | 1 | 4 | 3 |
|  | B | 1 | 3 | 2 | 5 | 4 |
|  | C | 1 | 5 | 2 | 4 | 3 |
| Office Bldg. | A | 3 | 1 | 2 | 4 | 5 |
|  | B | 1 | 3 | 2 | 4 | 5 |
|  | C | 1 | 3 | 2 | 4 | 5 |
| Commercial Stores | A | 3 | 1 | 2 | 4 | 5 |
|  | B | 2 | 3 | 1 | 4 | 5 |
|  | C | 1 | 3 | 2 | 4 | 5 |
| Religious Bldg. | A | 3 | 2 | 1 | 4 | 5 |
|  | B | 1 | 3 | 2 | 4 | 5 |
|  | C | 1 | 3 | 2 | 4 | 5 |
| Hospitals | A | 4 | 1 | 2 | 5 | 3 |
|  | B | 1 | 3 | 4 | 5 | 2 |
|  | C | 1 | 5 | 3 | 4 | 2 |

Climatic Zone A:   Fewer than 2500 degree days
Climate Zone B:    2500–5500 degree days
Climate Zone C:    5500–9500 degree days

Source:  Guidelines For Saving Energy In Existing Buildings ECM—1

**Note:** Numbers indicate energy consumption relative to each other
     (1) greatest consumption
     (5) least consumption

## ENERGY USE IN BUILDINGS[1]

More than 36% of the nation's primary energy was consumed in providing services to residential and commercial buildings in 1985 (26.9 quads). The residential sector consumed 15.3 quads; the commercial sector (including public buildings) consumed 11.6 quads.

The largest end use is space heating in both residential (40%) and commercial (34%) buildings. Water heating is second in residences (17%), followed by refrigerators (9%), air conditioning (7%), lighting (7%), and kitchen ranges and ovens (6%). In commercial buildings, lighting is second (25%), followed by ventilation (12%), air conditioning (10%) and water heating (6%). (See Table 3-12.)

**Table 3-12. Building Sector Energy Use**

| BUILDING SECTOR ENERGY USE BY END USE 1985-2010 FORECAST | | |
|---|---|---|
| | 1985 | 2010 |
| Space Heating | 38% | 34% |
| Air Conditioning and Ventilation | 15 | 16 |
| Lighting | 15 | 15 |
| Water Heating | 12 | 12 |
| Refrigeration | 7 | 5 |
| Other | 13 | 18 |

The forecast changes as the type of energy used becomes more significant. The analysis estimates a major increase in the use of electricity by the building sector, increasing from 17.0 quads in 1985 to 26.8 quads in 2010. These figures represent the source (primary) energy used to generate electricity.

During this period, natural gas use is forecast to increase slightly from 7.0 quads to 7.2 quads. The use of solar and renewable energy is expected to more than double from 1.2 quads to 2.9 quads; the use of oil is expected to remain flat at 2.6 quads.

---

[1] Source: Energy Conservation Goals for Buildings, A Report to the Congress of The United States, May 1988.

## THE ENERGY SURVEY

As part of the data acquisition phase, a detailed survey should be conducted. The various types of instrumentation commonly used in the survey are discussed in this section.

### Infrared Equipment

Some companies may have the wrong impression that infrared equipment can meet most of their instrumentation needs.

The primary use of infrared equipment in an energy utilization program is to detect building or equipment losses. Thus it is just one of the many options available.

Several energy managers find infrared in use in their plant prior to the energy utilization program. Infrared equipment, in many instances, was purchased by the electrical department and used to detect electrical hot spots.

Infrared energy is an invisible part of the electromagnetic spectrum. It exists naturally and can be measured by remote heat-sensing equipment. Within the last four years lightweight portable infrared systems became available to help determine energy losses. Differences in the infrared emissions from the surface of objects cause color variations to appear on the scanner. The hotter the object, the more infrared radiated. With the aid of an isotherm circuit, the intensity of these radiation levels can be accurately measured and quantified. In essence the infrared scanning device is a diagnostic tool which can be used to determine building heat losses. Equipment costs range from $400 to $25,000.

An overview energy scan of the plant can be made through an aerial survey using infrared equipment. Several companies offer aerial scan services starting at $1500. Aerial scans can determine underground stream pipe leaks, hot gas discharges, leaks, etc.

Since IR detection and measurement equipment have gained increased importance in the energy audit process, a summary of the fundamentals is reviewed in this section.

The visible portion of the spectrum runs from .4 to .75 micrometers ($\mu$m). The infrared or thermal radiation begins at this point and extends to approximately 1000 $\mu$m. Objects such as people, plants, or buildings will emit radiation with wavelengths around 10 $\mu$m. (See Figure 3-4.)

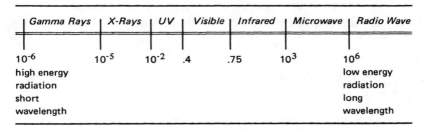

| Gamma Rays | X-Rays | UV | Visible | Infrared | Microwave | Radio Wave |

$10^{-6}$       $10^{-5}$    $10^{-2}$   .4      .75     $10^{3}$      $10^{6}$

high energy radiation short wavelength          low energy radiation long wavelength

Figure 3-4. Electromagnetic Spectrum

Infrared instruments are required to detect and measure the thermal radiation. To calibrate the instrument, a special "black body" radiator is used. A black body radiator absorbs all the radiation that impinges on it and has an absorbing efficiency or emissivity of 1.

The accuracy of temperature measurements by infrared instruments depends on the three processes which are responsible for an object acting like a black body. These processes—absorbed, reflected, and transmitted radiation—are responsible for the total radiation reaching an infrared scanner.

*The real temperature of the object is dependent only upon its emitted radiation.*

Corrections to apparent temperatures are made by knowing the emissivity of an object at a specified temperature.

The heart of the infrared instrument is the infrared detector. The detector absorbs infrared energy and converts it into electrical voltage or current. The two principal types of detectors are the thermal and photo type. The thermal detector generally requires a given period of time to develop an image on photographic film. The photo detectors are more sensitive and have a higher response time. Television-like displays on a cathode ray tube permit studies of dynamic thermal events on moving objects in real time.

There are various ways of displaying signals produced by infrared detectors. One way is by use of an isotherm contour. The lightest areas of the picture represent the warmest areas of the subject, and the darkest areas represent the coolest portions. These instruments can show thermal variations of less than 0.1°C and can cover a range of −30°C to over 2000°C.

The isotherm can be calibrated by means of a black body radiator so that a specific temperature is known. The scanner can then be moved and the temperatures of the various parts of the subject can be made.

## MEASURING ELECTRICAL SYSTEM PERFORMANCE

The ammeter, voltmeter, wattmeter, power factor meter, and footcandle meter are usually required to do an electrical survey. These instruments are described below.

### Ammeter and Voltmeter

To measure electrical currents, ammeters are used. For most audits, alternating currents are measured. Ammeters used in audits are portable and are designed to be easily attached and removed.

There are many brands and styles of snap-on ammeters commonly available that can read up to 1000 amperes continuously. This range can be extended to 4000 amperes continuously for some models with an accessory step-down current transformer.

The snap-on ammeters can be either indicating or recording with a printout. After attachment, the recording ammeter can keep recording current variations for as long as a full month on one roll of recording paper. This allows the study of current variations in a conductor for extended periods without constant operator attention.

The ammeter supplies a direct measurement of electrical current, which is one of the parameters needed to calculate electrical energy. The second parameter required to calculate energy is voltage, and it is measured by a voltmeter.

Several types of electrical meters can read the voltage or current. A voltmeter measures the difference in electrical potential between two points in an electrical circuit.

In series with the probes are the galvanometer and a fixed resistance (which determine the voltage scale). The current through this fixed resistance circuit is then proportional to the voltage, and the galvanometer deflects in proportion to the voltage.

The voltage drops measured in many instances are fairly constant and need only be performed once. If there are appreciable fluctuations, additional readings or the use of a recording voltmeter may be indicated.

Most voltages measured in practice are under 600 volts and there are many portable voltmeter/ammeter clamp-ons available for this and lower ranges.


## Wattmeter and Power Factor Meter

The portable wattmeter can be used to indicate by direct reading electrical energy in watts. It can also be calculated by measuring voltage, current and the angle between them (power factor angle).

The basic wattmeter consists of three voltage probes and a snap-on current coil which feeds the wattmeter movement.

The typical operating limits are 300 kilowatts, 650 volts, and 600 amperes. It can be used on both one- and three-phase circuits.

The portable power factor meter is primarily a three-phase instrument. One of its three voltage probes is attached to each conductor phase and a snap-on jaw is placed about one of the phases. By disconnecting the wattmeter circuitry, it will directly read the power factor of the circuit to which it is attached.

It can measure power factor over a range of 1.0 leading to 1.0 lagging with "ampacities" up to 1500 amperes at 600 volts. This range covers the large bulk of the applications found in light industry and commerce.

The power factor is a basic parameter whose value must be known to calculate electric energy usage. Diagnostically, it is a useful instrument to determine the sources of poor power factor in a facility.

Portable digital kWh and kW demand units are now available.

Digital read-outs of energy usage in both kWh and kW demand or in dollars and cents, including instantaneous usage, accumulated usage, projected usage for a particular billing period, alarms when over-target levels are desired for usage, and control-outputs for load-shedding and cycling are possible.

Continuous displays or intermittent alternating displays are available at the touch of a button for any information needed such as the cost of operating a production machine for one shift, one hour or one week.

### Footcandle Meter

Footcandle meters measure illumination in units of footcandles through a light-sensitive barrier layer of cells contained within them. They are usually pocket-size and portable and are meant to be used as field instruments to survey levels of illumination. Footcandle meters differ from conventional photographic lightmeters in that they are color and cosine corrected.

## TEMPERATURE MEASUREMENTS

To maximize system performance, knowledge of the temperature of a fluid, surface, etc. is essential. Several types of temperature devices are described in this section.

### Thermometer

There are many types of thermometers that can be used in an energy audit. The choice of what to use is usually dictated by cost, durability, and application.

For air-conditioning, ventilation and hot-water service applications (temperature ranges 50°F to 250°F), a multipurpose portable battery-operated thermometer is used. Three separate probes are usually provided to measure liquid, air or surface temperatures.

For boiler and oven stacks (1000°F) a dial thermometer is used. Thermocouples are used for measurements above 1000°F.

### Surface Pyrometer

Surface pyrometers are instruments which measure the temperature of surfaces. They are somewhat more complex than other temperature instruments because their probe must make intimate contact with the surface being measured.

Surface pyrometers are of immense help in assessing heat losses through walls and also for testing steam traps.

They may be divided into two classes: low-temperature (up to 250°F) and high-temperature (up to 600°F to 700°F). The low-temperature unit is usually part of the multipurpose thermometer kit. The high-temperature unit is more specialized but needed for evaluating fired units and general steam service.

There are also noncontact surface pyrometers which measure infrared radiation from surfaces in terms of temperature. These are suitable for general work and also for measuring surfaces which are visually but not physically accessible.

A more specialized instrument is the optical pyrometer. This is for high-temperature work (above 1500°F) because it measures the temperature of bodies which are incandescent because of their temperature.

## Psychrometer

A psychrometer is an instrument which measures relative humidity based on the relation of the dry-bulb temperature and the wet-bulb temperature.

Relative humidity is of prime importance in HVAC and drying operations. Recording psychrometers are also available. Above 200°F humidity studies constitute a specialized field of endeavor.

## Portable Electronic Thermometer

The portable electronic thermometer is an adaptable temperature measurement tool. The battery-powered basic instrument, when housed in a carrying case, is suitable for laboratory or industrial use.

A pocket-size digital, battery-operated thermometer is especially convenient for spot checks or where a number of rapid readings of process temperatures need to be taken.

## Thermocouple Probe

No matter what sort of indicating instrument is employed, the thermocouple used should be carefully selected to match the application and properly positioned if a representative temperature is to be measured. The same care is needed for all sensing devices—thermocouple, bimetals, resistance elements, fluid expansion, and vapor pressure bulbs.

## Suction Pyrometer

Errors arise if a normal sheathed thermocouple is used to measure gas temperatures, especially high ones. The suction pyrometer overcomes these by shielding the thermocouple from wall radiation and drawing gases over it at high velocity to ensure good convective heat transfer. The thermocouple thus produces a reading which approaches the true temperature at the sampling point rather than a temperature between that of the walls and the gases.

## MEASURING COMBUSTION SYSTEMS

To maximize combustion efficiency, it is necessary to know the composition of the flue gas. By obtaining a good air-fuel ratio, substantial energy will be saved.

## Combustion Tester

Combustion testing consists of determining the concentrations of the products of combustion in a stack gas. The products of combustion usually considered are carbon dioxide and carbon monoxide. Oxygen is tested to assure proper excess air levels.

The definitive test for these constituents is an Orsat apparatus. This test consists of taking a measured volume of stack gas and measuring successive volumes after intimate contact with selective absorbing solutions. The reduction in volume after each absorption is the measure of each constituent.

The Orsat has a number of disadvantages. The main ones are that it requires considerable time to set up and use and that its operator must have a good degree of dexterity and be in constant practice.

Instead of an Orsat, there are portable and easy to use absorbing instruments which can easily determine the concentrations of the constituents of interest on an individual basis. Setup and operating times are minimal and just about anyone can learn to use them.

The typical range of concentrations are $CO_2$, 0–20%; $O_2$, 0–21%; and CO, 0–0.5%. The $CO_2$ or $O_2$ content, along with knowledge of flue gas temperature and fuel type, allows the flue gas loss to be determined off standard charts.

## Boiler Test Kit

The boiler test kit contains the following:

$CO_2$  Gas analyzer
$O_2$  Gas analyzer
     Inclined monometer
CO  Gas analyzer.

The purpose of the components of the kit is to help evaluate fireside boiler operation. Good combustion usually means high carbon dioxide ($CO_2$), low oxygen ($O_2$), and little or no trace of carbon monoxide (CO).

### Gas Analyzers

The gas analyzers are usually of the Fyrite type. The Fyrite type differs from the Orsat apparatus in that it is more limited in application and less accurate. The chief advantages of the Fyrite are that it is simple and easy to use and is inexpensive. This device is used many times in an energy audit. Three readings using the Fyrite analyzer should be made and the results averaged.

### Draft Gauge

The draft gauge is used to measure pressure. It can be the pocket type or the inclined monometer type.

### Smoke Tester

To measure combustion completeness the smoke detector is used. Smoke is unburned carbon, which wastes fuel, causes air pollution, and fouls heat-exchanger surfaces. To use the instrument, a measured volume of flue gas is drawn through filter paper with the probe. The smoke spot is compared visually with a standard scale and a measure of smoke density is determined.

### Combustion Analyzer

The combustion electronic analyzer permits fast, close adjustments. The unit contains digital displays. A standard sampler assembly with probe allows for stack measurements through a single stack or breaching hole.

## MEASURING HEATING, VENTILATION
## AND AIR-CONDITIONING (HVAC)
## SYSTEM PERFORMANCE

### Air Velocity Measurement

The following suggests the preference, suitability, and approximate costs of particular equipment:

- *Smoke pellets*—limited use but very low cost. Considered to be useful if engineering staff has experience in handling.
- *Anemometer* (deflecting vane)—good indication of air movement with acceptable order of accuracy. Considered useful (approximately $50).
- *Anemometer* (revolving vane)—good indicator of air movement with acceptable accuracy. However, easily subject to damage. Considered useful (approximately $100).
- *Pitot tube*—a standard air measurement device with good levels of accuracy. Considered essential. Can be purchased in various lengths—12" about $20, 48" about $35. Must be used with a monometer. These vary considerably in cost, but could be on the order of $20 to $60.
- *Impact tube*—usually packaged air flow meter kits, complete with various jets for testing ducts, grills, open areas, etc. These units are convenient to use and of sufficient accuracy. The costs vary around $150 to $300, and therefore this order of cost could only be justified for a large system.
- *Heated thermocouple*—these units are sensitive and accurate but costly. A typical cost would be about $500 and can only be justified for regular use in a large plant.
- *Hot wire anemometer*—not recommended. Too costly and too complex.

### Temperature Measurement

The temperature devices most commonly used are as follows:
- *Glass thermometers*—considered to be the most useful to temperature measuring instruments—accurate and convenient but fragile. Cost runs from $5 each for 12" long mercury in glass. Engineers should have a selection of various ranges.

- *Resistance thermometers*—considered to be very useful for A/C testing. Accuracy is good and they are reliable and convenient to use. Suitable units can be purchased from $150 up, some with a selection of several temperature ranges.
- *Thermocouples*—similar to resistance thermocouple but do not require battery power source. Chrome-Alum or iron types are the most useful and have satisfactory accuracy and repeatability. Costs start from $50 up.
- *Bimetallic thermometers*—considered unsuitable.
- *Pressure bulb thermometers*—more suitable for permanent installation. Accurate and reasonable in cost—$40 up.
- *Optical pyrometers*—only suitable for furnace settings and therefore limited in use. Cost from $300 up.
- *Radiation pyrometers*—limited in use for A/C work and costs from $500 up.
- *Indicating crayons*—limited in use and not considered suitable for A/C testing—costs around $2/crayon.
- *Thermographs*—use for recording room or space temperature; gives a chart indicating variations over a 12- or 168-hour period. Reasonably accurate. Low cost at around $30 to $60. (Spring-wound drive.)

## Pressure Measurement (Absolute and Differential)

Common devices used for measuring pressure in HVAC applications (accuracy, range, application, and limitations are discussed in relation to HVAC work) are as follows:

- *Absolute pressure manometer*—not really suited to HVAC test work.
- *Diaphragm*—not really suited to HVAC test work.
- *Barometer (Hg manometer)*—not really suited to HVAC test work.
- *Micromanometer*—not usually portable, but suitable for fixed measurement of pressure differentials across filter, coils, etc. Cost around $30 up.
- *Draft gauges*—can be portable and used for either direct pressure or pressure differential. From $30 up.
- *Manometers*—can be portable. Used for direct pressure reading and with pitot tubes for air flows. Very useful. Costs from $20 up.

- *Swing Vane gauges*—can be portable. Usually used for air flow. Costs about $30.
- *Bourdon tube gauges*—very useful for measuring all forms of system fluid pressures from 5 psi up. Costs vary greatly, from $10 up. Special types for refrigeration plants.

### Humidity Measurement

The data given below indicate the type of instruments available for humidity measurement. The following indicates equipment suitable for HVAC applications:

- *Psychrometers*—basically these are wet and dry bulb thermometers. They can be fixed on a portable stand or mounted in a frame with a handle for revolving in air. Costs are low ($10 to $30) and they are convenient to use.
- *Dewpoint hygrometers*—not considered suitable for HVAC test work.
- *Dimensional change*—device usually consists of a "hair," which changes in length proportionally with humidity changes. Not usually portable, fragile, and only suitable for limited temperature and humidity ranges.
- *Electrical conductivity*—can be compact and portable but of a higher cost (from $200 up). Very convenient to use.
- *Electrolytic*—as above, but for very low temperature ranges. Therefore unsuitable for HVAC test work.
- *Gravemeter*—not suitable.

### IDENTIFYING STEAM AND UTILITY COSTS

Steam and utility costs are significant for most plants. It is important to quantify usage, fuel costs as a function of production. Figure 3-5 illustrates a typical Steam and Utility Cost Report. This report enables the plant manager to evaluate the total Btu's of fuel consumed, the total fuel cost, and the total steam generation cost as a function of production. This report is issued monthly.

Since each plant has the same report, plant to plant comparisons are made and the effectiveness of the energy use is measured.

**STEAM AND UTILITY COST REPORT**

Preparatory Production [＿＿＿＿＿] Lbs C & F        Month [＿＿＿＿＿] 19 [＿]

Tire Production ＿＿＿＿＿＿＿＿＿＿        Operating Days [＿＿]

---

### A. STEAM                    Quantities

**B. OPERATING CONDITIONS AND DATA IN BOILER PLANT**

1. Actual Steam Generated [＿＿＿] 000 lbs
   a) Press and Temp ＿ psi ＿°F ＿ 000 lbs
   b) Press and Temp ＿ psi ＿°F ＿ 000 lbs
   c) Press and Temp ＿ psi ＿°F ＿ 000 lbs

Quantity of Feedwater Makeup [＿＿＿ 000] (Gal.)

Water Treating Chemicals
- a) Phosphate ＿＿＿＿＿ lbs
- b) Salt ＿＿＿＿＿ lbs

2. Factor of Evaporation (ASME Standard)
   a) ＿＿＿＿
   b) ＿＿＿＿
   c) ＿＿＿＿

- c) ＿＿＿＿ ＿＿＿＿ lbs
- d) ＿＿＿＿ ＿＿＿＿ lbs
- e) ＿＿＿＿ ＿＿＿＿ lbs
- f) ＿＿＿＿ ＿＿＿＿ lbs
- g) ＿＿＿＿ ＿＿＿＿ lbs

3. Equivalent Steam (No. 1 x No. 2) [＿＿＿] 000 lbs

4. Blowdown [＿＿＿] lbs
   Heat Loss in BTU from Blowdown ＿＿＿ BTU
   BTU/Lb-Gal-Cu Ft

Ave. Feedwater Temp. ＿＿＿ °F

Max Steam Demand ＿＿＿ lbs/hr

Equivalent Steam Consumption Per 100 lbs C&F ＿＿＿ lbs

|  | Total $ | Total |  |
|---|---|---|---|
| 5. Coal | [＿＿] | [＿＿] | [＿＿] tons |
| 6. Oil | [＿＿] | [＿＿] | [＿＿] gals |
| 7. Gas | [＿＿] | [＿＿] | [＿＿] MCF |
| 8. Misc | [＿＿] | [＿＿] | [＿＿] Units |

Units ＿＿＿＿      Fuel ＿＿＿＿

9. Total BTU of Fuel Consumed [＿＿＿] 000,000 BTU

10. Boiler Efficiency $\frac{(970.3 \times No.\ 3) + No.\ 4}{No.\ 9}$ = [＿＿] %

Unit Cost $/1000 Lbs Equiv. Steam

Cost

| | Cost | Unit Cost $/1000 Lbs Equiv. Steam |
|---|---|---|
| 11. Total Fuel Cost (No. 5 + No. 6 + No. 7 + No. 8) | $ [＿＿] | $ ＿＿＿ |
| 12. Operating Labor (include wage benefits) | $ [＿＿] | $ ＿＿＿ |
| 13. Operating Supplies (water, chemicals, etc.) | $ ＿＿＿ | |
| 14. Electricity for Power Plant Auxiliaries ＿＿ KWH | $ ＿＿＿ | |
| 15. Maintenance Charges | $ ＿＿＿ | |
| 16. Other Miscellaneous Charges | $ ＿＿＿ | |
| 17. Total Operating Cost (No. 11 thru No. 16) | $ [＿＿] | $ ＿＿＿ |
| 18. Total Steam Generation Cost (No. 17 + fixed Cost) | $ [＿＿] | $ ＿＿＿ |

**Figure 3-5. Steam and Utility Cost Report**

## CALCULATING THE ENERGY
## CONTENT OF THE PROCESS

Knowing the energy content of the plant's process is the first step in understanding how to reduce its cost. Using energy more efficiently reduces the product cost, thus increasing profits. In order to account for the process energy content, all energy that enters and leaves a plant during a given period must be measured. Figure 3-6 illustrates Energy Content of a Process Report. The report applies to any manufacturing operation, whether it is a pulp mill, steel mill, or assembly line. This report enables one to quickly identify energy inefficient operations.

Attention can then be focused on which equipment should be replaced and what maintenance programs should be initiated. This report also focuses attention on the choice of raw materials. By using Btu's per unit of production, measurable goals can be set. This report will also identify opportunities where energy usage can be reduced. The energy content of raw materials can be estimated by using the heating values indicated in Table 3-13.

**Example Problem 3-1**

Comment on energy content by modifying process No. 1 as follows:

*Monthly Usage Rate*

|             | *Process No. 1*          | *Modified Process No. 1*  |
|-------------|--------------------------|---------------------------|
| Ethane      | 30,000 lb                | 50,000 lb                 |
| Steam       | $250 \times 10^6$ lb     | $200 \times 10^6$ lb      |
| Electricity | $0.5 \times 10^6$ kWh    | $0.8 \times 10^6$ kWh     |
| Natural gas | $350 \times 10^6$ ft$^3$ | $300 \times 10^6$ ft$^3$  |

Assuming a Btu content of steam of 1077 Btu/lb, compute the net energy content per process.

```
              ENERGY CONTENT OF PROCESS REPORT

Process_____ [1] Total Units Produced_____

      ENERGY OF RAW MATERIALS AND MISCELLANEOUS REQUIREMENTS

                        Total         Btu's          Total
                        Usage         Per Unit       Btu's
a. _____      ____    x   _____    =   _____
b. _____      ____        _____        _____
c. _____      ____        _____        _____
d. _____      ____        _____        _____

                    [2]   Total Raw Material _____Btu's

                    ENERGY OF MAJOR UTILITIES
         (ELECTRICAL, STEAM, COOLING WATER, NATURAL GAS, ETC.)

                        Total         Btu's          Total
                        Usage         Per Unit       Btu's
a. _____      ____    x   _____    =   _____
b. _____      ____        _____        _____
c. _____      ____        _____        _____

                    [3]   Total Energy Re-
                          quired from Utilities_____Btu's

           ENERGY CREDIT AS A RESULT OF BY-PRODUCTS

                        Total         Btu's          Total
                        Output        Per Unit       Btu's
a. _____      ____    x   _____    =   _____
b. _____      ____        _____        _____
c. _____      ____        _____        _____

                    [4]  Energy Credit         _____Btu's
[5]  NET ENERGY CONTENT OF PROCESS  [2] + [3] - [4]  _____Btu's
[6]  ENERGY CONTENT PER UNIT OF
                      PRODUCTION    [5] ÷ [1]       _____Btu's
                                                           Per
                                                           Unit
```

**Figure 3-6. Energy Content of Process Report**

**Table 3-13. Heat of Combustion for Raw Materials**

|  | Formula | Gross Heat of Combustion Btu/lb |
|---|---|---|
| Raw Material |  |  |
| Carbon | C | 14,093 |
| Hydrogen | $H_2$ | 61,095 |
| Carbon monoxide | CO | 4,347 |
| Paraffin Series |  |  |
| Methane | $CH_4$ | 23,875 |
| Ethane | $C_2H_4$ | 22,323 |
| Propane | $C_3H_8$ | 21,669 |
| n-Butane | $C_4H_{10}$ | 21,321 |
| Isobutane | $C_4H_{10}$ | 21,271 |
| n-Pentane | $C_5H_{12}$ | 21,095 |
| Isopentane | $C_5H_{12}$ | 21,047 |
| Neopentane | $C_5H_{12}$ | 20,978 |
| n-Hexane | $C_6H_{14}$ | 20,966 |
| Olefin Series |  |  |
| Ethylene | $C_2H_4$ | 21,636 |
| Propylene | $C_3H_6$ | 21,048 |
| n-Butene | $C_4H_8$ | 20,854 |
| Isobutene | $C_4H_8$ | 20,737 |
| n-Pentene | $C_5H_{10}$ | 20,720 |
| Aromatic Series |  |  |
| Benzene | $C_6H_6$ | 18,184 |
| Toluene | $C_7H_8$ | 18,501 |
| Xylene | $C_8H_{10}$ | 18,651 |
| Miscellaneous Gases |  |  |
| Acetylene | $C_2H_2$ | 21,502 |
| Naphthalene | $C_{10}H_8$ | 17,303 |
| Methyl alcohol | $CH_3OH$ | 10,258 |
| Ethyl alcohol | $C_2H_5OH$ | 13,161 |
| Ammonia | $NH_3$ | 9,667 |

*Source: NBS Handbook 115.*

**Answer**

*Process No. 1*
*Energy of Raw Materials*

|  | Total Usage | Btu's Per Unit | Total Btu's |
|---|---|---|---|
| Ethane | 30,000 lb | From Table 3-13 22,323 Btu/lb | $0.6 \times 10^9$ |

*Energy of Major Utilities*

|  |  |  |  |
|---|---|---|---|
| Steam | $250 \times 10^6$ lb | 1077 Btu/lb | $269.2 \times 10^9$ |
| Electricity | $0.5 \times 10^6$ kWh | From Table 1-2 10,000 Btu/kWh | $5 \times 10^9$ |
| Natural gas | $350 \times 10^6$ ft$^3$ | From Table 1-1 1000 Btu/ft$^3$ | $350 \times 10^9$ |
|  | Net energy content for Process No. 1 |  | $624 \times 10^9$ |

*Modified Process No. 1*
*Energy of Raw Materials*

| Ethane | 50,000 lb | From Table 3-13 22,323 Btu/lb | $1.1 \times 10^9$ |
|---|---|---|---|

*Energy of Major Utilities*

|  |  |  |  |
|---|---|---|---|
| Steam | $200 \times 10^6$ lb | 1077 Btu/lb | $215.4 \times 10^9$ |
| Electricity | $0.8 \times 10^6$ lb | From Table 1-2 10,000 Btu/kWh | $8 \times 10^9$ |
| Natural gas | $300 \times 10^6$ ft$^3$ | From Table 1-1 1000 Btu/ft$^3$ | $300 \times 10^9$ |
|  | Net energy content for Process No. 1 |  | $524 \times 10^9$ |

Modified Process No. 1 saves $100 \times 10^9$ Btu's per month.

# 4

# Electrical System Optimization

This chapter is divided into three main sections: *Power Distribution, Lighting Efficiency, and Energy Management Systems.*

By understanding the basic concepts of electrical systems it is possible to reduce energy consumption 25% or more. The first step is to analyze the billing structure. It may be possible to negotiate a better tariff rate with the local utility or modify the facility operation to qualify for a lower rate. In addition, specified charges or discounts for power factor, time of day or demand will determine if certain electrical efficiency measures are economically justified. This chapter reviews the basic parameters required to make sound energy engineering decisions.

## POWER DISTRIBUTION

### ELECTRICAL RATE TARIFF

The basic electrical rate charges contain the following elements:

*Billing Demand*—The maximum kilowatt requirement over a 15-, 30-, or 60-minute interval.

*Load Factor*—The ratio of the average load over a designated period to the peak demand load occurring in that period.

*Power Factor*—The ratio of resistive power to apparent power. Traditionally electrical rate tariffs have a decreasing kilowatt hour (kWh) charge with usage. This practice is likely to gradually phase out. New tariffs are containing the following elements:

*Time of Day*—Discounts are allowed for electrical usage during off-peak hours.

*Ratchet Rate*—The billing demand is based on 80-90% of peak demand for any one month. The billing demand will remain at that ratchet for 12 months even though the actual demand for the succeeding months may be less.

## THE POWER TRIANGLE

The total power requirement of a load is made up of two components, namely the resistive part and the reactive part. The resistive portion of a load cannot be added directly to the reactive component since it is essentially 90 degrees out of phase with the other. The pure resistive power is known as the watt, while the reactive power is referred to as the reactive volt amperes. To compute the total volt ampere load, it is necessary to analyze the power triangle indicated below.

$$KVA = \frac{KW}{Cos\Theta}$$

$$KVAR = KVA \; Sin\Theta$$
(REACTIVE)

$\Theta$ = POWER FACTOR ANGLE

$$KW = KVA \times Cos\Theta$$
(RESISTIVE)

$$
\begin{aligned}
K &= 1000 \\
W &= \text{Watts} \\
VA &= \text{Volt Amperes} \\
VAR &= \text{Volt Amperes Reactive} \\
\Theta &= \text{Angle Between KVA and KW} \\
Cos\Theta &= \text{Power Factor} \\
Tan \; \Theta &= \frac{KVAR}{KW}
\end{aligned}
$$

The windings of transformers and motors are usually connected in a wye or delta configuration. The relationships for line and phase voltages and currents are illustrated by Figure 4-1.

WYE CONNECTIONS

$$V_L = \sqrt{3} \;\; V_p$$
$$I_L = I_p$$

$V_L$ = VOLTAGE BETWEEN HOT LEGS

$V_p$ = VOLTAGE FROM LINE TO GROUND

DELTA CONNECTIONS

$$V_L = V_p$$
$$I_L = \sqrt{3} \cdot I_p$$

Figure 4-1. Three-Phase Windings

For a balanced 3-phase load

$$\text{Power} = \underbrace{\sqrt{3} \; V_L \; I_L}_{} \; \text{Cos}\Theta \qquad \qquad \textit{Formula (4-1)}$$

$$\text{Watts} = \quad \begin{matrix} \text{Volt} \\ \text{Amperes} \end{matrix} \quad \begin{matrix} \text{Power} \\ \text{Factor} \end{matrix}$$

For a balanced 1-phase load

$$P = V_L \; I_L \; \text{Cos}\Theta \qquad \qquad \textit{Formula (4-2)}$$

The primary windings of 13.8 Kv — 480-volt unit substations are usually delta-connected with the secondary wye-connected.

## MOTOR HORSEPOWER

The standard power rating of a motor is referred to as a horsepower. In order to relate the motor horsepower to a kilowatt (kW), multiply the horsepower by .746 (Conversion Factor) and divide by the motor efficiency.

$$\text{KVA} = \frac{\text{HP} \times .746}{\eta \times \text{P.F.}} \qquad \qquad \textit{Formula (4-3)}$$

$$\begin{aligned}
\text{HP} &= \text{Motor Horsepower} \\
\eta &= \text{Efficiency of Motor} \\
\text{P.F.} &= \text{Power Factor of Motor}
\end{aligned}$$

Motor efficiencies and power factors vary with load. Typical values are shown in Table 4-1. Values are based on totally enclosed fan-cooled motors (TEFC) running at 1800 rpm "T" frame.

## POWER FLOW CONCEPT

Power flowing is analogous to water flowing in a pipe. To supply several small water users, a large pipe services the plant at a high pressure. Several branches from the main pipe service various loads. Pressure reducing stations lower the main pressure to meet the requirements of each user. Similarly, a large feeder at a high voltage

Table 4-1.

| HP RANGE | 3-30 | 40-100 |
|---|---|---|
| η% at | | |
| ½ Load | 83.3 | 89.2 |
| ¾ Load | 85.8 | 90.7 |
| Full Load | 86.2 | 90.9 |
| P.F. at | | |
| ½ Load | 70.1 | 79.2 |
| ¾ Load | 79.2 | 85.4 |
| Full Load | 83.5 | 87.4 |

services a plant. Through switchgear breakers, the main feeder is distributed into smaller feeders. The switchgear breakers serve as a protector for each of the smaller feeders. Transformers are used to lower the voltage to the nominal value needed by the user.

## ELECTRICAL EQUIPMENT

Electrical equipment commonly specified is as follows:

- *Switchgear-Breakers*—used to distribute power.
- *Unit Substation*—used to step down voltage. Consists of a high voltage disconnect switch, transformer and low-voltage breakers. Typical 480-volt transformer sizes are 300 KVA, 500 KVA, 750 KVA, 1000 KVA, 2500 KVA and 3000 KVA.
- *Motor Control Center (M.C.C.)*—a structure which houses starters and circuit breakers or fuses for motor control. It consists of the following:
  - (1) Thermal overload relays which guard against motor overloads,
  - (2) Fuse disconnect switches or breakers which protect the cable and motor and can be used as a disconnecting means,
  - (3) Contactors (relays) whose contacts are capable of opening and closing the power source to the motor.

## MOTORS

•   *Squirrel Cage Induction Motors* are commonly used. These motors require three power loads. For two-speed applications several different types of motors are available. Depending on the process requirements such as constant horsepower or constant torque, the windings of the motor are connected differently. The theory of two-speed operation is based on Formula 4-4.

$$\text{Frequency} = \frac{\text{No. of poles} \times \text{speed}}{120} \qquad \textit{Formula (4-4)}$$

Thus, if the frequency is fixed, the effective number of motor poles should be changed to change the speed. This can be accomplished by the manner in which the windings are connected. Two-speed motors require six power leads.

•   *D. C. Motors* are used where speed control is essential. The speed of a D.C. Motor is changed by varying the field voltage through a rheostat. A  D.C. Motor requires two power wires to the armature and two smaller cables for the field.

•   *Synchronous Motors* are used when constant speed operation is essential. Synchronous motors are sometimes cheaper in the large horsepower categories when slow speed operation is required. Synchronous motors also are considered for power factor correction. A .8 P.F. synchronous motor will supply corrective KVARs to the system. A synchronous motor requires A.C. for power and D.C. for the field. Since many synchronous motors are self-excited, only the power cables are required to the motor.

## IMPORTANCE OF POWER FACTOR

Transformer size is based on KVA. The closer $\Theta$ equals $0°$ or power factor approaches unity, the smaller the KVA. Many times utility companies have a power factor clause in their contract with the customer. The statement usually causes the customer to pay an additional power rate if the power factor of the plant deviates substantially from unity. The utility company wishes to maximize the efficiency of its transformers and associated equipment.

## POWER FACTOR CORRECTION

One problem facing the energy engineer is to estimate the power factor of a new plant and to install equipment such as capacitor banks or synchronous motors so that the overall power factor will meet the utility company's objectives.

Capacitor banks lower the total reactive KVAR by the value of the capacitors installed.

A second problem is to retrofit an existing plant so that the overall power factor desired is obtained.

### Example Problem 4-1

A total motor horsepower load of 854 is made up of motors ranging from 40–100 horsepower. Calculate the connected KVA.

It is desired to operate the plant at a power factor of .95. What approximate capacitor bank is required?

### Answer

$$KVA = \frac{HP \times .746}{Motor\ Eff. \times Motor\ P.F.}$$

From Table 4-1 at full load

$$Eff. = .909 \text{ and } P.F. = .87$$

$$KVA = \frac{854 \times .746}{.909 \times .87} = 806$$

The plant is operating at a power factor of .87. The power factor of .87 corresponds to an angle of 29°.

KVAR = 806 Sin 29°
= 806 x .48 = 386

KVAR = 386

806 KVA

Θ = 29°

KW = 806 CoSΘ = 806 x .87 = 701

A power factor of .95 is required.

$$CoS\Theta = .95$$
$$\Theta = 18°$$

The KVAR of 386 needs to be reduced by adding capacitors.

Remember kW does not change with different power factors, but KVA does.

Thus, the desired power triangle would look as follows:

$$CoS\Theta = .95$$
$$\Theta = 18° \quad Sin\ 18° = .31$$
$$KVA_c = \frac{701}{CoS\Theta} = \frac{701}{.95} = 738$$

Note: Power factor correction reduces total KVA.

$$KVAR_c = 738\ Sin\ 18° = 738 \times .31 = 228$$
$$Capacitance\ Bank = 386-228$$
$$158\ KVAC$$

**Example Problem 4-2**

The client wishes to know the expected power factor for a new plant. The lighting load is 40 kW. The plant is composed of two identical modules (2 motors for each equipment number listed). Remember that KVAs at different power factors cannot be added directly.

## Motor List — Module 1

| Motor No. | Description | HP | Voltage | Phase |
|-----------|-------------|-----|---------|-------|
| AG-1 | Agitator Motor | 60 | 460 | 3 |
| CF-3 | Centrifuge Motor | 100 | 460 | 3 |
| FP-4 | Feed Pump Motor | 30 | 460 | 3 |
| TP-5 | Transfer Pump Motor | 10 | 460 | 3 |
| CTP-6 | Cooling Tower | 25 | 460 | 3 |
| CT-9 | Cooling Tower Motor | 20 | 460 | 3 |
| HF-10 | H&V Supply Fan Motor | 40 | 460 | 3 |
| HF-11 | H&V Exhaust Fan Motor | 20 | 460 | 3 |
| BC-13 | Brine Compressor Motor | 50 | 460 | 3 |
| C-16 | Conveyor Motor | 20 | 460 | 3 |
| H-17 | Hoist Motor | 5 | 460 | 3 |

## Answer

Based on the motor list, the plant power factor is estimated as follows:

### Module 1

Lighting $kW_3$ = 40 Total

| Motors 3-30 | Motors 40-100 |
|:-----------:|:-------------:|
| 30 | 60 |
| 10 | 100 |
| 25 | 40 |
| 20 | 50 |
| 20 | 250 |
| 20 | |
| 5 | |
| 130 | |

At Full Load:

| | |
|---|---|
| P.F. = 83.5 | P.F. = 87.4 |
| $\eta$ = 86.2 | $\eta$ = 90.9 |

$$KVA_1 = \frac{130 \times .746}{.83 \times .86} = 135 \qquad KVA_2 = \frac{250 \times .746}{.90 \times .87} = 238$$

$$kW_1 = KVA \, Cos\Theta$$
$$= KVA \, .83$$
$$= 112$$

$$kW_2 = KVA_2 \, Cos\Theta$$
$$= KVA \, .87$$
$$= 207$$

$$KVAR_1 = KVA_1 \, Sin\Theta$$

$$\Theta = 33°$$

$$KVAR_2 = KVA_2 \, Sin\Theta$$

$$\Theta = 29°$$

$$KVAR_1 = KVA_1 \times .54$$
$$= 135 \times .54$$
$$= 73$$

$$KVAR_2 = KVA_2 \times .48$$
$$KVAR_2 = 115$$

$$kW_{total} = \underbrace{kW_1 + kW_2}_{\text{Module 1}} + \underbrace{kW_1 + kW_2}_{\text{Module 2}} + kW_3 = 678 \, KW$$

$$KVAR_{total} = \underbrace{KVAR_1 + KVAR_2}_{\text{Module 1}} + \underbrace{KVAR_1 + KVAR_2}_{\text{Module 2}}$$
$$= 73 + 115 + 73 + 115 = 376$$

$$KVA_{total} = \sqrt{(678)^2 + (376)^2} = 775 \, KVA$$

$$Cos\Theta = \frac{kW_{total}}{KVA_{total}} = \frac{678}{775} = .87$$

**Shortcut Methods**

A handy shortcut table which can be used to find the value of the capacitor required to improve the plant power factor is illustrated by Table 4-2.

### Table 4-2. Shortcut Method — Power Factor Correction.

**KW MULTIPLIERS FOR DETERMINING CAPACITOR KILOVARS**

**DESIRED POWER-FACTOR IN PERCENTAGE**

| ORIG. PF | 80 | 81 | 82 | 83 | 84 | 85 | 86 | 87 | 88 | 89 | 90 | 91 | 92 | 93 | 94 | 95 | 96 | 97 | 98 | 99 | 100 |
|---|---|---|---|---|---|---|---|---|---|---|---|---|---|---|---|---|---|---|---|---|---|
| 50 | .982 | 1.008 | 1.034 | 1.060 | 1.086 | 1.112 | 1.139 | 1.165 | 1.192 | 1.220 | 1.248 | 1.276 | 1.303 | 1.337 | 1.369 | 1.403 | 1.441 | 1.481 | 1.529 | 1.590 | 1.732 |
| 51 | .936 | .962 | .988 | 1.014 | 1.040 | 1.066 | 1.093 | 1.119 | 1.146 | 1.174 | 1.202 | 1.230 | 1.257 | 1.291 | 1.323 | 1.357 | 1.395 | 1.435 | 1.483 | 1.544 | 1.688 |
| 52 | .894 | .920 | .946 | .972 | .998 | 1.024 | 1.051 | 1.077 | 1.104 | 1.132 | 1.160 | 1.188 | 1.215 | 1.249 | 1.281 | 1.315 | 1.353 | 1.393 | 1.441 | 1.502 | 1.644 |
| 53 | .850 | .876 | .902 | .928 | .954 | .980 | 1.007 | 1.033 | 1.060 | 1.088 | 1.116 | 1.144 | 1.171 | 1.205 | 1.237 | 1.271 | 1.309 | 1.349 | 1.397 | 1.458 | 1.600 |
| 54 | .809 | .835 | .861 | .887 | .913 | .939 | .966 | .992 | 1.019 | 1.047 | 1.075 | 1.103 | 1.130 | 1.164 | 1.196 | 1.230 | 1.268 | 1.308 | 1.356 | 1.417 | 1.559 |
| 55 | .769 | .795 | .821 | .847 | .873 | .899 | .926 | .952 | .979 | 1.007 | 1.035 | 1.063 | 1.090 | 1.124 | 1.156 | 1.190 | 1.228 | 1.268 | 1.316 | 1.377 | 1.519 |
| 56 | .730 | .756 | .782 | .808 | .834 | .860 | .887 | .913 | .940 | .968 | .996 | 1.024 | 1.051 | 1.085 | 1.117 | 1.151 | 1.189 | 1.229 | 1.277 | 1.338 | 1.480 |
| 57 | .692 | .718 | .744 | .770 | .796 | .822 | .849 | .875 | .902 | .930 | .958 | .986 | 1.013 | 1.047 | 1.079 | 1.113 | 1.151 | 1.191 | 1.239 | 1.300 | 1.442 |
| 58 | .655 | .681 | .707 | .733 | .759 | .785 | .812 | .838 | .865 | .893 | .921 | .949 | .976 | 1.010 | 1.042 | 1.076 | 1.114 | 1.154 | 1.202 | 1.263 | 1.405 |
| 59 | .618 | .644 | .670 | .696 | .722 | .748 | .775 | .801 | .828 | .856 | .884 | .912 | .939 | .973 | 1.005 | 1.039 | 1.077 | 1.117 | 1.165 | 1.226 | 1.368 |
| 60 | .584 | .610 | .636 | .662 | .688 | .714 | .741 | .767 | .794 | .822 | .849 | .878 | .905 | .939 | .971 | 1.005 | 1.043 | 1.083 | 1.131 | 1.192 | 1.334 |
| 61 | .549 | .575 | .601 | .627 | .653 | .679 | .706 | .732 | .759 | .787 | .815 | .843 | .870 | .904 | .936 | .970 | 1.008 | 1.048 | 1.096 | 1.157 | 1.299 |
| 62 | .515 | .541 | .567 | .593 | .619 | .645 | .672 | .698 | .725 | .753 | .781 | .809 | .836 | .870 | .902 | .936 | .974 | 1.014 | 1.062 | 1.123 | 1.265 |
| 63 | .483 | .509 | .535 | .561 | .587 | .613 | .640 | .666 | .693 | .721 | .749 | .777 | .804 | .838 | .870 | .904 | .942 | .982 | 1.030 | 1.091 | 1.233 |
| 64 | .450 | .476 | .502 | .528 | .554 | .580 | .607 | .633 | .660 | .688 | .716 | .744 | .771 | .805 | .837 | .871 | .909 | .949 | .997 | 1.058 | 1.200 |
| 65 | .419 | .445 | .471 | .497 | .523 | .549 | .576 | .602 | .629 | .657 | .685 | .713 | .740 | .774 | .806 | .840 | .878 | .918 | .966 | 1.027 | 1.169 |
| 66 | .388 | .414 | .440 | .466 | .492 | .518 | .545 | .571 | .598 | .626 | .654 | .682 | .709 | .743 | .775 | .809 | .847 | .887 | .935 | .996 | 1.138 |
| 67 | .358 | .384 | .410 | .436 | .462 | .488 | .515 | .541 | .568 | .596 | .624 | .652 | .679 | .713 | .745 | .779 | .817 | .857 | .905 | .966 | 1.108 |
| 68 | .329 | .355 | .381 | .407 | .433 | .459 | .486 | .512 | .539 | .567 | .595 | .623 | .650 | .684 | .716 | .750 | .788 | .828 | .876 | .937 | 1.079 |
| 69 | .299 | .325 | .351 | .377 | .403 | .429 | .456 | .482 | .509 | .537 | .565 | .593 | .620 | .654 | .686 | .720 | .758 | .798 | .840 | .907 | 1.049 |
| 70 | .270 | .296 | .322 | .348 | .374 | .400 | .427 | .453 | .480 | .508 | .536 | .564 | .591 | .625 | .657 | .691 | .729 | .769 | .811 | .878 | 1.020 |
| 71 | .242 | .268 | .294 | .320 | .346 | .372 | .399 | .425 | .452 | .480 | .508 | .536 | .563 | .597 | .629 | .663 | .701 | .741 | .783 | .850 | .992 |
| 72 | .213 | .239 | .265 | .291 | .317 | .343 | .370 | .396 | .423 | .451 | .479 | .507 | .534 | .568 | .600 | .634 | .672 | .712 | .754 | .821 | .963 |
| 73 | .186 | .212 | .238 | .264 | .290 | .316 | .343 | .369 | .396 | .424 | .452 | .480 | .507 | .541 | .573 | .607 | .645 | .685 | .727 | .794 | .936 |
| 74 | .159 | .185 | .211 | .237 | .263 | .289 | .316 | .342 | .369 | .397 | .425 | .453 | .480 | .514 | .546 | .580 | .618 | .658 | .700 | .767 | .909 |
| 75 | .132 | .158 | .184 | .210 | .236 | .262 | .289 | .315 | .342 | .370 | .398 | .426 | .453 | .487 | .519 | .553 | .591 | .631 | .673 | .740 | .882 |
| 76 | .105 | .131 | .157 | .183 | .209 | .235 | .262 | .288 | .315 | .343 | .371 | .399 | .426 | .460 | .492 | .526 | .564 | .604 | .652 | .713 | .855 |
| 77 | .079 | .105 | .131 | .157 | .183 | .209 | .236 | .262 | .289 | .317 | .345 | .373 | .400 | .434 | .466 | .500 | .538 | .578 | .620 | .687 | .829 |
| 78 | .053 | .079 | .105 | .131 | .157 | .183 | .210 | .236 | .263 | .291 | .319 | .347 | .374 | .408 | .440 | .474 | .512 | .552 | .594 | .661 | .803 |
| 79 | .026 | .052 | .078 | .104 | .130 | .156 | .183 | .209 | .236 | .264 | .292 | .320 | .347 | .381 | .413 | .447 | .485 | .525 | .567 | .634 | .776 |
| 80 | .000 | .026 | .052 | .078 | .104 | .130 | .157 | .183 | .210 | .238 | .266 | .294 | .321 | .355 | .387 | .421 | .450 | .499 | .541 | .608 | .750 |
| 81 | — | .000 | .026 | .052 | .078 | .104 | .131 | .157 | .184 | .212 | .240 | .268 | .295 | .329 | .361 | .395 | .433 | .473 | .515 | .582 | .724 |
| 82 | — | — | .000 | .026 | .052 | .078 | .105 | .131 | .158 | .186 | .214 | .242 | .269 | .303 | .335 | .369 | .407 | .447 | .489 | .556 | .698 |
| 83 | — | — | — | .000 | .026 | .052 | .079 | .105 | .132 | .160 | .188 | .216 | .243 | .277 | .309 | .343 | .381 | .421 | .463 | .530 | .672 |
| 84 | — | — | — | — | .000 | .026 | .053 | .079 | .106 | .134 | .162 | .190 | .217 | .251 | .283 | .317 | .355 | .395 | .437 | .504 | .645 |
| 85 | — | — | — | — | — | .000 | .027 | .053 | .080 | .108 | .136 | .164 | .191 | .225 | .257 | .291 | .329 | .369 | .417 | .478 | .620 |

*(Row labels at left = ORIGINAL POWER FACTOR IN PERCENTAGE)*

Example: Total kw input of load from wattmeter reading 100 kw at a power factor of 60%. The leading reactive kvar necessary to raise the power factor to 90% is found by multiplying the 100 kw by the factor found in the table, which is .849. Then 100 kw × 0.849 = 84.9 kvar. Use 85 kvar.

Reprinted by permission of Federal Pacific Electric Company.

## WHERE TO LOCATE CAPACITORS

As indicated, the primary purpose of capacitors is to reduce the power consumption. Additional benefits are derived by capacitor location. Figure 4-2 indicates typical capacitor locations. Maximum benefit of capacitors is derived by locating them as close as possible to the load. At this location, its kilovars are confined to the smallest possible segment, decreasing the load current. This, in turn, will reduce power losses of the system substantially. Power losses are proportional to the square of the current. When power losses are reduced, voltage at the motor increases; thus, motor performance also increases.

**Figure 4-2. Power Distribution Diagram Illustrating Capacitor Locations**

Locations $C1A$, $C1B$ and $C1C$ of Figure 4-2 indicate three different arrangements at the load. Note that in all three locations extra switches are not required, since the capacitor is either switched with the motor starter or the breaker before the starter. Case $C1A$ is recommended for new installation, since the maximum benefit is derived and the size of the motor thermal protector is reduced. In Case $C1B$, as in Case $C1A$, the capacitor is energized only when the motor is in operation. Case $C1B$ is recommended in cases where the installation is existing and the thermal protector does not need to be re-sized. In position $C1C$, the capacitor is permanently connected to the circuit but does not require a separate switch, since it can be disconnected by the breaker before the starter.

It should be noted that the rating of the capacitor should *not* be greater than the no-load magnetizing KVAR of the motor. If this condition exists, damaging overvoltage or transient torques can occur. This is why most motor manufacturers specify maximum capacitor ratings to be applied to specific motors.

The next preference for capacitor locations as illustrated by Figure 4-2 is at locations *C2* and *C3*. In these locations, a breaker or switch will be required. Location *C4* requires a high voltage breaker. The advantage of locating capacitors at power centers or feeders is that they can be grouped together. When several motors are running intermittently, the capacitors are permitted to be on-line all the time, reducing the total power regardless of load.

## ENERGY EFFICIENT MOTORS

Energy efficient motors are now available. These motors are approximately 30% more expensive than their standard counterpart. Based on the energy cost, it can be determined if the added investment is justified. With the emphasis on energy conservation, new lines of energy efficient motors are being introduced. Figures 4-3 and 4-4 illustrate a typical comparison between energy efficient and standard motors.

**Figure 4-3. Efficiency vs Horsepower Rating (Dripproof Motors)**

Figure 4-4. Power Factor vs Horsepower Rating (Dripproof Motors)

## LIGHTING EFFICIENCY

### LIGHTING BASICS

About 20 percent of all electricity generated in the United States today is used for lighting.

By understanding the basics of lighting design, several ways to improve the efficiency of lighting systems will become apparent.

There are two common lighting methods used. One is called the "Lumen" method, while the other is the "Point by Point" method. The Lumen method assumes an equal footcandle level throughout the area. This method is used frequently by lighting designers since it is simplest; however, it wastes energy, since it is the light "at the task" which must be maintained and not the light in the surrounding areas. The "Point by Point" method calculates the lighting requirements for the task in question.

The "Point by Point" method makes use of the inverse-square law, which states that the illuminance at a point on a surface perpendicular to the light ray is equal to the luminous intensity of the source at that point divided by the square of the distance between the source and the point of calculation, as illustrated in Formula 4-5.

$$E = \frac{I}{D^2} \qquad \qquad Formula\ (4\text{-}5)$$

Where

- $E$ = Illuminance in footcandles
- $I$ = Luminous intensity in candles
- $D$ = Distance in feet between the source and the point of calculation.

If the surface is not perpendicular to the light ray, the appropriate trigonometrical functions must be applied to account for the deviation.

## Lumen Method

A footcandle is the illuminance on a surface of one square foot in area having a uniformly distributed flux of one lumen. From this definition, the "Lumen Method" is developed and illustrated by Formula 4-6.

$$N = \frac{F_1 \times A}{Lu \times L_1 \times L_2 \times Cu} \qquad Formula\ (4\text{-}6)$$

Where

- $N$   is the number of lamps required.
- $F_1$   is the required footcandle level at the task. A footcandle is a measure of illumination; one standard candle power measured one foot away.
- $A$   is the area of the room in square feet.
- $Lu$   is the Lumen output per lamp. A Lumen is a measure of lamp intensity: its value is found in the manufacturer's catalogue.
- $Cu$   is the coefficient of utilization. It represents the ratio of the Lumens reaching the working plane to the total Lumens generated by the lamp. The coefficient of utilization makes allowances for light absorbed or reflected by walls, ceilings, and the fixture itself. Its values are found in the manufacturer's catalogue.
- $L_1$   is the lamp depreciation factor. It takes into account that the lamp Lumen depreciates with time. Its value is found in the manufacturer's catalogue.

$L_2$     is the luminaire (fixture) dirt depreciation factor. It takes into account the effect of dirt on a luminaire and varies with type of luminaire and the atmosphere in which it is operated.

The Lumen method formula illustrates several ways lighting efficiency can be improved.

Faced with the desire to reduce their energy use,[1] lighting consumers have four options: i) reduce light levels, ii) purchase more efficient equipment, iii) provide light when needed at the task at the required level, and iv) add control and reduce lighting loads automatically. The multitude of equipment options to meet one or more of the above needs permits the consumer and the lighting designer-engineer to consider the trade-offs between the initial and operating costs based upon product performance (life, efficacy, color, glare, and color rendering).

Some definitions and terms used in the field of lighting will be presented to help consumers evaluate and select lighting products best suiting their needs. Then, some state-of-the-art advances will be characterized so that their benefits and limitations are explicit.

## Lighting Terminology

*Efficacy* — Is the amount of visible light (lumens) produced for the amount of power (watts) expended. It is a measure of the efficiency of a process but is a term used in place of efficiency when the input (W) has different units than the output (lm) and expressed in lm/W.

*Color Temperature* — A measure of the color of a light source relative to a black body at a particular temperature expressed in degrees Kelvin (°K). Incandescents have a low color temperature (~2800°K) and have a red-yellowish tone; daylight has a high color temperature (~6000°K) and appears bluish. Today, the phosphors used in fluorescent lamps can be blended to provide any desired color temperature in the range from 2800°K to 6000°K.

---

[1] Source: *Lighting Systems Research*, R.R. Verderber

*Color Rendering* — A parameter that describes how a light source renders a set of colored surfaces with respect to a black body light source at the same color temperature. The color rendering index (CRI) runs from 0 to 100. It depends upon the specific wavelengths of which the light is composed. A black body has a continuous spectrum and contains all of the colors in the visible spectrum. Fluorescent lamps and high intensity discharge lamps (HID) have a spectrum rich in certain colors and devoid in others. For example, a light source that is rich in blues and low in reds could appear white, but when reflected from a substance, it would make red materials appear faded. The same material would appear different when viewed with an incandescent lamp, which has a spectrum that is rich in red.

## LIGHT SOURCES[2]

Figure 4-5 indicates the general lamp efficiency ranges for the generic families of lamps most commonly used for both general and supplementary lighting systems. Each of these sources is discussed briefly here. It is important to realize that in the case of fluorescent and high intensity discharge lamps, the figures quoted for "lamp efficacy" are for the lamp only and do not include the associated ballast losses. To obtain the total system efficiency, ballast input watts must be used rather than lamp watts to obtain an overall system lumen per watt figure. This will be discussed in more detail in a later section.

**Incandescent lamps** have the lowest lamp efficacies of the commonly used lamps. This would lead to the accepted conclusion that incandescent lamps should generally not be used for large-area, general lighting systems where a more efficient source could serve satisfactorily. However, this does not mean that incandescent lamps should never be used. There are many applications where the size, convenience, easy control, color rendering, and relatively low cost of incandescent lamps are suitable for a specific application.

General service incandescent lamps do not have good lumen maintenance throughout their lifetime. This is the result of the tungsten's evaporation off the filament during heating as it deposits on the bulb wall, thus darkening the bulb and reducing the lamp lumen output.

---

[2] **Source:** *Selection Criteria for Lighting Energy Management*, Roger L. Knott

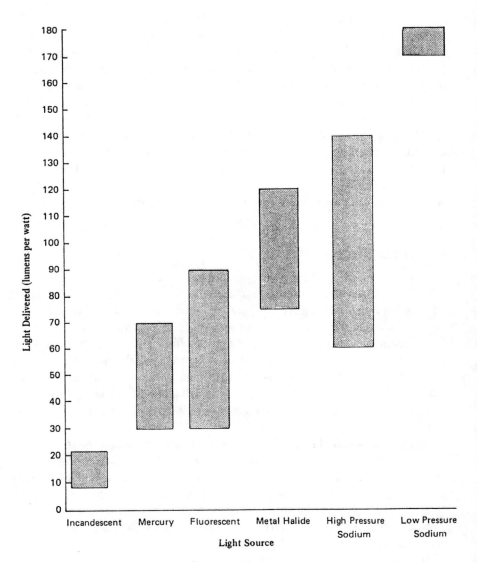

Figure 4-5. Efficiency of Various Light Sources

## Efficient Types of Incandescents
## for Limited Use

Attempts to increase the efficiency of incandescent lighting while maintaining good color rendition have led to the manufacture of a number of energy-saving incandescent lamps for limited residential use.

*Tungsten Halogen*—These lamps vary from the standard incandescent by the addition of halogen gases to the bulb. Halogen gases keep the glass bulb from darkening by preventing the filament's evaporation, thereby increasing lifetime up to four times that of a standard bulb. The lumen-per-watt rating is approximately the same for both types of incandescents, but tungsten halogen lamps average 94% efficiency throughout their extended lifetime, offering significant energy and operating cost savings. However, tungsten halogen lamps require special fixtures, and during operation the surface of the bulb reaches very high temperatures, so they are not commonly used in the home.

*Reflector or R-Lamps*—Reflector lamps are incandescents with an interior coating of aluminum that directs the light to the front of the bulb. Certain incandescent light fixtures, such as recessed or directional fixtures, trap light inside. Reflector lamps project a cone of light out of the fixture and into the room, so that more light is delivered where it is needed. In these fixtures, a 50-watt reflector bulb will provide better lighting and use less energy when substituted for a 100-watt standard incandescent bulb.

Reflector lamps are an appropriate choice for task lighting (because they directly illuminate a work area) and for accent lighting. Reflector lamps are available in 25, 30, 50, 75, and 150 watts. While they have a lower initial efficiency (lumens per watt) than regular incandescents, they direct light more effectively, so that more light is actually delivered than with regular incandescents. (See Figure 4-6.)

*PAR Lamps*— Parabolic aluminized reflector (PAR) lamps are reflector lamps with a lens of heavy, durable glass, which makes them an appropriate choice for outdoor flood and spot lighting. They are available in 75, 150, and 250 watts. They have longer lifetimes with less depreciation than standard incandescents.

*ER Lamps*—Ellipsoidal reflector (ER) lamps are ideally suited for recessed fixtures, because the beam of light produced is focused two inches ahead of the lamp to reduce the amount of light trapped in the fixture. In a directional fixture, a 75-watt ellipsoidal reflector lamp delivers more light than a 150-watt R-lamp. (See Figure 4-6.)

**Standard Incandescent**                                          **R-Lamp**

A high percentage                                          An aluminum
of light output                                          coating directs light
is trapped in fixture                                          out of the fixture

**ER Lamp**

The beam is focused 2 inches
ahead of the lamp, so that very
little light is trapped in the fixture

**Figure 4-6. Comparison of Incandescent Lamps**

**Mercury vapor lamps** find limited use in today's lighting systems because fluorescent and other high intensity discharge (HID) sources have surpassed them in both lamp efficacy and system efficiency. Typical ratings for mercury vapor lamps range from about 30 to 70 lumens per watt. The primary advantages of mercury lamps are a good range of color, availability in sizes as low as 30 watts, long life and relatively low cost. However, fluorescent systems are available today which can do many of the jobs mercury used to do and they do it more efficiently. There are still places for mercury vapor lamps in lighting system design, but they are becoming fewer as technology advances in fluorescent and higher efficacy HID sources.

**Fluorescent lamps** have made dramatic advances in the last 10 years. From the introduction of reduced wattage lamps in the mid-1970s, to the marketing of several styles of low wattage, compact lamps recently, there has been a steady parade of new products. Lamp efficacy now ranges from about 30 lumens per watt to near 90 lumens per watt. The range of colors is more complete than mercury vapor, and lamp manufacturers have recently made significant progress in developing fluorescent and metal halide lamps which have much more consistent color rendering properties allowing greater flexibility in mixing these two sources without creating disturbing color mismatches. The recent compact fluorescent lamps open up a whole new market for fluorescent sources. These lamps permit design of much smaller luminaries which can compete with incandescent and mercury vapor in the low cost, square or round fixture market which the incandescent and mercury sources have dominated for so long. While generally good, lumen maintenance throughout the lamp lifetime is a problem for some fluorescent lamp types.

## Energy Efficient "Plus" Fluorescents[3]

The energy efficient "plus" fluorescents represent the second generation of improved fluorescent lighting. These bulbs are available for replacement of standard 4-foot, 40-watt bulbs and require only 32 watts of electricity to produce essentially the same light levels. To the authors' knowledge, they are not available for 8-foot fluores-

---

[3] **Source:** *Fluorescent Lighting – An Expanding Technology*, R.E. Webb, M.G. Lewis, W.C. Turner.

cent bulb retrofit. The energy efficient plus fluorescents require a ballast change. The light output is similar to the energy efficient bulbs, and the two types may be mixed in the same area if desired.

Examples of energy efficient plus tubes include the Super-Saver Plus by Sylvania and General Electric's Watt Mizer Plus.

### Energy Efficient Fluorescents System Change

The third generation of energy efficient fluorescents requires both a ballast and a fixture replacement. The standard 2-foot by 4-foot fluorescent fixture, containing four bulbs and two ballasts, requires approximately 180 watts (40 watts per tube and 20 watts per ballast). The new-generation fluorescent manufacturers claim the following:

- General Electric — "Optimizer" requires only 116 watts with a slight reduction in light output.

- Sylvania — "Octron" requires only 132 watts with little reduction in light level.

- General Electric — "Maximizer" requires 169 watts but supplies 22 percent more light output.

The fixtures and ballasts designed for the third-generation fluorescents are not interchangeable with earlier generations.

**Metal halide lamps** fall into a lamp efficacy range of approximately 75-125 lumens per watt. This makes them more energy efficient than mercury vapor but somewhat less so than high pressure sodium. Metal halide lamps generally have fairly good color rendering qualities. While this lamp displays some very desirable qualities, it also has some distinct drawbacks including relatively short life for an HID lamp, long restrike time to restart after the lamp has been shut off (about 15-20 minutes at 70°F) and a pronounced tendency to shift colors as the lamp ages. In spite of the drawbacks, this source deserves serious consideration and is used very successfully in many applications.

**High pressure sodium lamps** introduced a new era of extremely high efficacy (60-140 lumens/watt) in a lamp which operates in fixtures having construction very similar to those used for mercury vapor and metal halide. When first introduced, this lamp suffered from ballast problems. These have now been resolved and luminaries employing high quality lamps and ballasts provide very satisfactory service. The 24,000-hour lamp life, good lumen maintenance and high efficacy of these lamps make them ideal sources for industrial and outdoor applications where discrimination of a range of colors is not critical.

The lamp's primary drawback is the rendering of some colors. The lamp produces a high percentage of light in the yellow range of the spectrum. This tends to accentuate colors in the yellow region. Rendering of reds and greens shows a pronounced color shift. This can be compensated for in the selection of the finishes for the surrounding areas, and, if properly done, the results can be very pleasing. In areas where color selection, matching and discrimination are necessary, high pressure sodium should not be used as the only source of light. It is possible to gain quite satisfactory color rendering by mixing high pressure sodium and metal halide in the proper proportions. Since both sources have relatively high efficacies, there is not a significant loss in energy efficiency by making this compromise.

High pressure sodium has been used quite extensively in outdoor applications for roadway, parking and façade or security lighting. This source will yield a high efficiency system; however, it should be used only with the knowledge that foliage and landscaping colors will be severely distorted where high pressure sodium is the only, or predominant, illuminant. Used as a parking lot source, there may be some difficulty in identification of vehicle colors in the lot. It is necessary for the designer or owner to determine the extent of this problem and what steps might be taken to alleviate it.

Recently lamp manufacturers have introduced high pressure sodium lamps with improved color rendering qualities. However, the improvement in color rendering was not gained without cost—the efficacy of the color-improved lamps is somewhat lower approximately 90 lumens per watt.

**Low pressure sodium lamps** provide the highest efficacy of any of the sources for general lighting with values ranging up to 180 lumens per watt. Low pressure sodium produces an almost pure yellow light with very high efficacy, and renders all colors gray except yellow or near yellow. This effect results in no color discrimination under low pressure sodium lighting; it is suitable for use in a very limited number of applications. It is an acceptable source for warehouse lighting where it is only necessary to read labels but not to choose items by color. This source has application for either indoor or outdoor safety or security lighting as long as color rendering is not important.

In addition to these primary sources, there are a number of retrofit lamps which allow use of higher efficacy sources in the sockets of existing fixtures. Therefore, metal halide or high pressure sodium lamps can be retrofitted into mercury vapor fixtures, or self-ballasted mercury lamps can replace incandescent lamps. These lamps all make some compromises in operating characteristics, life and/or efficacy.

Figure 4-7 presents data on the efficacy of each of the major lamp types in relation to the wattage rating of the lamps. Without exception, the efficacy of the lamp increases as the lamp wattage rating increases.

The lamp efficacies discussed here have been based on the lumen output of a new lamp after 100 hours of operation or the "initial lumens." Not all lamps age in the same way. Some lamp

Figure 4-7. Lamp Efficacy
(Does Not Include Ballast Losses)

types, such as lightly loaded fluorescent and high pressure sodium, hold up well and maintain their lumen output at a relatively high level until they are into, or past, middle age. Others, as represented by heavily loaded fluorescent, mercury vapor and metal halide, decay rapidly during their early years and then coast along at a relatively lower lumen output throughout most of their useful life. These factors must be considered when evaluating the various sources for overall energy efficiency.

## Incandescent Replacement

The most efficacious lamps that can be used in incandescent sockets are the compact fluorescent lamps. The most popular systems are the twin tubes and double twin tubes. These are closer to the size and weight of the incandescent lamp than the earlier type of fluorescent (circline) replacements.

Twin tubes with lamp wattages from 5 to 13 watts provide amounts of light ranging from 240 to 850 lumens. Table 4-3 lists the characteristics of various types of incandescent and compact fluorescent lamps that can be used in the same type sockets.

### Table 4-3. Lamp Characteristics

| Lamp Type (Total Input Power)* | Lamp Power (W) | Light Output (lumens) | Lamp Life (hour) | Efficacy (lm/W) |
|---|---|---|---|---|
| 100 W (Incandescent) | 100 | 1750 | 750 | 18 |
| 75 W (Incandescent) | 75 | 1200 | 750 | 16 |
| 60 W (Incandescent) | 60 | 890 | 1000 | 15 |
| 40 W (Incandescent) | 40 | 480 | 1500 | 12 |
| 25 W (Incandescent) | 25 | 238 | 2500 | 10 |
| 22 W (Fl. Circline) | 18 | 870 | 9000 | 40 |
| 44 W (Fl. Circline) | 36 | 1750 | 9000 | 40 |
| 7 W (Twin) | 5 | 240 | 10000 | 34 |
| 10 W (Twin) | 7 | 370 | 10000 | 38 |
| 13 W (Twin) | 9 | 560 | 10000 | 43 |
| 19 W (Twin) | 13 | 850 | 10000 | 45 |
| 18 W (Solid-State)** | — | 1100 | 7500 | 61 |

*Includes ballast losses.
**Operated at high frequency.

The advantages of the compact fluorescent lamps are larger and increased efficacy, longer life and reduced total cost. The cost per $10^6$ lumen hours of operating the 75-watt incandescent and the 18-watt fluorescent is $5.47 and $3.29, respectively. This is based upon an energy cost of $0.075 per kWh and lamp costs of $0.70 and $17 and the 75-watt incandescent and 18-watt fluorescent lamps, respectively. The circline lamps were much larger and heavier than the incandescents and would fit in a limited number of fixtures. The twin tubes are only slightly heavier and larger than the equivalent incandescent lamp. However, there are some fixtures that are too small for them to be employed.

The narrow tube diameter compact fluorescent lamps are now possible because of the recently developed rare earth phosphors. These phosphors have an improved lumen depreciation at high lamp power loadings. The second important characteristic of these narrow band phosphors is their high efficiency in converting the ultraviolet light generated in the plasma into visible light. By proper mixing of these phosphors, the color characteristics (color temperature and color rendering) are similar to the incandescent lamp.

There are two types of compact fluorescent lamps. In one type of lamp system, the ballast and lamp are integrated into a single package; in the second type, the lamp and ballast are separate, and when a lamp burns out it can be replaced. In the integrated system, both the lamp and the ballast are discarded when the lamp burns out.

It is important to recognize when purchasing these compact fluorescent lamps that they provide the equivalent light output of the lamps being replaced. The initial lumen output for the various lamps is shown in Table 4-3.

## Lighting Efficiency Options

Several lighting efficiency options are illustrated below: (Refer to Formula 4-6.)

*Footcandle Level*—The footcandle level required is that level at the task. Footcandle levels can be lowered to one third of the levels for surrounding areas such as aisles. (A minimum 20-footcandle level should be maintained.)

The placement of the lamp is also important. If the luminaire can be lowered or placed at a better location, the lamp wattage may be reduced.

*Coefficient of Utilization (Cu)*—The color of the walls, ceiling, and floors, the type of luminaire, and the characteristics of the room determine the *Cu*. This value is determined based on manufacturer's literature. The *Cu* can be improved by analyzing components such as lighter colored walls and more efficient luminaires for the space.

*Lamp Depreciation Factor* and *Dirt Depreciation Factor*—These two factors are involved in the maintenance program. Choosing a luminaire which resists dirt build-up, group relamping and cleaning the luminaire will keep the system in optimum performance. Taking these factors into account can reduce the number of lamps initially required.

The light loss factor *(LLF)* takes into account that the lamp lumen depreciates with time $(L_1)$, that the lumen output depreciates due to dirt build-up $(L_2)$, and that lamps burn out $(L_3)$. Formula 4-7 illustrates the relationship of these factors.

$$LLF = L_1 \times L_2 \times L_3 \qquad \textit{Formula (4-7)}$$

To reduce the number of lamps required which in turn reduces energy consumption, it is necessary to increase the overall light loss factor. This is accomplished in several ways. One is to choose the luminaire which minimizes dust build-up. The second is to improve the maintenance program to replace lamps prior to burn-out. Thus if it is known that a group relamping program will be used at a given percentage of rated life, the appropriate lumen depreciation factor can be found from manufacturer's data. It may be decided to use a shorter relamping period in order to increase $(L_1)$ even further. If a group relamping program is used, $(L_3)$ is assumed to be unity.

Figure 4-6 illustrates the effect of dirt build-up on $(L_2)$ for a dustproof luminaire. Every luminaire has a tendency for dirt build-up. Manufacturer's data should be consulted when estimating $(L_2)$ for the luminaire in question.

## Solid-State Ballasts

After more than 10 years of development and 5 years of manufacturing experience, operating fluorescent lamps at high frequency (20 to 30 kHz) with solid-state ballasts has achieved credibility. The fact that all of the major ballast manufacturers offer solid-state

**Figure 4-6.** Effect of Dirt Build-Up on Dustproof Luminaires for
Various Atmospheric Conditions

ballasts and the major lamp companies have designed new lamps to be operated at high frequency is evidence that the solid-state high frequency ballast is now state-of-the-art.

It has been shown that fluorescent lamps operated at high frequency are 10 to 15 percent more efficacious than 60 Hz operation. In addition, the solid-state ballast is more efficient than conventional ballasts in conditioning the input power for the lamps such that the total system efficacy increase is between 20 and 25 percent. That is, for a standard two-lamp, 40-watt F40 T-12 rapid-start system, overall efficacy is increased from 63 lm/W to over 80 lm/W.

In the past few years, continued development of the product has improved reliability and reduced cost. Today, solid-state ballasts can be purchased for less than $30, and, in sufficiently large quantities, some bids have been less than $20. The industry's growth is evidenced by the availability of ballasts for the 8-foot fluorescent lamp, both slimline and high power, as well as the more common F40 (4-ft.) size. In order to be more competitive with initial costs, there are three- and four-lamp ballasts for the F40-type lamps. These multi-lamp ballasts reduce the initial cost per lamp, as well as the installation cost, and are even more efficient than the one- and two-lamp ballast system.

The American National Standards Institute (ANSI) ballast committee has been developing standards for solid-state ballasts for the past few years. The ballast factor is one parameter that will be specified by ANSI. However, solid-state ballasts are available with different ballast factors. The ballast factor is the light output provided by the ballast-lamp system compared to the light output of the lamp specified by the lamp manufacturer. The ANSI ballast factor standard for 40-watt F40 fluorescent lamps is 95 ± 2.5 percent. Because most solid-state ballasts were initially sold in the retrofit market, their ballasts were designed to have a lower ballast factor. Thus, energy was saved not only by the increased efficacy but also by reducing the light output. The thrust was to reduce illumination levels in overlit spaces.

Today, there are solid-state ballasts with a ballast factor exceeding 100 percent. These ballasts are most effectively used in new installations. In these layouts, more light from each luminaire will reduce the number of luminaires, ballasts and lamps, hence reducing both initial and operating costs. It is essential that the lighting designer-engineer and consumer know the ballast factor for the lamp-ballast system. The ballast factor for a ballast also depends upon the lamp. For example, a core-coil ballast will have a ballast factor of 95 ± 2 percent when operating a 40-watt F40 argon-filled lamp and less when operating an "energy saving" 34-watt F40 Krypton-filled lamp. The ballast factor instead will be about 87 ± 2.5 percent with the 34-watt energy saving lamp. Because of this problem, the ANSI standard for the ballast factor for the 34-watt lamps has recently been reduced to 85 percent. Table 4-4 provides some data for several types of solid-state ballasts operating 40-watt and 34-watt F40 lamps and lists some parameters of concern for the consumer.

The table compares several types of solid-state ballasts with a standard core-coil ballast that meets the ANSI standard with the two-lamp, 40-watt F40 lamp. Notice that the system efficacy of any ballast system is about the same operating a 40-watt or a 34-watt F40 lamp. Although the 34-watt "lite white" lamp is about 6 percent more efficient than the 40-watt "cool white" lamp, the ballast losses are greater with the 34-watt lamp due to an increased lamp current. The lite white phosphor is more efficient than the cool white phosphor but has poorer color rendering characteristics.

Note that the percent flicker is drastically reduced when the lamps are operated at high frequency with solid-state ballasts. A recent scientific field study of office workers in the U.K. has shown that complaints of headaches and eye-strain are 50 percent less under high frequency lighting when compared to lamps operating at 50 cycles, the line frequency in the U.K.

**Table 4-4. Performance of F40 Fluorescent Lamp Systems**

| Characteristic | Core-Coil 2 Lamps, T-12 40W | 34W | — Solid-State Ballasts — 2 Lamps, T-12 40W | 34W | 4 Lamps, T-12 40W | 34W | 2 Lamps, T-8 32W |
|---|---|---|---|---|---|---|---|
| Power (W) | 96 | 79 | 72 | 63 | 136 | 111 | 65 |
| Power Factor (%) | 98 | 92 | 95 | 93 | 94 | 94 | 89 |
| Filament Voltage (V) | 3.5 | 3.6 | 3.1 | 3.1 | 2.0 | 1.6 | 0 |
| Light Output (lm) | 6050 | 5060 | 5870 | 5060 | 11,110 | 9250 | 5820 |
| Ballast Factor | .968 | .880 | .932 | .865 | .882 | .791 | 1.003 |
| Flicker (%) | 30 | 21 | 15 | 9 | 1 | 0 | 1 |
| System Efficacy (lm/W) | 63 | 64 | 81 | 81 | 82 | 83 | 90 |

Each of the above ballasts has different factors, which are lower when operating the 34-watt Krypton-filled lamp. Table 4-4 also lists the highest system efficacy of 90 lumens per watt for the solid-state ballast and T-8, 32-watt lamp.

All of the above solid-state ballasts can be used in place of core-coil ballasts specified to operate the same lamps. To determine the illumination levels, or the change in illumination levels, the manufacturer must supply the ballast factor for the lamp type employed. The varied light output from the various systems allows the lighting designer-engineer to precisely tailor the lighting level.

## CONTROL EQUIPMENT

Table 4-5 lists various types of equipment that can be components of a lighting control system, with a description of the predominant characteristic of each type of equipment. Static equipment can alter light levels semipermanently. Dynamic equipment can alter light levels automatically over short intervals to correspond to the activities in a space. Different sets of components can be used to

### Table 4-5. Lighting Control Equipment

| System | Remarks |
|---|---|
| STATIC: | |
| Delamping | Method for reducing light level 50%. |
| Impedance Monitors | Method for reducing light level 30, 50%. |
| DYNAMIC: | |
| Light Controllers | |
| Switches/Relays | Method for on-off switching of large banks of lamps. |
| Voltage/Phase Control | Method for controlling light level continuously 100 to 50%. |
| Solid-State Dimming Ballasts | Ballasts that operate fluorescent lamps efficiently and can dim them continuously (100 to 10%) with low voltage. |
| SENSORS: | |
| Clocks | System to regulate the illumination distribution as a function of time. |
| Personnel | Sensor that detects whether a space is occupied by sensing the motion of an occupant. |
| Photocell | Sensor that measures the illumination level of a designated area. |
| COMMUNICATION: | |
| Computer/Microprocessor | Method for automatically communicating instructions and/or input from sensors to commands to the light controllers. |
| Power-Line Carrier | Method for carrying information over existing power lines rather than dedicated hard-wired communication lines. |

form various lighting control systems in order to accomplish different combinations of control strategies.

## FLUORESCENT LIGHTING CONTROL SYSTEMS

The control of fluorescent lighting systems is receiving increased
attention. Two major categories of lighting control are available—
personnel sensors and lighting compensators.

### Personnel Sensors

There are three classifications of personnel sensors—ultrasonic,
infrared and audio.

*Ultrasonic sensors* generate sound waves outside the human
hearing range and monitor the return signals. Ultrasonic sensor
systems are generally made up of a main sensor unit with a network of satellite sensors providing coverage throughout the lighted
area. Coverage per sensor is dependent upon the sensor type and
ranges between 500 and 2,000 square feet. Sensors may be mounted
above the ceiling, suspended below the ceiling or mounted on the
wall. Energy savings are dependent upon the room size and occupancy. Advertised savings range from 20 to 40 percent.

Several companies manufacture ultrasonic sensors including
Novita and Unenco.

*Infrared sensor* systems consist of a sensor and control unit.
Coverage is limited to approximately 130 square feet per sensor.
Sensors are mounted on the ceiling and usually directed towards
specific work stations. They can be tied into the HVAC control
and limit its operation also. Advertised savings range between 30 and
50 percent. (See Figure 4-9.)

*Audio sensors* monitor sound within a working area. The coverage of the sensor is dependent upon the room shape and the mount-
ing height. Some models advertise coverage of up to 1,600 square
feet. The first cost of the audio sensors is approximately one half
that of the ultrasonic sensors. Advertised energy savings are approximately the same as the ultrasonic sensors. Advertised energy savings
are approximately the same as the ultrasonic sensors. Several restrictions apply to the use of the audio sensors. First, normal background
noise must be less than 60 dB. Second, the building should be at
least 100 feet from the street and may not have a metal roof.

**Figure 4-9. Transformer, relay and wide view infrared sensor to control lights.** *(Photograph courtesy of Sensorswitch)*

## Lighting Compensators

Lighting compensators are divided into two major groups—switched and sensored.

*Switched compensators* control the light level using a manually operated wall switch. These particular systems are used frequently in residential settings and are commonly known as "dimmer switches." Based on discussions with manufacturers, the switched controls are available for the 40-watt standard fluorescent bulbs only. The estimated savings are difficult to determine, as usually switched control systems are used to control room mood. The only restriction to their use is that the luminaire must have a dimming ballast.

*Sensored compensators* are available in three types. They may be very simple or very complex. They may be integrated with the building's energy management system or installed as a stand-alone system. The first type of system is the Excess Light Turn-Off (ELTO) system. This sytem senses daylight levels and automatically turns off lights as the sensed light level approaches a programmed upper limit. Advertised paybacks for these types of systems range from 1.8 to 3.8 years.

The second type of system is the Daylight Compensator (DAC) system. This system senses daylight levels and automatically dims lights to achieve a programmed room light level. Advertised savings range from 40 to 50 percent. The primary advantage of this system is it maintains a uniform light level across the controlled system area. The third system type is the Daylight Compensator + Excess Light Turn-Off system. As implied by the name, this system is a combination of the first two systems. It automatically dims light outputs to achieve a designated light level and, as necessary, automatically turns off lights to maintain the desired room conditions.

*Specular reflectors*: Fluorescent fixtures can be made more efficient by the insertion of a suitably shaped specular reflector. The specular reflector material types are aluminum, silver and multiple dielectric film mirrors. The latter two have the highest reflectivity while the aluminum reflectors are less expensive.

Measurements show the fixture efficiency with higher reflectance specular reflectors (silver or dielectric films) is improved by 15 percent compared to a new fixture with standard diffuse reflectors.

Specular reflectors tend to concentrate more light downward with reduced light at high exit angles. This increases the light modulation in the space, which is the reason several light readings at different sites around the fixture are required for determining the average illuminance. The increased downward component of candle power may increase the potential for reflected glare from horizontal surfaces.

When considering reflectors, information should be obtained on the new candle power characteristics. With this information a lighting designer or engineer can estimate the potential changes in modulation and reflected glare.


## ENERGY MANAGEMENT

The availability of computers at moderate costs and the concern for reducing energy consumption have resulted in the application of computer-based controllers to more than just industrial process applications. These controllers, commonly called Energy Management Systems (EMS), can be used to control virtually all non-process energy using pieces of equipment in buildings and industrial plants. Equipment controlled can include fans, pumps, boilers, chillers and lights. This section will investigate the various types of Energy Management Systems which are available and illustrate some of the methods used to reduce energy consumption.

## THE TIMECLOCK

One of the simplest and most effective methods of conserving energy in a building is to operate equipment only when it is needed. If, due to time, occupancy, temperature or other means, it can be determined that a piece of equipment does not need to operate, energy savings can be achieved without affecting occupant comfort by turning off the equipment.

One of the simplest devices to schedule equipment operation is the mechanical timeclock. The timeclock consists of a rotating disk which is divided into segments corresponding to the hour of the day and the day of the week. This disk makes one complete revolution in, depending on the type, a 24-hour or a 7-day period. (See Figure 4-10.)

On and off "lugs" are attached to the disk at appropriate positions corresponding to the schedule for the piece of equipment. As the disk rotates, the lugs cause a switch contact to open and close, thereby controlling equipment operation.

A common application of timeclocks is scheduling office building HVAC equipment to operate during business hours Monday through Friday and to be off all other times. As is shown in the following problem, significant savings can be achieved through the correct application of timeclocks.

### Example Problem 4-3

An office building utilizes two 50 hp supply fans and two 15 hp return fans which operate continuously to condition the building. What are the annual savings that result from installing a timeclock to operate these fans from 7:00 a.m. to 5:00 p.m., Monday through Friday? Assume an electrical rate of $0.08/kWh.

Figure 4-10. Mechanical Timeclock

**Answer**

Annual Operation Before Timeclock =
52 weeks X 7 days/week X 24 hours/day = 8736 hours

Annual Operation After Timeclock =
52 X (5 days/week X 10 hours/day) = 2600 hours

Savings = 130 hp X 0.746 kW/hp X (8736−2600) hours X
$0.08/kWh = $47,600

Although most buildings today utilize some version of a time-clock, the magnitude of the savings value in this example illustrates the importance of correct timeclock operation and the potential for additional costs if this device should malfunction or be adjusted inaccurately. Note that the above example also ignores heating and cooling savings which would result from the installation of a time-clock.

## PROBLEMS WITH MECHANICAL TIMECLOCKS

Although the use of mechanical timeclocks in the past has resulted in significant energy savings, they are being replaced by Energy Management Systems because of problems that include the following:

- The on/off lugs sometimes loosen or fall off.

- Holidays, when the building is unoccupied, cannot easily be taken into account.

- Power failures require the timeclock to be reset or it is not synchronized with the building schedule.

- Inaccuracies in the mechanical movement of the timeclock prevent scheduling any closer than ±15 minutes of the desired times.

- There are a limited number of on and off cycles possible each day.

- It is a time-consuming process to change schedules on multiple timeclocks.

Energy Management Systems, or sometimes called electronic timeclocks, are designed to overcome these problems plus provide increased control of building operations.

### ENERGY MANAGEMENT SYSTEMS

Recent advances in digital technology, dramatic decreases in the cost of this technology and increased energy awareness have resulted in the increased application of computer-based controllers (i.e., Energy Management Systems) in commercial buildings and industrial plants. These devices can control anywhere from one to a virtually unlimited number of items of equipment.

By concentrating the control of many items of equipment at a single point, the EMS allows the building operator to tailor building operation to precisely satisfy occupant needs. This ability to maximize energy conservation, while preserving occupant comfort, is the ultimate goal of an energy engineer.

### Microprocessor Based

Energy Management Systems can be placed in one of two broad, and sometimes overlapping, categories referred to as microprocessor-based and mini-computer based.

Microprocessor-based systems can control from 1 to 40 input/outpoints and can be linked together for additional loads. Programming is accomplished by a keyboard or hand-held console, and an LED display is used to monitor/review operation of the unit. A battery maintains the programming in the event of power failure. (See Figures 4-11 and 4-12.)

Capabilities of this type of EMS are generally pre-programmed so that operation is relatively straightforward. Programming simply involves entering the appropriate parameters (e.g., the point number and the on and off times) for the desired function. Microprocessor-based EMS can have any or all of the following capabilities:

- Scheduling
- Duty Cycling
- Demand Limiting
- Optimal Start
- Monitoring
- Direct Digital Control

### Scheduling

Scheduling with an EMS is very much the same as it is with a timeclock. Equipment is started and stopped based on the time of day and the day of week. Unlike a timeclock, however, multiple start/stops can be accomplished very easily and accurately (e.g.,

**Figure 4-11. Microprocessor-Based Programmable EMS**
*(Photograph Courtesy Control Systems International)*

**Figure 4-12. Keyboard Used to Access EMS**
*(Photo Courtesy Control Systems International)*

in a classroom, lights can be turned off during morning and afternoon break periods and during lunch). It should be noted that this single function, if accurately programmed and depending on the type of facility served, can account for the largest energy savings attributable to an EMS.

Additionally, holiday dates can be entered into the EMS a year in advance. When the holiday occurs, regular programming is overridden and equipment can be kept off.

## Duty Cycling

Most HVAC fan systems are designed for peak load conditions, and consequently these fans are usually moving much more air than is needed. Therefore, they can sometimes be shut down for short periods each hour, typically 15 minutes, without affecting occupant comfort. Turning equipment off for pre-determined periods of time during occupied hours is referred to as duty cycling, and can be accomplished very easily with an EMS. Duty cycling saves fan and pump energy but does not reduce the energy required for space heating or cooling since the thermal demand must still be met.

The more sophisticated EMS's monitor the temperature of the conditioned area and use this information to automatically modify the duty cycle length when temperatures begin to drift. If, for example, the desired temperature in an area is 70° and at this temperature equipment is cycled 50 minutes on and 10 minutes off, a possible temperature-compensated EMS may respond as shown in Figure 4-13. As the space temperature increases above (or below if so programmed) the setpoint, the equipment off time is reduced until, at 80° in this example, the equipment operates continuously.

Duty cycling is best applied in large, open-space offices which are served by a number of fans. Each fan could be programmed so that the off times do not coincide, thereby assuring adequate air flow to the offices at all times.

Duty cycling of fans which provide the only air flow to an area should be approached carefully to insure that ventilation requirements are maintained and that varying equipment noise does not annoy the occupants. Additionally, duty cycling of equipment imposes extra stress on motors and associated equipment. Care should be taken, particularly with motors over 20 hp, to prevent

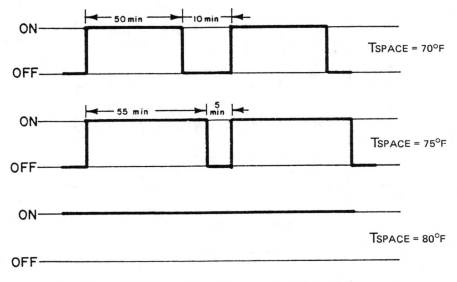

Figure 4-13. Temperature Compensated Duty Cycling

starting and stopping of equipment in excess of what is recommended by the manufacturer.

## Demand Charges

Electrical utilities charge commercial customers based not only on the amount of energy used (kWh) but also on the peak demand (kW) for each month. Peak demand is very important to the utility so that they may properly size the required electrical service and insure that sufficient peak generating capacity is available to that given facility.

In order to determine the peak demand during the billing period, the utility establishes short periods of time called the demand interval (typically 15, 30, or 60 minutes). The billing demand is defined as the highest average demand recorded during any one demand interval within the billing period. (See Figure 4-14.) Many utilities now utilize "ratchet" rate charges. A "ratchet" rate means that the billed demand for the month is based on the highest demand in the previous 12 months, or an average of the current month's peak demand and the previous highest demand in the past year.

T - DEMAND INTERVAL

**Figure 4-14. Peak Demand**

Depending on the facility, the demand charge can be a significant portion, as much as 20%, of the utility bill. The user will get the most electrical energy per dollar if the load is kept constant, thereby minimizing the demand charge. The objective of demand control is to even out the peaks and valleys of consumption by deferring or rescheduling the use of energy during peak demand periods.

A measure of the electrical efficiency of a facility can be found by calculating the load factor. The load factor is defined as the ratio of energy usage (kWh) per month to the peak demand (kW) X the facility operating hours.

### Example Problem 4-4

What is the load factor of a continuously operating facility that consumed 800,000 kWh of energy during a 30-day billing period and established a peak demand of 2000 kW?

### Answer

$$\text{Load Factor} = \frac{800,000 \text{ kWh}}{2000 \text{ kW X 30 days X 24 hours/day}} = 0.55$$

The ideal load factor is 1.0, at which demand is constant; therefore, the difference between the calculated load factor and 1.0 gives an indication of the potential for reducing peak demand (and demand charges) at a facility.

## Demand Limiting

Energy Management Systems with demand limiting capabilities utilize either pulses from the utility meter or current transformers to predict the facility demand during any demand interval. If the facility demand is predicted to exceed the user-entered setpoint, equipment is "shed" to control demand. Figure 4-15 illustrates a typical demand chart before and after the actions of a demand limiter.

Electrical load in a facility consists of two major categories: essential loads which include most lighting, elevators, escalators, and most production machinery; and non-essential ("sheddable") loads such as electric heaters, air conditioners, exhaust fans, pumps, snow melters, compressors and water heaters. Sheddable loads will not,

Typical Demand Chart (Full Scale = 10,000 KW)

Demand Chart After Installation Of Demand Control
(Full Scale = 10,000 KW)

**Figure 4-15. Demand Limiting Comparison**

when turned off for short periods of time to control demand, affect productivity or comfort.

To prevent excessive cycling of equipment, most Energy Management Systems have a deadband that demand must drop below before equipment operation is restored (See figure 4-16). Additionally, minimum on and maximum off times and shed priorities can be entered for each load to protect equipment and insure that comfort is maintained.

**Figure 4-16. Demand Limiting Actions**

It should be noted that demand shedding of HVAC equipment in commercial office buildings should be applied with caution. Since times of peak demand often occur during times of peak air conditioning loads, excessive demand limiting can result in occupant discomfort.

### Time of Day Billing

Many utilities are beginning to charge their larger commercial users based on the time of day that consumption occurs. Energy and demand during peak usage periods (i.e., summer weekday afternoons and winter weekday evenings) are billed at much higher rates than consumption during other times. This is necessary because utilities must augment the power production of their large power plants during periods of peak demand with small generators which are expensive to operate. Some of the more sophisticated Energy Management Systems can now account for these peak billing periods with different demand setpoints based on the time of day and day of week.

## Optimal Start

External building temperatures have a major influence on the amount of time it takes to bring the building temperature up to occupied levels in the morning. Buildings with mechanical time clocks usually start HVAC equipment operation at an early enough time in the morning (as much as 3 hours before occupancy time) to bring the building up to temperature on the coldest day of the year. During other times of the year when temperatures are not as extreme, building temperatures can be up to occupied levels several hours before it is necessary, and consequently unnecessary energy is used. (See Figure 4-17.)

Energy Management Systems with optimal start capabilities, however, utilize indoor and outdoor temperature information, along with learned building characteristics, to vary start time of HVAC equipment so that building temperatures reach desired values just as occupancy occurs. Consequently, if a building is scheduled to be occupied at 8:00 a.m., on the coldest day of the year, the HVAC equipment may start at 5:00 a.m. On milder days, however, equipment may not be started until 7:00 a.m. or even later, thereby saving significant amounts of energy.

Most Energy Management Systems have a "self-tuning" capability to allow them to learn the building characteristics. If the building is heated too quickly or too slowly on one day, the start time is adjusted the next day to compensate.

## Monitoring

Microprocessor-based EMS can usually accomplish a limited amount of monitoring of building conditions including the following:

- Outside air temperature

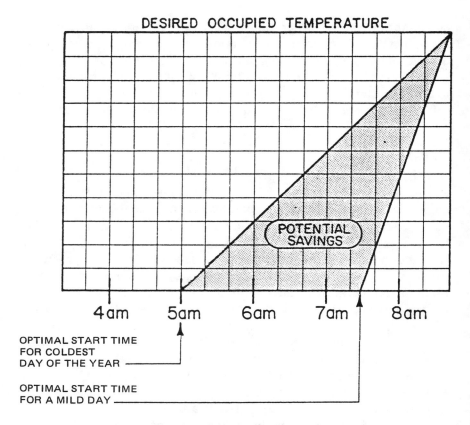

**Figure 4-17. Typical Variation in Building Warm-Up Times**

- Several indoor temperature sensors
- Facility electrical energy consumption and demand
- Several status input points

The EMS can store this information to provide a history of the facility. Careful study of these trends can reveal information about facility operation that can lead to energy conservation strategies that might not otherwise be apparent.

### Direct Digital Control

The most sophisticated of the microprocessor-based EMSs provide a function referred to as direct digital control (DDC). This capability allows the EMS to provide not only sophisticated energy management but also basic temperature control of the building's HVAC systems.

Direct digital control has taken over the majority of all process control application and is now becoming an important part of the HVAC industry. Traditionally, pneumatic controls were used in most commercial facilities for environmental control.

The control function in a traditional facility is performed by a pneumatic controller which receives its input from pneumatic sensors (i.e., temperature, humidity) and sends control signals to pneumatic actuators (valves, dampers, etc.). Pneumatic controllers typically perform a single, fixed function which cannot be altered unless the controller itself is changed or other hardware is added. (See Figure 4-18 for a typical pneumatic control configuration.)

With direct digital control, the microprocessor functions as the primary controller. Electronic sensors are used to measure variables such as temperature, humidity and pressure. This information is used, along with the appropriate application program, by the microprocessor to determine the correct control signal, which is then sent directly to the controlled device (valve or damper actuator). (See Figure 4-18 for a typical DDC configuration.)

Direct digital control (DDC) has the following advantage over pneumatic controls:

- Reduces overshoot and offset errors, thereby saving energy
- Flexibility to easily and inexpensively accomplish changes of control strategies
- Calibration is maintained more accurately, thereby saving energy and providing better performance.

To program the DDC functions, a user programming language is utilized. This programming language uses simple commands in English to establish parameters and control strategies.

## Mini-Computer Based

Mini-computer based EMS can provide all the functions of the microprocessor based EMS, as well as the following:

- Extensive graphics

- Special reports and studies
- Fire and security monitoring and detection
- Custom programs

These devices can control and monitor from 50 to an unlimited number of points and form the heart of a building's (or complex's) operations.

CONVENTIONAL PNEUMATIC CONTROL SYSTEM

DIRECT DIGITAL CONTROL SYSTEM

**Figure 4-18. Comparison of Pneumatic and DDC Controls**

Figure 4-19 shows a typical configuration for this type of system.

The "central processing unit" (CPU) is the heart of the EMS. It is a mini-computer with memory for the operating system and applications software. The CPU performs arithmetic and logical decisions necessary to perform central monitoring and control.

Data and programs are stored or retrieved from the memory or mass storage devices (generally a disk storage system). The CPU has programmed I/O ports for specific equipment, such as printers and cathode ray tube (CRT) consoles. During normal operation, it coordinates operation of all other EMS components.

**Figure 4-19. Mini-Computer Based EMS**

A cathode ray tube console (CRT), either color and/or black and white, with a keyboard is used for operator interaction with the EMS. It accepts operator commands, displays data and graphically displays systems controlled or monitored by the EMS. A "printer," (or printers) provides a permanent copy of system operations and historical data.

A "field interface device" (FID) provides an interface to the points which are monitored and controlled, performs engineering conversions to or from a digital format, performs calculations and logical operations, accepts and processes CPU commands and is capable, in some versions, of stand-alone operations in the event of CPU or communications link failure.

The FID is essentially a microprocessor based EMS as described in the previous section. It may or may not have a keyboard/display unit on the front panel.

The FIDs are generally located in the vicinity of the points to be monitored and/or controlled and are linked together and to the CPU by a single twisted pair of wires which carries multiplexed data (i.e., data from a number of sources combined on a single channel) from the FID to the CPU and back. In some versions, the FIDs can communicate directly with each other.

Early versions of mini-computer based EMS used the CPU to perform all of the processing with the FID used merely for input and output. A major disadvantage of this type of "centralized" system is that the loss of the CPU disables the entire control system. The development of "intelligent" FIDs in a configuration known as "distributed processing" helped to solve this problem. This system, which is becoming prevalent today, utilizes microprocessor-based FIDs to function as remote CPUs. Each panel has its own battery pack to insure continued operation should the main CPU fail.

Each intelligent panel sends signals back to the main CPU only upon a change of status rather than continuously transmitting the same value as previous "centralized" systems have done. This streamlining of data flow to the main CPU frees it to perform other functions such as trend reporting. The CPU's primary function becomes one of directing communications between various FID panels, generating reports and graphics and providing operator interface for programming and monitoring.

## Features

The primary difference in operating functions of the mini-computer based EMS is its increased capability to monitor building operations. For this reason, these systems are sometimes referred to as Energy Monitoring and Control Systems (EMCS). Analog inputs such as temperature and humidity can be monitored, as well as digital inputs such as pump or valve status.

The mini-computer based EMS is also designed to make operator interaction very easy. Its operation can be described as "user friendly" in that the operator, working through the keyboard, enters information in English in a question and response format. In addition, custom programming languages are available so that powerful programs can be created specifically for the building through the use of simplified English commands.

The graphics display CRT can be used to create HVAC schematics, building layouts, bar charts, etc. to better understand building systems operation. These graphics can be "dynamic" so that values and statuses are continuously updated.

Many mini-computer based EMS can also easily incorporate fire and security monitoring functions. Such a configuration is sometimes referred to as a Building Automation System (BAS). By combining these functions with energy management, savings in initial equipment costs can be achieved. Reduced operating costs can be achieved as well by having a single operator for these systems.

The color graphics display can be particularly effective in pinpointing alarms as they occur within a building and guiding quick and appropriate response to that location. In addition, management of fan systems to control smoke in a building during a fire is facilitated with a system that combines energy management and fire monitoring functions.

Note, however, that the incorporation of fire, security and energy management functions into a single system increases the complexity of that system. This can result in longer start-up time for the initial installation and more complicated troubleshooting if problems occur. Since the function of fire monitoring is critical to building operation, these disadvantages must be weighed against the previously mentioned advantages to determine if a combined BAS is desired.

## DATA TRANSMISSION METHODS

A number of different transmission systems can be used in an EMS for communications between the CPU and FID panels. These transmission systems include telephone lines, coaxial cables, electrical power lines, radio frequency, fiber optics and microwave. Table 4-6 compares the various transmission methods.

**Table 4-6. Transmission Method Comparisons**

| Method | First Cost | Scan Rates | Reliability | Maint. Effort | Expandibility | Compatibility with Future Requirements |
|---|---|---|---|---|---|---|
| Coaxial | high | fast | excellent | min. | unlimited | unlimited |
| Twisted pair | low | med. | very good | min. | unlimited | unlimited |
| RF | med. | fast but limited | low | high | very limited | very limited |
| Micro-wave | very high | very fast | excellent | high | unlimited | unlimited |
| Tele-phone | very low | slow | low to high | min. | limited | limited |
| Fiber optics | high | very fast | excellent | min. | unlimited | unlimited |
| Power Line Carrier | med. | med. | med. | high | limited | limited |

### Twisted Pair

One of the most common data transmission methods for an EMS is a twisted pair of wires. A twisted pair consists of two insulated conductors twisted together to minimize interference from unwanted signals.

Twisted pairs are permanently hardwired lines between the equipment sending and receiving data that can carry information over a wide range of speeds, depending on line characteristics. To maintain a particular data communication rate, the line bandwidth, time delay or the signal-to-noise ratio may require adjustment by conditioning the line.

Data transmission in twisted pairs, in most cases, is limited to 1200 bps (bits per second) or less. By using signal conditioning, operating speeds up to 9600 bps may be obtained.

### Voice Grade Telephone Lines

Voice grade lines used for data transmission are twisted pair circuits defined as type 3002 by the Bell Telephone Company. The local telephone company charges a small connection fee for this service, plus a monthly equipment lease fee. Maintenance is included in the monthly lease fee, with a certain level of service guaranteed.

Two of the major problems involve the quality of telephone pairs provided to the installer and the transmission rate. The minimum quality of each line intended for use in the trunk wiring system must be clearly defined.

The most common voice grade line used for data communication is the unconditioned type 3002 allowing transmission rates up to 1200 bps. The 3002 type line may be used for data transmission up to 9600 bps with the proper line conditioning. Voice grade lines must be used with the same constraints and guidelines for twisted pairs.

### Coaxial Cable

Coaxial cable consists of a center conductor surrounded by a shield. The center conductor is separated from the shield by a dielectric. The shield protects against electromagnetic interference. Coaxial cables can operate at data transmission rates in the megabits per second range, but attenuation becomes greater as the data transmission rate increases.

The transmission rate is limited by the data transmission equipment and not by the cable. Regenerative repeaters are required at specific intervals depending on the data rate, nominally every 2000 feet to maintain the signal at usable levels.

### Power Line Carrier

Data can be transmitted to remote locations over electric power lines using carrier current transmission that superimposes a low power RF (radio frequency) signal, typically 100 kHz, onto the 60 Hz power distribution system. Since the RF carrier signal cannot operate across transformers, all communicating devices must be connected to the same power circuit (same transformer secondary and phase) unless RF couplers are installed across transformers permitting the transmitters and receivers to be connected over a wider area of the power system. Transmission can be either one-way or two-way.

Note that power line carrier technology is sometimes used in microprocessor based EMS retrofit applications to control single loads in a facility where hardwiring would be difficult and expensive (e.g., wiring between two buildings). Figure 4-20 shows a basic power line carrier system configuration.

### Radio Frequency

Modulated RF signals, usually VHF or FM radio, can be used as a data transmission method with the installation of radio receivers and transmitters. RF systems can be effectively used for two-way communication between CPU and FID panels where other data transmission methods are not available or suitable for the application. One-way RF systems can be effectively used to control loads at remote locations such as warehouses and unitary heaters and for family housing projects.

The use of RF at a facility, however, must be considered carefully to avoid conflict with other existing or planned facility RF systems. Additionally, there may be a difficulty in finding a frequency on which to transmit, since there are a limited number available.

The kinds of signals sent over an FM radio system are also limited as are the distances over which the signals can be transmitted. The greater the distance, the greater the likelihood that erroneous signals will be received.

### Fiber Optics

Fiber optics uses the wideband properties of infrared light traveling through transparent fibers. Fiber optics is a reliable communications media which is rapidly becoming cost competitive when compared to other high-speed data transmission methods.

The bandwidth of this media is virtually unlimited, and extremely high transmission rates can be obtained. The signal attenuation of high-quality fiber optic cable is lower than the best coaxial cables.

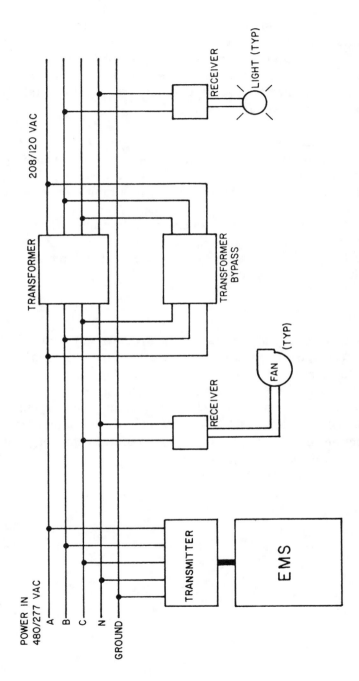

Figure 4-20. Power Line Carrier System Configuration

Repeaters required nominally every 2000 feet, for coaxial cable, are 3 to 6 miles apart in fiber optics systems.

Fiber optics terminal equipment selection is limited to date, and there is a lack of skilled installers and maintenance personnel familiar with this media. Fiber optics must be carefully installed and cannot be bent at right angles.

## Microwave Transmission

For long distance transmission, a microwave link can be used. The primary drawback of microwave links is first cost. Receivers/transmitters are needed at each building in a multi-facility arrangement.

Microwave transmission rates are very fast and are compatible with present and future data requirements. Reliability is excellent, too, but knowledgeable maintenance personnel are required. The only limit on expansion is cost.

## SUMMARY

The term Energy Management System denotes equipment whose functions can range from simple timeclock control to sophisticated building automation. Two broad and overlapping categories of these systems are microprocessor and mini-computer based.

Capabilities of EMS can include scheduling, duty cycling, demand limiting, optimal start, monitoring, direct digital control, fire detection and security. Direct digital control capability enables the EMS to replace the environmental control system so that it directly manages HVAC operations.

# 5

# Waste Heat Recovery

## INTRODUCTION

Waste heat is heat which is generated in a process but then "dumped" to the environment even though it could still be reused for some useful and economic purpose.

The essential quality of heat is not the amount but rather its "value."

The strategy of how to recover this heat depends in part on the temperature of the waste heat gases and the economics involved.

This chapter will present the various methods involved in traditionally recovering waste heat.

Portions of material used in Chapters 5 and 6 are based upon the *Waste Heat Management Guidebooks* published by the U.S. Department of Commerce/National Bureau of Standards. The authors express appreciation to Kenneth G. Kreider; Michael B. McNeil; W. M. Rohrer, Jr.; R. Ruegg; B. Leidy and W. Owens who have contributed extensively to this publication.

## SOURCES OF WASTE HEAT

Sources of waste energy can be divided according to temperature into three temperature ranges. The high temperature range refers to temperatures above 1200°F. The medium temperature range is between 450°F and 1200°F, and the low temperature range is below 450°F.

High and medium temperature waste heat can be used to produce process steam. If high temperature waste heat exists, instead of producing steam directly, consider the possibility of using the high temperature energy to do useful work before the waste heat is extracted. Both gas and steam turbines are useful and fully developed heat engines.

In the low temperature range, waste energy which would be otherwise useless can sometimes be made useful by application of mechanical work through a device called the heat pump.

## HIGH TEMPERATURE HEAT RECOVERY

The combustion of hydrocarbon fuels produces product gases in the high temperature range. The maximum theoretical temperature possible in atmospheric combustors is somewhat under 3500°F, while measured flame temperatures in practical combustors are just under 3000°F. Secondary air or some other dilutant is often admitted to the combustor to lower the temperature of the products to the required process temperature (for example, to protect equipment) thus lowering the practical waste heat temperature.

Table 5-1 gives temperatures of waste gases from industrial process equipment in the high temperature range. All of these result from direct fuel fired processes.

**Table 5-1**

| Type of Device | Temperature F |
|---|---|
| Nickel refining furnace | 2500–3000 |
| Aluminum refining furnace | 1200–1400 |
| Zinc refining furnace | 1400–2000 |
| Copper refining furnace | 1400–1500 |
| Steel heating furnaces | 1700–1900 |
| Copper reverberatory furnace | 1650–2000 |
| Open hearth furnace | 1200–1300 |
| Cement kiln (Dry process) | 1150–1350 |
| Glass melting furnace | 1800–2800 |
| Hydrogen plants | 1200–1800 |
| Solid waste incinerators | 1200–1800 |
| Fume incinerators | 1200–2600 |

## MEDIUM TEMPERATURE HEAT RECOVERY

Table 5-2 gives the temperatures of waste gases from process equipment in the medium temperature range. Most of the waste heat in this temperature range comes from the exhausts of directly fired process units. Medium temperature waste heat is still hot enough to allow consideration of the extraction of mechanical work from the waste heat, by a steam or gas turbine. Gas turbines can be economically utilized in some cases at inlet pressures in the range of 15 to 30 lb/in² g. Steam can be generated at almost any desired pressure and steam turbines used when economical.

**Table 5-2**

| Type of Device | Temperature F |
|---|---|
| Steam boiler exhausts | 450–900 |
| Gas turbine exhausts | 700–1000 |
| Reciprocating engine exhausts | 600–1100 |
| Reciprocating engine exhausts (turbocharged) | 450–700 |
| Heat treating furnaces | 800–1200 |
| Drying and baking ovens | 450–1100 |
| Catalytic crackers | 800–1200 |
| Annealing furnace cooling systems | 800–1200 |

## LOW TEMPERATURE HEAT RECOVERY

Table 5-3 lists some heat sources in the low temperature range. In this range it is usually not practical to extract work from the source, though steam production may not be completely excluded if there is a need for low pressure steam. Low temperature waste heat may be useful in a supplementary way for preheating purposes. Taking a common example, it is possible to use economically the energy from an air-conditioning condenser operating at around 90°F to heat the domestic water supply. Since the hot water must be heated to about 160°F, obviously the air-conditioner waste heat is not hot enough. However, since the cold water enters the domestic water system at about 50°F, energy interchange can take place to raise the water to something less than 90°F. Depending upon the relative air-conditioning load and hot water requirements, any excess condenser heat can be rejected, and the additional energy required by the hot water can be provided by the usual electrical or fired heater.

**Table 5-3**

| Source | Temperature F |
|---|---|
| Process steam condensate | 130–190 |
| Cooling water from: | |
|     Furnace doors | 90–130 |
|     Bearings | 90–190 |
|     Welding machines | 90–190 |
|     Injection molding machines | 90–190 |
|     Annealing furnaces | 150–450 |
|     Forming dies | 80–190 |
|     Air compressors | 80–120 |
|     Pumps | 80–190 |
|     Internal combustion engines | 150–250 |
|     Air conditioning and | |
|       refrigeration condensers | 90–110 |
|     Liquid still condensers | 90–190 |
| Drying, baking and curing ovens | 200–450 |
| Hot processed liquids | 90–450 |
| Hot processed solids | 200–450 |

## WASTE HEAT RECOVERY APPLICATIONS

To use waste heat from sources such as those above, one often wishes to transfer the heat in one fluid stream to another (e.g., from flue gas to feedwater or combustion air). The device which accomplishes the transfer is called a heat exchanger. In the discussion immediately below is a listing of common uses for waste heat energy and, in some cases, the name of the heat exchanger that would normally be applied in each particular case.

The equipment that is used to recover waste heat can range from something as simple as a pipe or duct to something as complex as a waste heat boiler.

Some applications of waste heat are as follows:

• Medium to high temperature exhaust gases can be used to preheat the combustion air for:

    Boilers using air preheaters

    Furnaces using recuperators

Ovens using recuperators

Gas turbines using regenerators

• Low to medium temperature exhaust gases can be used to preheat boiler feedwater or boiler makeup water using *economizers,* which are simply gas-to-liquid water heating devices.

• Exhaust gases and cooling water from condensers can be used to preheat liquid and/or solid feedstocks in industrial processes. Finned tubes and tube-in-shell *heat exchangers* are used.

• Exhaust gases can be used to generate steam in *waste heat boilers* to produce electrical power, mechanical power, process steam, and any combination of above.

• Waste heat may be transferred to liquid or gaseous process units directly through pipes and ducts or indirectly through a secondary fluid such as steam or oil.

• Waste heat may be transferred to an intermediate fluid by heat exchangers or waste heat boilers, or it may be used by circulating the hot exit gas through pipes or ducts. Waste heat can be used to operate an absorption cooling unit for air conditioning or refrigeration.

## THE WASTE HEAT RECOVERY SURVEY

In order to identify sources of waste heat, a survey is usually made. Figure 5-1 illustrates a survey form which can be used for the Waste Heat Audit. It is important to record flow and temperature of waste gases.

Composition data is required for heat recovery and system design calculations. Be sure to note contaminants since this factor could limit the type of heat recovery equipment to apply. Contaminants can foul or plug heat exchangers.

Operation schedule affects the economics and type of equipment to be specified. For example, an incinerator that is only used one shift per day may require a different method of recovering discharges than if it were used three shifts a day. A heat exchanger used for waste heat recovery in this service would soon deteriorate due to metal fatigue. A different type of heat recovery incinerator utilizing heat storage materials such as rock or ceramic would be more suitable.

SURVEY FORM FOR INDUSTRIAL PROCESS UNITS

NAME OF PROCESS UNIT _____ INVENTORY NUMBER _____

LOCATION OF PROCESS UNIT, PLANT NAME _____ BUILDING _____

MANUFACTURER _____ MODEL _____ SERIAL NUMBER _____

| | NAME | FIRING RATE | HHV | TEMPERATURE OF | | | FLUE GAS COMPOSITION % VOLUME | | | | |
|---|---|---|---|---|---|---|---|---|---|---|---|
| | | | | COMB. AIR | FUEL | STACK | $CO_2$ | $O_2$ | CO | CH | $N_2$ |
| PRIMARY FUEL | | | | | | | | | | | |
| FIRST ALTERNAT. | | | | | | | | | | | |
| SECOND ALTERNAT. | | | | | | | | | | | |

| | FLOW PATH 1 | FLOW PATH 2 | FLOW PATH 3 | FLOW PATH 4 |
|---|---|---|---|---|
| FLUID COMPOSITION | | | | |
| FLOW RATE | | | | |
| INLET TEMPERATURE | | | | |
| OUTLET TEMPERATURE | | | | |
| DESCRIPTION | | | | |

ANNUAL HOURS OPERATION _____ ANNUAL CAPACITY FACTOR, % _____

ANNUAL FUEL CONSUMPTION: PRIMARY FUEL _____; FIRST ALTERN. _____ SEC. ALTERN. _____

PRESENT FUEL COST:    PRIMARY FUEL _____; FIRST ALTERN. _____ SEC. ALTERN. _____

ANNUAL ELECTRICAL ENERGY CONSUMPTION, KWHR. _____

PRESENT ELECTRICAL ENERGY RATE _____

CONTAMINANTS _____

Figure 5-1.  Waste Heat Survey

## WASTE HEAT RECOVERY CALCULATIONS

From the heat balance (Chapter 6), the heat recovered from the source is determined by Formula 5-1.

$$q = m\ c_p\ \Delta T \qquad \text{\textit{Formula (5-1)}}$$

Where $q$ = heat recovered, Btu/hr
$m$ = mass flow rate lb/hr
$c_p$ = specific heat of fluid, Btu/lb°F
$\Delta T$ = temperature change of gas or liquid during heat recovery °F

If the flow is air, then Formula 5-1 can be expressed as

$$q = 1.08\ \text{cfm}\ \ \Delta T \qquad \text{\textit{Formula (5-2)}}$$

Where cfm = volume flow rate in standard cubic feet per minute

If the flow is water, then Formula 5-1 can be expressed as

$$q = 500\ \text{gpm}\ \Delta T \qquad \text{\textit{Formula (5-3)}}$$

Where gpm = volume flow rate in gallons per minute

### Example Problem 5-1

A Waste Heat Audit survey indicates 10,000 lb/hr of water at 190°F is discharged to the sewer. How much heat can be saved by utilizing this fluid as makeup to the boiler instead of the 70°F feedwater supply? Fuel cost is $6 per million Btu, boiler efficiency .8, and hours of operation 4000.

### Analysis

$q = mc_p\ \Delta T = 10{,}000 \times (190{-}70) = 1.2 \times 10^6$ Btu/hr
Savings $= 1.2 \times 10^6 \times 4000 \times \$6/10^6 /.8 = \$36{,}000$

### Heat Transfer by Convection

Convection is the transfer of heat to or from a fluid, gas, or liquid. Formula 5-4 is indicative of the basic form of convective heat transfer. $U_0$, in this case, represents the convection film conductance, Btu/ft$^2 \cdot$ hr $\cdot$ °F.

Heat transferred for heat exchanger applications is predominantly a combination of conduction and convection expressed as

$$q = U_0 \, A \Delta T_m \qquad \qquad \textit{Formula (5-4)}$$

Where  $q$   = rate of heat flow by convection, Btu/hr
       $U_0$  = is the overall heat transfer coefficient Btu/ft$^2$ · hr · °F
       $A$   = is the area of the tubes in square feet
       $\Delta T_m$ = is the logarithmic mean temperature difference and represents the situation where the temperature of two fluids change as they transverse the surface.

$$\Delta T_m = \frac{\Delta T_1 - \Delta T_2}{\text{Log}_e \, [\Delta T_1/\Delta T_2]} \qquad \qquad \textit{Formula (5-5)}$$

To understand the different logarithmic mean temperature relationships, Figure 5-2 should be used. Referring to Figure 5-2, the $\Delta T_m$ for the counterflow heat exchanger is

$$\Delta T_m = \frac{(t_1 - t_2') - (t_2 - t_1')}{\text{Log}_e \, [t_1 - t_2'/t_2 - t_1']} \qquad \qquad \textit{Formula (5-6)}$$

The $\Delta T_m$ for the parallel flow heat exchanger is

$$\Delta T_m = \frac{(t_1 - t_1') - (t_2 - t_2')}{\text{Log}_e \, [t_1 - t_1'/t_2 - t_2']} \qquad \qquad \textit{Formula (5-7)}$$

## HEAT TRANSFER BY RADIATION

Radiation is the transfer of heat energy by electromagnetic means between two materials whose surfaces "see" each other. The governing equation is known as the Stefan-Boltzmann equation, and is written

$$q = \sigma F_e F_a A \, (T^4_{aba_1} - T^4_{aba_2}) \qquad \qquad \textit{Formula (5-8)}$$

Again, $q$ and $A$ are as defined for conduction and convection, and $T_{aba_1}$ and $T_{aba_2}$ are the absolute temperatures of the two surfaces involved. The factor $F_e$ is a function of the condition of the radiation surfaces and in some cases the areas of the surfaces. The factor $F_a$ is

A. COUNTERFLOW

B. PARALLEL FLOW

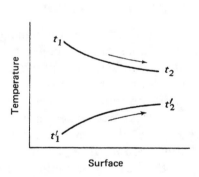

**Figure 5-2. Temperature Relationships for Heat Exchangers**

the configuration factor and is a function of the areas and their positions. Both $F_e$ and $F_a$ are dimensionless. The Stefan-Boltzmann constant $\sigma$ is equal to 0.1714 But/h·ft$^2$·$R^4$. Although radiation is usually associated with solid surfaces, certain gases can emit and absorb radiation. These include the so-called nonpolar molecular gases such as $H_2O$, $CO_2$, CO, $SO_2$, $NH_3$ and the hydrocarbons. Some of these gases are present in every combustion process.

It is convenient in certain calculations to express the heat transferred by radiation in the form of

$$q = h_r A (T_1 - T_2) \qquad \text{Formula (5-9)}$$

where the coefficient of radiation, $h_r$, is defined as

$$h_r = \frac{\sigma F_e F_a (T^4 aba_1 - T^4 aba_2)}{(T_1 - T_1)} \qquad \text{Formula (5-10)}$$

Note that $h_r$ is still dependent on the factors $F_e$ and $F_a$ and also on the absolute temperatures to the fourth power. The main advantage is that in Formula 5-10 the rate of heat flow by radiation is a function of the temperature difference and can be combined with the coefficient of convection to determine the total heat flow to or from a surface.

## WASTE HEAT RECOVERY EQUIPMENT

Industrial heat exchangers have many pseudonyms. They are sometimes called recuperators, regenerators, waste heat steam generators, condensers, heat wheels, temperature and moisture exchangers, etc. Whatever name they may have, they all perform one basic function: the transfer of heat.

Heat exchangers are characterized as single or multipass gas to gas, liquid to gas, liquid to liquid, evaporator, condenser, parallel flow, counterflow, or crossflow. The terms single or multipass refer to the heating or cooling media passing over the heat transfer surface once or a number of times. Multipass flow involves the use of internal baffles. The next three terms refer to the two fluids between which heat is transferred in the heat exchanger and imply that no phase changes occur in those fluids. Here the term "fluid" is used in the most general sense. Thus, we can say that these terms apply to non-evaporator and noncondensing heat exchangers. The term evaporator applies to a heat exchanger in which heat is transferred to an evaporating (boiling) liquid, while a condenser is a heat exchanger in which heat is removed from a condensing vapor. A parallel flow heat exchanger is one in which both fluids flow in approximately the same direction, whereas in counterflow the two fluids move in opposite directions. When the two fluids move at right angles to each other, the heat exchanger is considered to be of the crossflow type.

The principal methods of reclaiming waste heat in industrial plants make use of heat exchangers. The heat exchanger is a system which separates the stream containing waste heat and the medium which is to absorb it but which allows the flow of heat across the separation boundaries. The reasons for separating the two streams may be any of the following:

(1) A pressure difference may exist between the two streams of fluid. The rigid boundaries of the heat exchanger can be designed to withstand the pressure difference.

(2) In many, if not most, cases the one stream would contaminate the other, if they were permitted to mix. The heat exchanger prevents mixing.

(3) Heat exchangers permit the use of an intermediate fluid better suited than either of the principal exchange media for transporting waste heat through long distances. The secondary fluid is often steam, but another substance may be selected for special properties.

(4) Certain types of heat exchangers, specifically the heat wheel, are capable of transferring liquids as well as heat. Vapors being cooled in the gases are condensed in the wheel and later re-evaporated into the gas being heated. This can result in improved humidity and/or process control, abatement of atmospheric air pollution and conservation of valuable resources.

The various names or designations applied to heat exchangers are partly an attempt to describe their function and partly the result of tradition within certain industries. For example, a recuperator is a heat exchanger which recovers waste heat from the exhaust gases of a furnace to heat the incoming air for combustion. This is the name used in both the steel and the glass making industries. The heat exchanger performing the same function in the steam generator of an electric power plant is termed an air preheater, and in the case of a gas turbine plant, a regenerator.

However, in the glass and steel industries the word regenerator refers to two chambers of brick checkerwork which alternately absorb heat from the exhaust gases and then give up part of that heat to the incoming air. The flows of flue gas and of air are periodically reversed by valves so that one chamber of the regenerator is being heated by the products of combustion while the other is being cooled by the incoming air. Regenerators are often more expensive to buy and more expensive to maintain than are recuperators, and their application is primarily in glass melt tanks and in open hearth steel furnaces.

It must be pointed out, however, that although their functions are similar, the three heat exchangers mentioned above may be structurally quite different as well as different in their principal modes of heat transfer. A more complete description of the various industrial heat exchangers follows later in this chapter, and details of their differences will be clarified.

The specification of an industrial heat exchanger must include the heat exchange capacity, the temperatures of the fluids, the allowable pressure drop in each fluid path, and the properties and volumetric flow of the fluids entering the exchanger. These specifications will determine construction parameters and thus the cost of the heat exchanger. The final design will be a compromise between pressure drop, heat exchanger effectiveness, and cost. Decisions leading to that final design will balance out the cost of maintenance and operation of the overall system against the fixed costs in such a way as to minimize the total. Advice on selection and design of heat exchangers is available from vendors.

The essential parameters which should be known in order to make an optimum choice of waste heat recovery devices are

- Temperature of waste heat fluid
- Flow rate of waste heat fluid
- Chemical composition of waste heat fluid
- Minimum allowable temperature of waste heat fluid
- Temperature of heated fluid
- Chemical composition of heated fluid
- Maximum allowable temperature of heated fluid
- Control temperature, if control required

In the rest of this chapter, some common types of waste heat recovery devices are discussed in some detail.

## GAS TO GAS HEAT EXCHANGERS

### Recuperators

The simplest configuration for a heat exchanger is the metallic radiation recuperator which consists of two concentric lengths of metal tubing as shown in Figure 5-3.

The inner tube carries the hot exhaust gases while the external annulus carries the combustion air from the atmosphere to the air inlets of the furnace burners. The hot gases are cooled by the incoming combustion air which now carries additional energy into the combustion chamber. This is energy which does not have to be supplied by the fuel; consequently, less fuel is burned for a given furnace loading. The saving in fuel also means a decrease in combustion air and therefore stack losses are decreased not only by lowering the stack gas temperatures but also by discharging smaller quantities of exhaust gas. This particular recuperator gets its name from the fact that a substantial portion of the heat transfer from the hot gases to

**Figure 5-3. Diagram of Metallic Radiation Recuperator**

the surface of the inner tube takes place by radiative heat transfer. The cold air in the annulus, however, is almost transparent to infrared radiation so that only convection heat transfer takes place to the incoming air. As shown in the diagram, the two gas flows are usually parallel, although the configuration would be simpler and the heat transfer more efficient if the flows were opposed in direction (or counterflow). The reason for the use of parallel flow is that recuperators frequently serve the additional function of cooling the duct carrying away the exhaust gases and consequently extending its service life.

The inner tube is often fabricated from high temperature materials such as stainless steels of high nickel content. The large temperature differential at the inlet causes differential expansion, since the outer shell is usually of a different and less expensive material. The mechanical design must take this effect into account. More elaborate designs of radiation recuperators incorporate two sections: the bottom operating in parallel flow and the upper section using the more efficient counterflow arrangement. Because of the large axial expansions experienced and the stress conditions at the bottom of the recuperator, the unit is often supported at the top by a free-standing support frame with an expansion joint between the furnace and recuperator.

A second common configuration for recuperators is called the tube type or convective recuperator. As seen in the schematic diagram of Figure 5-4, the hot gases are carried through a number of parallel small diameter tubes, while the incoming air to be heated enters a shell surrounding the tubes and passes over the hot tubes one or more times in a direction normal to their axes.

If the tubes are baffled to allow the gas to pass over them twice, the heat exchanger is termed a two-pass recuperator; if two baffles are used, a three-pass recuperator, etc. Although baffling increases both the cost of the exchanger and the pressure drop in the combustion air path, it increases the effectiveness of heat exchange. Shell- and tube-type recuperators are generally more compact and have a higher effectiveness than radiation recuperators, because of the larger heat transfer area made possible through the use of multiple tubes and multiple passes of the gases.

The principal limitation on the heat recovery of metal recuperators is the reduced life of the liner at inlet temperatures exceeding 2000F. At this temperature, it is necessary to use the less efficient arrangement of parallel flows of exhaust gas and coolant in order to maintain sufficient cooling of the inner shell. In addition, when furnace combustion air flow is dropped back because of reduced load, the heat transfer rate from hot waste gases to preheat combustion air becomes excessive, causing rapid surface deterioration. Then, it is usually necessary to provide an ambient air bypass to cool the exhaust gases.

**Figure 5-4. Diagram of Convective-Type Recuperator**

In order to overcome the temperature limitations of metal recuperators, ceramic tube recuperators have been developed whose materials allow operation on the gas side to 2800°F and on the preheated air side to 2200°F on an experimental basis and to 1500°F on a more or less practical basis. Early ceramic recuperators were built of tile and joined with furnace cement, and thermal cycling caused cracking of joints and rapid deterioration of the tubes. Later developments introduced various kinds of short silicon carbide tubes which can be joined by flexible seals located in the air headers. This kind of patented design illustrated in Figure 5-5 maintains the seals at comparatively low temperatures and has reduced the seal leakage rates to a few percent.

Figure 5-5.  Ceramic Recuperator

Earlier designs had experienced leakage rates from 8 to 60 percent. The new designs are reported to last two years with air preheat
temperatures as high as 1300°F, with much lower leakage rates.

An alternative arrangement for the convective type recuperator, in which the cold combustion air is heated in a bank of parallel
vertical tubes that extend into the flue gas stream, is shown sche-
matically in Figure 5-6. The advantage claimed for this arrangement
is the ease of replacing individual tubes, which can be done during
full capacity furnace operation. This minimizes the cost, the inconvenience  and possible furnace damage due to a shutdown forced by
recuperator failure.

For maximum effectiveness of heat transfer, combinations of
radiation type and convective type recuperators are used, with the
convective type always following the high temperature radiation recuperator. A schematic diagram of this arrangement is seen in
Figure 5-7.

Although the use of recuperators conserves fuel in industrial
furnaces  and although their original cost is relatively modest, the
purchase of the unit is often just the beginning of a somewhat more

**Figure 5-6. Diagram of Vertical Tube-Within-Tube Recuperator**

extensive capital improvement program. The use of a recuperator, which raises the temperature of the incoming combustion air, may require purchase of high temperature burners, larger diameter air lines with flexible fittings to allow for expansion, cold air lines for cooling the burners, modified combustion controls to maintain the required air/fuel ratio despite variable recuperator heating, stack dampers, cold air bleeds, controls to protect the recuperator during blower failure or power failures and larger fans to overcome the additional pressure drop in the recuperator. It is vitally important to protect the recuperator against damage due to excessive temperatures, since the cost of rebuilding a damaged recuperator may be as high as 90 percent of the initial cost of manufacture, and the drop in efficiency of a damaged recuperator may easily increase fuel costs by 10 to 15 percent.

Figure 5-8 shows a schematic diagram of one radiant tube burner fitted with a radiation recuperator. With such a short stack, it is necessary to use two annuli for the incoming air to achieve reasonable heat exchange efficiencies.

**Figure 5-7. Diagram of Combined Radiation and Convective Type Recuperator**

Recuperators are used for recovering heat from exhaust gases to heat other gases in the medium to high temperature range. Some typical applications are in soaking ovens, annealing ovens, melting furnaces, afterburners and gas incinerators, radiant-tube burners, reheat furnaces, and other gas to gas waste heat recovery applications in the medium to high temperature range.

**Heat Wheels**

A rotary regenerator (also called an air preheater or a heat wheel) is finding increasing applications in low to medium temperature waste heat recovery. Figure 5-9 is a sketch illustrating the application of a heat wheel. It is a sizable porous disk, fabricated from some material having a fairly high heat capacity, which rotates be-

**Figure 5-8. Diagram of a Small Radiation-Type Recuperator Fitted to
a Radiant Tube Burner**

tween two side-by-side ducts: one a cold gas duct, the other a hot gas
duct. The axis of the disk is located parallel to, and on the partition
between, the two ducts. As the disk slowly rotates, sensible heat (and,
in some cases, moisture that contains latent heat) is transferred to the
disk by the hot air and, as the disk rotates, from the disk to the cold
air. The overall efficiency of sensible heat transfer for this kind of
regenerator can be as high as 85 percent. Heat wheels have been built
as large as 70 feet in diameter with air capacities up to 40,000 ft³/
min. Multiple units can be used in parallel. This may help to prevent
a mismatch between capacity requirements and the limited number
of sizes available in packaged units. In very large installations such as

**Figure 5-9.** Heat and Moisture Recovery Using a
Heat Wheel Type Regenerator

those required for preheating combustion air in fixed station electrical generating stations, the units are custom designed.

The limitation on temperature range for the heat wheel is primarily due to mechanical difficulties introduced by uneven expansion of the rotating wheel when the temperature differences mean large differential expansion, causing excessive deformations of the wheel and thus difficulties in maintaining adequate air seals between duct and wheel.

Heat wheels are available in four types. The first consists of a metal frame packed with a core of knitted mesh stainless steel or aluminum wire, resembling that found in the common metallic kitchen pot scraper; the second, called a laminar wheel, is fabricated from corrugated metal and is composed of many parallel flow passages; the third variety is also a laminar wheel but is constructed from a ceramic matrix of honeycomb configuration. This type is used for higher temperature applications with a present-day limit of about 1600°F. The fourth variety is of laminar construction, but the flow passages are coated with a hygroscopic material so that latent heat may be recovered. The packing material of the hygroscopic wheel may be any of a number of materials. The hygroscopic material is often termed a desiccant.

Most industrial stack gases contain water vapor (since water vapor is a product of the combustion of all hydrocarbon fuels and since water is introduced into many industrial processes) and part of the process water evaporates as it is exposed to the hot gas stream. Each pound of water requires approximately 1000 Btu for its evaporation at atmospheric pressure; thus each pound of water vapor leaving in the exit stream will carry 1000 Btu of energy with it. This latent heat may be a substantial fraction of the sensible energy in the exit gas stream. A hygroscopic material is one such as lithium chloride (LiCl) which readily absorbs water vapor. Lithium chloride is a solid which absorbs water to form a hydrate, $LiCl \cdot H_2O$, in which one molecule of lithium chloride combines with one molecule of water. Thus, the ratio of water to lithium chloride in $LiCl \cdot H_2O$ is 3/7 by weight. In a hygroscopic heat wheel, the hot gas stream gives up part of its water vapor to the coating; the cool gases which enter the wheel to be heated are drier than those in the inlet duct, and part of the absorbed water is given up to the incoming gas stream. The latent heat of the water adds directly to the total quantity of recovered waste heat. The efficiency of recovery of water vapor can be as high as 50 percent.

Since the pores of heat wheels carry a small amount of gas from the exhaust to the intake duct, cross contamination can result. If this contamination is undesirable, the carryover of exhaust gas can be partially eliminated by the addition of a purge section where a small amount of clean air is blown through the wheel and then exhausted to the atmosphere, thereby clearing the passages of exhaust gas. Figure 5-10 illustrates the features of an installation using a purge section. Note that additional seals are required to separate the purge ducts. Common practice is to use about six air changes of clean air for purging. This limits gas contamination to as little as 0.04 percent and particle contamination to less than 0.2 percent in laminar wheels, and cross contamination to less than 1 percent in packed wheels. If inlet gas temperature is to be held constant, regardless of heating loads and exhaust gas temperatures, then the heat wheel must be driven at variable speed. This requires a variable speed drive and a speed control system using an inlet air temperature sensor as the control element. This feature, however, adds considerably

Figure 5-10. Heat Wheel Equipped with Purge Section to
Clear Contaminants from the Heat Transfer Surface

to the cost and complexity of the system. When operating with outside air in periods of high humidity and sub-zero temperatures, heat
wheels may require preheat systems to prevent frost formation.
When handling gases which contain water-soluble, greasy or adhesive
contaminants or large concentrations of process dust, air filters may
be required in the exhaust system upstream from the heat wheel.

One application of heat wheels is in space heating situations
where unusually large quantities of ventilation air are required for
health or safety reasons. As many as 20 or 30 air changes per hour
may be required to remove toxic gases or to prevent the accumulation of explosive mixtures. Comfort heating for that quantity of ventilation air is frequently expensive enough to make the use of heat
wheels economical. In the summer season the heat wheel can be
used to cool the incoming air from the cold exhaust air, reducing the
air-conditioning load by as much as 50 percent. It should be pointed
out that in many circumstances where large ventilating requirements
are mandatory, a better solution than the installation of heat wheels
may be the use of local ventilation systems to reduce the hazards
and/or the use of infrared comfort heating at principal work areas.

Heat wheels are finding increasing use for process heat recovery
in low and moderate temperature environments. Typical applications
would be curing or drying ovens and air preheaters in all sizes for
industrial and utility boilers.

## Air Preheaters

Passive gas to gas regenerators, sometimes called air preheaters, are available for applications which cannot tolerate any cross contamination. They are constructed of alternate channels (see Figure 5-11) which put the flows of the heating and the heated gases in close contact with each other, separated only by a thin wall of conductive metal. They occupy more volume and are more expensive to construct than are heat wheels, since a much greater heat transfer surface area is required for the same efficiency. An advantage, besides the absence of cross contamination, is the decreased mechanical complexity since no drive mechanism is required. However, it becomes more difficult to achieve temperature control with the passive regeneration, and, if this is a requirement, some of the advantages of its basic simplicity are lost.

Figure 5-11. A Passive Gas to Gas Regenerator

Gas to gas regenerators are used for recovering heat from exhaust gases to heat other gases in the low to medium temperature range. A list of typical applications follows:

- Heat and moisture recovery from building heating and ventilation systems
- Heat and moisture recovery from moist rooms and swimming pools

- Reduction of building air-conditioner loads
- Recovery of heat and water from wet industrial processes
- Heat recovery from steam boiler exhaust gases
- Heat recovery from gas and vapor incinerators
- Heat recovery from baking, drying, and curing ovens
- Heat recovery from gas turbine exhausts
- Heat recovery from other gas to gas applications in the low through high temperature range.

### Heat-Pipe Exchangers

The heat pipe is a heat transfer element that has only recently become commercial, but it shows promise as an industrial waste heat recovery option because of its high efficiency and compact size. In use, it operates as a passive gas to gas finned-tube regenerator. As can be seen in Figure 5-12, the elements form a bundle of heat pipes which extend through the exhaust and inlet ducts in a pattern that resembles the structured finned coil heat exchangers. Each pipe, however, is a separate sealed element consisting of an annular wick on the inside of the full length of the tube, in which an appropriate heat transfer fluid is entrained.

Figure 5-13 shows how the heat absorbed from hot exhaust gases evaporates the entrained fluid, causing the vapor to collect in the center core. The latent heat of vaporization is carried in the vapor to the cold end of the heat pipe located in the cold gas duct. Here the vapor condenses, giving up its latent heat. The condensed liquid is then carried by capillary (and/or gravity) action back to the hot end where it is recycled. The heat pipe is compact and efficient because: (1) the finned-tube bundle is inherently a good configuration for convective heat transfer in both gas ducts, and (2) the evaporative-condensing cycle within the heat tubes is a highly efficient way of transferring the heat internally. It is also free from cross contamination. Possible applications include

- Drying, curing and baking ovens
- Waste steam reclamation
- Air preheaters in steam boilers
- Air dryers

- Brick kilns (secondary recovery)
- Reverberatory furnaces (secondary recovery)
- Heating, ventilating and air-conditioning systems

**Figure 5-12.  Heat Pipe Bundle Incorporated in Gas to Gas Regenerator**

**Figure 5-13.  Heat Pipe Schematic**

## GAS OR LIQUID TO LIQUID REGENERATORS

### Finned-Tube Heat Exchangers

When waste heat in exhaust gases is recovered for heating liquids for purposes such as providing domestic hot water, heating the feedwater for steam boilers, or for hot water space heating, the finned-tube heat exchanger is generally used. Round tubes are connected together in bundles to contain the heated liquid, and fins are welded or otherwise attached to the outside of the tubes to provide additional surface area for removing the waste heat in the gases.

Figure 5-14 shows the usual arrangement for the finned-tube exchanger positioned in a duct and details of a typical finned-tube construction. This particular type of application is more commonly known as an economizer. The tubes are often connected all in series

**Figure 5-14. Finned-Tube Gas to Liquid Regenerator (Economizer)**

but can also be arranged in series-parallel bundles to control the liquid side pressure drop. The air side pressure drop is controlled by the spacing of the tubes and the number of rows of tubes within the duct.

Finned-tube exchangers are available prepackaged in modular sizes or can be made up to custom specifications very rapidly from standard components. Temperature control of the heated liquid is usually provided by a bypass duct arrangement which varies the flow rate of hot gases over the heat exchanger. Materials for the tubes and the fins can be selected to withstand corrosive liquids and/or corrosive exhaust gases.

Finned-tube heat exchangers are used to recover waste heat in the low to medium temperature range from exhaust gases for heating liquids. Typical applications are domestic hot water heating, heating boiler feedwater, hot water space heating, absorption-type refrigeration or air conditioning and heating process liquids.

### Shell and Tube Heat Exchanger

When the medium containing waste heat is a liquid or a vapor which heats another liquid, then the shell and tube heat exchanger must be used since both paths must be sealed to contain the pressures of their respective fluids. The shell contains the tube bundle, and usually internal baffles, to direct the fluid in the shell over the tubes in multiple passes. The shell is inherently weaker than the tubes so that the higher pressure fluid is circulated in the tubes while the lower pressure fluid flows through the shell. When a vapor contains the waste heat, it usually condenses, giving up its latent heat to the liquid being heated. In this application, the vapor is almost invariably contained within the shell. If the reverse is attempted, the condensation of vapors within small diameter parallel tubes causes flow instabilities. Tube and shell heat exchangers are available in a wide range of standard sizes with many combinations of materials for the tubes and shells.

Typical applications of shell and tube heat exchangers include heating liquids with the heat contained by condensates from refrigeration and air-conditioning systems; condensate from process steam; coolants from furnace doors, grates, and pipe supports; coolants from engines, air compressors, bearings, and lubricants; and the condensates from distillation processes.

## Waste Heat Boilers

Waste heat boilers are ordinarily water tube boilers in which the hot exhaust gases from gas turbines, incinerators, etc., pass over a number of parallel tubes containing water. The water is vaporized in the tubes and collected in a steam drum from which it is drawn off for use as heating or processing steam.

Figure 5-15 indicates one arrangement that is used, where the exhaust gases pass over the water tubes twice before they are exhausted to the air. Because the exhaust gases are usually in the medium temperature range and in order to conserve space, a more compact boiler can be produced if the water tubes are finned in order to increase the effective heat transfer area on the gas side. The diagram shows a mud drum, a set of tubes over which the hot gases make a double pass, and a steam drum which collects the steam generated above the water surface. The pressure at which the steam is generated and the rate of steam production depend on the temperature of the hot gases entering the boiler, the flow rate of the hot gases, and the efficiency of the boiler. The pressure of a pure vapor in the presence of its liquid is a function of the temperature of the liquid from which it is evaporated. The steam tables tabulate this relationship between saturation pressure and temperature. Should the waste heat in the exhaust gases be insufficient for generating the required amount of process steam, it is sometimes possible to add auxiliary burners which burn fuel in the waste heat boiler or to add an afterburner to the exhaust gas duct just ahead of the boiler. Waste heat boilers are built in capacities from less than a thousand to almost a million ft³/min. of exhaust gas.

Typical applications of waste heat boilers are to recover energy from the exhausts of gas turbines, reciprocating engines, incinerators, and furnaces.

## Gas and Vapor Expanders

Industrial steam and gas turbines are in an advanced state of development and readily available on a commercial basis. Recently,

**Figure 5-15. Waste Heat Boiler for Heat Recovery from
Gas Turbines or Incinerators**

special gas turbine designs for low pressure waste gases have become
available; for example, a turbine is available for operation from the
top gases of a blast furnace. In this case, as much as 20 MW of power
could be generated, representing a recovery of 20 to 30 percent of
the available energy of the furnace exhaust gas stream. Maximum top
pressures are of the order of 40 lb/in$^2$ g.

Perhaps of greater applicability than the last example are steam turbines used for producing mechanical work or for driving electrical generators. After removing the necessary energy for doing work, the steam turbine exhausts partially spent steam at a lower pressure than the inlet pressure. The energy in the turbine exhaust stream can then be used for process heat in the usual ways. Steam turbines are classified as back-pressure turbines, available with allowable exit pressure operation above 400 lb/in² g, or condensing turbines which operate below atmospheric exit pressures. The steam used for driving the turbines can be generated in direct fired or waste heat boilers. A list of typical applications for gas and vapor expanders follows:

- Electrical power generation
- Compressor drives
- Pump drives
- Fan drives

### Heat Pumps

In the commercial options previously discussed in this chapter, we find waste heat being transferred from a hot fluid to a fluid at a lower temperature. Heat must flow spontaneously "downhill", that is, from a system at high temperature to one at a lower temperature. This can be expressed scientifically in a number of ways—all the variations of the statement of the second law of thermodynamics. The practical impact of these statements is that energy, as it is transformed again and again and transferred from system to system, becomes less and less available for use. Eventually that energy has such low intensity (resides in a medium at such low temperature) that it is no longer available at all to perform a useful function. It has been taken as a general rule of thumb in industrial operations that fluids with temperatures less than 250°F are of little value for waste heat extraction; flue gases should not be cooled below 250°F (or, better, 300°F to provide a safe margin), because of the risk of condensation of corrosive liquids. However, as fuel costs continue to rise, such waste heat can be used economically for space heating and other low temperature applications. It is possible to reverse the direction of spontaneous energy flow by the use of a thermodynamic system known as a heat pump.

This device consists of two heat exchangers, a compressor and an expansion device. A liquid or a mixture of liquid and vapor of a pure chemical species flows through an evaporator, where it absorbs heat at low temperature and, in doing so, is completely vaporized. The low temperature vapor is compressed by a compressor which requires external work. The work done on the vapor raises its pressure and temperature to a level where its energy becomes available for use. The vapor flows through a condenser where it gives up its energy as it condenses to a liquid. The liquid is then expanded through a device back to the evaporator where the cycle repeats. The heat pump was developed as a space heating system where low temperature energy from the ambient air, water, or earth is raised to heating system temperatures by doing compression work with an electric motor-driven compressor. The performance of the heat pump is ordinarily described in terms of the coefficient of performance or COP, which is defined as

$$\text{COP} = \frac{\text{Heat transferred in condenser}}{\text{Compressor work}} \qquad \textit{Formula (5-11)}$$

which in an ideal heat pump is found as

$$\text{COP} = \frac{T_H}{T_H - T_L} \qquad \textit{Formula (5-12)}$$

where $T_L$ is the temperature at which waste heat is extracted from the low temperature medium and $T_H$ is the high temperature at which heat is given up by the pump as useful energy. The coefficient of performance expresses the economy of heat transfer.

In the past, the heat pump has not been applied generally to industrial applications. However, several manufacturers are now redeveloping their domestic heat pump systems as well as new equipment for industrial use. The best applications for the device in this new context are not yet clear, but it may well make possible the use of large quantities of low-grade waste heat with relatively small expenditures of work.

## Summary

Table 5-4 presents the collation of a number of significant attributes of the most common types of industrial heat exchangers in matrix form. This matrix allows rapid comparisons to be made in selecting competing types of heat exchangers. The characteristics given in the table for each type of heat exchanger are allowable temperature range, ability to transfer moisture, ability to withstand large temperature differentials, availability as packaged units, suitability for retrofitting, and compactness and the allowable combinations of heat transfer fluids.

Table 5-4. Operation and Application Characteristics of
Industrial Heat Exchangers

| COMMERCIAL HEAT TRANSFER EQUIPMENT \ SPECIFICATIONS FOR WASTE RECOVERY UNIT | Low Temperature Sub-Zero – 250°F | Intermediate Temp. 250°F – 1200°F | High Temperature 1200°F – 2000°F | Recovers Moisture | Large Temperature Differentials Permitted | Packaged Units Available | Can Be Retrofit | No Cross-Contamination | Compact Size | Gas-to-Gas Heat Exchange | Gas-to-Liquid Heat Exchanger | Liquid-to-Liquid Heat Exchanger | Corrosive Gases Permitted with Special Construction |
|---|---|---|---|---|---|---|---|---|---|---|---|---|---|
| Radiation Recuperator | | | • | | • | 1 | • | • | | • | | | • |
| Convection Recuperator | | • | • | | • | • | • | • | | • | | | • |
| Metallic Heat Wheel | • | • | | 2 | • | • | | 3 | • | • | | | • |
| Hygroscopic Heat Wheel | • | | | • | • | • | | 3 | • | • | | | |
| Ceramic Heat Wheel | | • | • | | • | • | | | • | • | | | • |
| Passive Regenerator | • | • | | | • | • | • | | | • | | | • |
| Finned-Tube Heat Exchanger | • | • | | | • | • | • | • | | | • | | 4 |
| Tube Shell-and-Tube Exchanger | • | • | | | • | • | • | • | | | • | • | |
| Waste Heat Boilers | • | • | | | • | • | • | | | | | • | 4 |
| Heat Pipes | • | • | | | 5 | • | • | • | • | • | | | • |

1. Off-the-shelf items available in small capacities only.
2. Controversial subject. Some authorities claim moisture recovery. Do not advise depending on it.
3. With a purge section added, cross-contamination can be limited to less than 1% by mass.
4. Can be constructed of corrosion-resistant materials, but consider possible extensive damage to equipment caused by leaks or tube ruptures.
5. Allowable temperatures and temperature differential limited by the phase equilibrium properties of the internal fluid.

# 6

# Utility System Optimization

## BASIS OF THERMODYNAMICS

Thermodynamics deals with the relationships between heat and work. It is based on two basic laws of nature: the first and second laws of thermodynamics. The principles are used in the design of equipment such as steam engines, turbines, pumps, and refrigerators, and in practically every process involving a flow of heat or a chemical equilibrium.

First Law: The first law states that energy can neither be created nor destroyed, thus, it is referred to as the law of conservation of energy. Formula 6-1 expresses the first law for the steady state condition.

$$E_2 - E_1 = Q - W \qquad \qquad Formula\ (6\text{-}1)$$

Where

$E_2 - E_1$ is the change in stored energy at the boundary states 1 and 2 of the system

$Q$ is the heat added to the system

$W$ is the work done by the system

Figure 6-1 illustrates a thermodynamic process where mass enters and leaves the system. The potential energy $(Z)$ and the kinetic energy $(V^2/64.2)$ plus the enthalpy represent the stored energy of the mass. Note, $Z$ is the elevation above the reference point in feet, and $V$ is the velocity of the mass in ft/sec. In the case of the steam turbine, the change in $Z$, $V$, and $Q$ are small in comparison to the change in enthalpy. Thus, the energy equation reduces to

$$W/778 = h_1 - h_2 \qquad \qquad Formula\ (6\text{-}2)$$

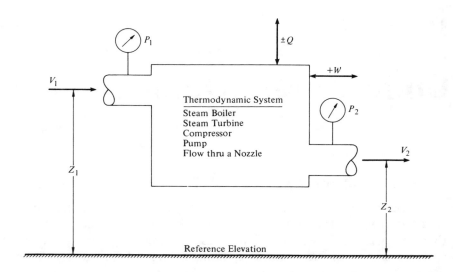

**Figure 6-1.  System Illustrating Conservation of Energy**

Where

$W$   is the work done in ft · lb/lb
$h_1$   is the enthalpy of the entering steam, Btu/lb
$h_2$   is the enthalpy of the exhaust steam, Btu/lb
And 1 Btu equals 778 ft · lb

Second Law: The second law qualifies the first law by discussing the conversion between heat and work. All forms of energy, including work, can be converted to heat, but the converse is not generally true. The Kelvin-Planck statement of the second law of thermodynamics says essentially the following: Only a portion of the heat from a heat work cycle, such as a steam power plant, can be converted to work. The remaining heat must be rejected as heat to a sink of lower temperature (to the atmosphere, for instance).

The Clausius statement, which also deals with the second law, states that heat, in the absence of some form of external assistance, can only flow from a hotter to a colder body.

## THE CARNOT CYCLE

The Carnot cycle is of interest because it is used as a comparison of the efficiency of equipment performance. The Carnot cycle offers the maximum thermal efficiency attainable between any given temperatures of heat source and sink. A thermodynamic cycle is a series of processes forming a closed curve on any system of thermodynamic coordinates. The Carnot cycle is illustrated on a temperature-entropy diagram, Figure 6-2A, and on the Mollier Diagram for superheated steam, Figure 6-2B.

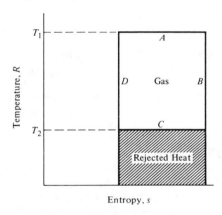

**Figure 6-2A. Temperature—entropy diagram; gas.**

**Figure 6-2B. Mollier diagram; superheated vapor.**

**Figure 6-2. Carnot Cycles**

The cycle consists of the following:

1. Heat addition at constant temperature, resulting in expansion work and changes in enthalpy.
2. Adiabatic isentropic expansion (change in entropy is zero) with expansion work and an equivalent decrease in enthalpy.
3. Constant temperature heat rejection to the surroundings, equal to the compression work and any changes in enthalpy.
4. Adiabatic isentropic compression returning to the starting temperature with compression work and an equivalent increase in enthalpy.

The Carnot cycle is an example of a reversible process and has no counterpart in practice. Nevertheless, this cycle illustrates the principles of thermodynamics. The thermal efficiency for the Carnot cycle is illustrated by Formula (6-3).

$$\text{Thermal efficiency} = \frac{T_1 - T_2}{T_1} \qquad \textit{Formula (6-3)}$$

Where

$T_1$ = Absolute temperature of heat source, °R (Rankine)
$T_2$ = Absolute temperature of heat sink, °R

and absolute temperature is given by Formula 6-4.

Absolute temperature = 460 + temperature in Fahrenheit.

*Formula (6-4)*

$T_2$ is usually based on atmospheric temperature, which is taken as 500°R.

## PROPERTIES OF STEAM PRESSURE
## AND TEMPERATURE

Water boils at 212°F when it is in an open vessel under atmospheric pressure equal to 14.7 psia (pounds per square inch, absolute). Absolute pressure is the amount of pressure exerted by a system on its boundaries and is used to differentiate it from gage pressure. A pressure gage indicates the difference between the pressure of the system and atmospheric pressure.

psia = psig + atmospheric pressure in psia     *Formula (6-5)*

Changing the pressure of water changes the boiling temperature. Thus, water can be vaporized at 170°F, at 300°F, or any other temperature, as long as the applied pressure corresponds to that boiling point.

## SOLID, LIQUID AND
## VAPOR STATES OF A LIQUID

Water, as well as other liquids, can exist in three states: solid, liquid, and vapor. In order to change the state from ice to water or from water to steam, heat must be added. The heat required to change a solid to liquid is called the *latent heat of fusion.* The heat required to change a liquid to a vapor is called the *latent heat of vaporization.*

In condensing steam, heat must be removed. The quantity is exactly equal to the latent heat that went into the water to change it to steam.

Heat supplied to a fluid, during the change of state to a vapor, will not cause the temperature to rise; thus, it is referred to as the latent heat of vaporization. Heat given off by a substance when it condenses from steam to a liquid is called *sensible* heat. Physical properties of water, such as the latent heat of vaporization, also change with variations in pressure.

## USE OF THE STEAM TABLES

Steam properties are illustrated in Chapter 15 by Tables 15-13, 15-14 and 15-15 (ASME Steam Tables). Table 15-13 is referred to as the steam table for saturated steam. Steam properties are shown and correlated to temperature. Table 15-14 is another form of the steam table, where steam properties are shown and correlated to pressure. (Pressure must be converted to psia.) The properties of superheated steam are indicated by Table 15-15.

The term $h_{fg}$ is the latent heat or enthalpy of vaporization. Thus, from Table 15-14, at atmospheric pressure the latent heat of vaporization is 970.3 Btu/lb. At 200 psia the latent heat of vaporization is 842.8 Btu/lb.

The enthalpy $h_f$ represents the amount of heat required to raise one pound of water from 32°F to a liquid state at another temperature. As an example, from Table 15-13, to raise water from 32°F to 170°F will require 137.9 Btu/lb.

### Example Problem 6-1

How much heat is required to raise 100 pounds of water at 126°F to 170°F?

**Answer**

From Table 15-13:
At $126°F - h_f = 93.9$ Btu/lb
At $170°F - h_f = 137.9$ Btu/lb
$Q = 100(137.9 - 93.9) = 44 \times 10^2$ Btu.

## USE OF THE SPECIFIC HEAT CONCEPT

Another physical property of a material is the *specific heat*. The specific heat is defined as the amount of heat required to raise a unit of mass of a substance one degree. For water, it can be seen from the previous example that one Btu of heat is required to raise one lb water 1°F; thus, the specific heat of water $C_p = 1$. Specific heats for other materials are illustrated in Table 6-1. This leads to two equations.

$$Q = wCp \, \Delta T \qquad\qquad Formula\ (6\text{-}6)$$

**Table 6-1. Specific Heat of Various Substances**

| SUBSTANCE | SPECIFIC HEAT | SUBSTANCE | SPECIFIC HEAT |
|---|---|---|---|
| **SOLIDS** | | **LIQUIDS** | |
| ALUMINUM | 0.230 | ALCOHOL | 0.600 |
| ASBESTOS | 0.195 | AMMONIA | 1.100 |
| BRASS | 0.086 | BRINE, CALCIUM (20% SOLUTION) | 0.730 |
| BRICK | 0.220 | BRINE, SODIUM (20% SOLUTION) | 0.810 |
| BRONZE | 0.086 | CARBON TETRACHLORIDE | 0.200 |
| CHALK | 0.215 | CHLOROFORM | 0.230 |
| CONCRETE | 0.270 | ETHER | 0.530 |
| COPPER | 0.093 | GASOLINE | 0.700 |
| CORK | 0.485 | GLYCERINE | 0.576 |
| GLASS, CROWN | 0.161 | KEROSENE | 0.500 |
| GLASS, FLINT | 0.117 | MACHINE OIL | 0.400 |
| GLASS, THERMOMETER | 0.199 | MERCURY | 0.033 |
| GOLD | 0.030 | PETROLEUM | 0.500 |
| GRANITE | 0.192 | SULPHURIC ACID | 0.336 |
| GYPSUM | 0.259 | TURPENTINE | 0.470 |
| ICE | 0.480 | WATER | 1.000 |
| IRON, CAST | 0.130 | WATER, SEA | 0.940 |
| IRON, WROUGHT | 0.114 | | |
| LEAD | 0.031 | **GASES** | |
| LEATHER | 0.360 | AIR | 0.240 |
| LIMESTONE | 0.216 | AMMONIA | 0.520 |
| MARBLE | 0.210 | BROMINE | 0.056 |
| MONEL METAL | 0.128 | CARBON DIOXIDE | 0.200 |
| PORCELAIN | 0.255 | CARBON MONOXIDE | 0.243 |
| RUBBER | 0.481 | CHLOROFORM | 0.144 |
| SILVER | 0.055 | ETHER | 0.428 |
| STEEL | 0.118 | HYDROGEN | 3.410 |
| TIN | 0.045 | METHANE | 0.593 |
| WOOD | 0.330 | NITROGEN | 0.240 |
| ZINC | 0.092 | OXYGEN | 0.220 |
| | | SULPHUR DIOXIDE | 0.154 |
| | | STEAM (SUPERHEATED, 1 PSI) | 0.450 |

Reprinted by permission of The Trane Company.

Where

$Q$   = quantity of heat, Btu
$w$   = weight of substance, lb
$Cp$  = specific heat of substance, Btu per lb °F
$\Delta T$  = temperature change of substance °F

$$Q = MCp\ \Delta T \qquad\qquad\qquad \text{Formula (6-7)}$$

Where

$Q$ = quantity of heat, Btu/hr (Btu/hr is sometimes abbreviated as Btuh)
$M$ = flow rate, lbs/hr
$Cp$ = specific heat, Btu per lb °F
$\Delta T$ = temperature change of substance°F


**Example Problem 6-2**

Check answer to Example Problem 6-1 using Formula 6-6.

**Answer**

$Q$   = $WCp\ \Delta T$
   = $100 \times 1 \times (170 - 126) = 44 \times 10^2$ Btu.

The saturated vapor enthalpy $b_g$ represents the amount of heat necessary to change water at 32°F to steam at a specified temperature and pressure.

**Example Problem 6-3**

How much heat is required to raise 100 pounds of water at 126°F to 15 psig steam?

**Answer**

At 126°F $- b_f = 93.9$ Btu/lb
15 psig corresponds to 30 psia at 250°F
At 30 psia, $b_g = 1164.1$ Btu/lb
$Q = 100(1164.1 - 93.9) = 1070.2 \times 10^2$ Btu.

## USE OF THE LATENT HEAT CONCEPT

When water is raised from $32°F$ to $212°F$ steam, the total heat required is composed of two components:

  (a)  The heat required to raise the temperature of water from $32°F$ to $212°F$; $h_f$ = 180.17 Btu/lb from Table 15-13.
  (b)  The heat required to evaporate the water at $212°F$; $h_{fg}$ = 970.3 Btu/lb from Table 15-13.

Thus, $h_g = h_f + h_{fg}$ = 180.17 + 970.3 = 1150.5 Btu/lb, which agrees with the value of $h_g$ found in Steam Table 15-13.

### Example Problem 6-4

30 psig steam is used for a heat exchanger and returns to the system as 30 psig condensate. What amount of heat is given off to the process fluid?

### Answer

30 psig = 45 psia. This corresponds to an $h_{fg}$ of approximately 928 Btu/lb.

## THE USE OF THE SPECIFIC VOLUME CONCEPT

Another property of water is the specific volume $v_f$ of water and the specific volume $v_g$ of steam. *Specific volume* is defined as the space occupied by one pound of a material. For example, at 50 psia, water occupies 0.01727 ft$^3$ per lb and steam occupies 8.515 ft$^3$ per lb.

The *specific weight* is simply the weight of one cubic foot of a material and is the reciprocal of the specific volume.

## THE MOLLIER DIAGRAM

A visual tool for understanding and using the properties of steam is illustrated by the Mollier Diagram, Figure 6-3. The Mollier Diagram enables one to find the relationship between temperature, pressure, enthalpy, and entropy for steam. Constant temperature and pressure curves illustrate the effect of various processes on steam.

MOLLIER CHART

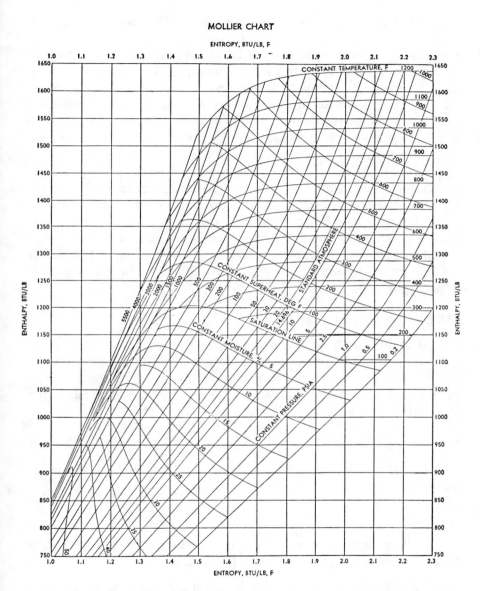

**Figure 6-3. Mollier Diagram** *(Courtesy Babcock & Wilcox Company and ASME)*

For a constant temperature process (isothermal), the change in entropy is equal to the heat added (or subtracted) divided by the temperature at which the process is carried out. This is a simple way of explaining the physical meaning of entropy. The *change* in entropy is of interest. Referring to Table 15-13, the value of entropy at 32°F is zero. Increases in entropy are a measure of the portion of heat in a process which is unavailable for conversion to work. Entropy has a close relationship to the second law of thermodynamics, discussed earlier in this chapter.

Another item of interest from the Mollier Diagram is the saturation line. The saturation line indicates temperature and pressure relationships corresponding to saturated steam. Below this curve, steam contains a % moisture, as indicated by the % moisture curves. Steam at temperatures above the saturation curve is referred to as superheated. As an example, this chart indicates that at 212°F and 14.696 psia, the enthalpy $h_g$ is 1150 Btu/lb, which agrees with Steam Table 15-14.

## SUPERHEATED STEAM

If additional heat is added to raise the temperature of steam over the point at which it was evaporated, the steam is termed superheated. Thus, steam at the same temperature as boiling water is saturated steam. Steam at temperatures higher than boiling water, at the same pressure, is superheated steam.

### Example Problem 6-5

Using the Mollier Diagram, find the enthalpy of steam at 14.696 psia and 300°F.

### Answer

Follow the constant pressure line until it intersects the 300°F curve. The answer to the right is 1192 Btu/lb.

In addition to the Mollier Diagram, the Superheated Steam Table 15-15 is helpful.

Steam cannot be superheated in the presence of water because the heat supplied will only evaporate the water. Thus, the water will be evaporated prior to becoming superheated. Superheated steam is condensed by first cooling it down to the boiling point corresponding to its pressure. When the steam has been de-superheated, further removal of heat will result in its condensation.

In the generation of power, superheated steam has many uses.

## HEAT BALANCE

A heat balance is an analysis of a process which shows where all the heat comes from and where it goes. This is a vital tool in assessing the profit implications of heat losses and proposed waste heat utilization projects. The heat balance for a steam boiler, process furnace, air conditioner, etc. must be derived from measurements made during actual operating periods. Chapter 3 provides information on the instrumentation available to make these measurements. The measurements that are needed to get a complete heat balance involve energy inputs, energy losses to the environment and energy discharges.

### Energy Input

Energy enters most process equipment either as chemical energy in the form of fossil fuels, of sensible enthalpy of fluid streams, of latent heat in vapor streams, or as electrical energy.

For each input it is necessary to meter the quantity of fluid flowing or the electrical current. This means that if accurate results are to be obtained, submetering for each flow is required (unless all other equipment served by a main meter can be shut down so that the main meter can be used to measure the inlet flow to the unit). It is not necessary to continuously submeter every flow since temporary installations can provide sufficient information. In the case of furnaces and boilers that use pressure ratio combustion controls, the control flow meters can be utilized to yield the correct information. It should also be pointed out that for furnaces and boilers only the fuel should be metered. Tests of the exhaust products provide sufficient information to derive the oxidant (usually air) flow if accurate fuel flow data are available. For electrical energy inflows, the current is measured with an ammeter, or a kilowatt hour meter may be installed as a submeter. Ammeters using split core transformers are available for measuring alternating current flow without opening the line. These are particularly convenient for temporary installations.

In addition to measuring the flow for each inlet stream, it is necessary to know the chemical composition of the stream. For air, water, and other pure substances no tests for composition are required, but for fossil fuels the composition must be determined by chemical analysis or secured from the fuel supplier. For vapors one should know the quality—this is the mass fraction of vapor present in the mixture of vapor and droplets. Measurement of quality is made with a vapor calorimeter which requires only a small sample of the vapor stream.

Other measurements that are required are the entering temperatures of the inlet stream of fluid and the voltages of the electrical energy entering (unless kilowatt-hour meters are used).

The testing routines discussed above involve a good deal of time, trouble and expense. However, they are necessary for accurate analyses and may constitute the critical element in the engineering and economic analyses required to support decisions to expend capital on waste heat recovery equipment.

**Energy Losses**

Energy loss from process equipment to the ambient environment is usually by radiative and convective heat transfer. Radiant heat transfer, that is, heat transfer by light or other electromagnetic radiation, is discussed in the section of Chapter 3 dealing with infrared thermography. Convective heat transfer, which takes place by hot gas at the surface of the hot material being displaced by cooler gas, may be analyzed using Newton's law of cooling.

$$Q = UA \, (T_s - T_o) \qquad\qquad Formula \ (6\text{-}8)$$

Where

$Q$ = rate of heat loss in energy units Btu/h
$U$ = heat transfer coefficient in Btu/h$\cdot$ft$^2 \cdot$F
$A$ = area of surface losing heat in ft$^2$
$T_s$ = surface temperature
$T_o$ = ambient temperature

Although heat flux meters are available, it is usually easier to measure the quantities above and derive the heat loss from the equation. The problems encountered in using the equation involve the measurement of surface temperatures and the finding of accurate values for the heat transfer coefficient.

Unfortunately, the temperature distribution over the surface of a process unit can be very nonuniform so that an estimate of the overall average is quite difficult. New infrared measurement techniques, which are discussed in Chapter 3, make the determination somewhat more accurate. The heat transfer coefficient is not only a strong function of surface and ambient temperatures but also depends on geometric considerations and surface conditions. Thus for given surface and ambient temperatures a flat, vertical plate will have a different $b_{cr}$ value than will a horizontal or inclined plate.

### Energy Discharges

The composition, discharge rate and temperature of each outflow from the process unit are required in order to complete the heat balance. For a fuel-fired unit, only the composition of the exhaust products, the flue gas temperature and the fuel input rate to the unit are required to derive

(1)   air input rate
(2)   exhaust gas flow rate
(3)   energy discharge rate from exhaust stack

The composition of the exhaust products can be determined from an Orsat analysis, a chromatographic test or, less accurately, from a determination of the volumetric fraction of oxygen or $CO_2$. Figure 6-4 can be used for determining the quantity of excess or deficiency of air in the combustible mixture. It is based on the fact that chemical reactions occur with fixed ratios of reactants to form given products.

For example, natural gas with the following composition:

$CO_2$  —  0.7% volume
$O_2$  —  0.0
$CH_4$  —  92.0
$C_2H_6$  —   6.8
$N_2$  —   0.5
              100.00

is burned to completion with the theoretical amount of air indicated in the volume equation below:

FUEL:        $0.92\ CH_4 + 0.068\ C_2H_6$
(1 ft³)       $+ 0.007\ CO_2 + 0.005\ N_2$
plus
AIR:         $2.078\ O_2 + 7.813\ N_2 \rightarrow$
(9.891 ft³)
yields
DRY PRODUCTS:   $1.063\ CO_2 + 7.818\ N_2$
(10.925 ft³)      $+ 2.044\ H_2O$                    *Formula (6-9)*

The equation is based upon the laws of conservation of mass and elemental chemical species. The ratio of $N_2$ and $O_2$ in the combustible mixture results from the approximate volumetric ratio of $N_2$ to $O_2$ in air, i.e.,

$$20.9\%\ O_2,\ 79.1\%\ N_2\ \text{ or }\ \frac{79.1}{20.9} = 3.76 = \frac{\text{Volume } N_2}{\text{Volume } O_2}.$$

For gases the coefficients of the chemical equation represent relative volumes of each species reacting. Ordinarily, excess air is provided to

Figure 6-4. Natural Gas Combustion Chart

the fuel so that every fuel molecule will react with the necessary number of oxygen molecules even though the physical mixing process is imperfect. If 10 percent excess air were supplied, this mixture of reactants and products would give a chemical equation appropriately modified as given below.

FUEL: $0.92\ CH_4 + 0.068\ C_2H_6$
(1 ft³) $+ 0.007\ CO_2 + 0.005\ N_2$
plus
AIR: $2.286\ O_2 + 8.594\ N_2 \rightarrow$
(10.880 ft³)
yields
DRY PRODUCTS: $1.063\ CO_2 + 0.208\ O_2 + 8.599\ N_2$
(11.914 ft³) $+ 2.044\ H_2O$      *Formula (6-10)*

where an additional term representing the excess oxygen appears in the products, along with a corresponding increase in the nitrogen.

As an example let us assume that an oxygen meter has indicated a reading of 7% for the products of combustion from the natural gas whose composition was given previously and that the exhaust gas temperature was measured as 700°F. Figure 6-4 is used as indicated to determine that 45% excess air is mixed with fuel. The combustion equation then becomes:

FUEL: $0.92\ CH_4 + 0.068\ C_2H_6$
(1 ft³) $+ 0.007\ CO_2 + 0.005\ N_2$
plus
AIR: $3.013\ O_2 + 11.329\ N_2 \rightarrow$
(14.342 ft³)
yields
DRY PRODUCTS: $1.063\ CO_2 + 0.935\ O_2 + 11.334\ N_2$
(15.376 ft³) $+ 2.044\ H_2O$

For each 1 ft³ of fuel, 14.342 ft³ of air is supplied and 15.376 ft³ of exhaust products (at mixture temperature) is formed. Each cubic foot of fuel contains 1055 Btu of energy, so the fuel energy input is 250,000ft³/h X 1055 = 263,750,000 Btu/h. From Figure 6-5 we compute the exhaust gas losses at 700°F as

$$CO_2: 250,000 \times 1.063 \times 17.5 = 4,651,000 \text{ Btu/h}$$
$$H_2O: 250,000 \times 2.044 \times 14.0 = 7,154,000$$
$$O_2: 250,000 \times 0.935 \times 12.3 = 2,875,000$$
$$N_2: 250,000 \times 11.334 \times 11.5 = \underline{32,585,000}$$
$$47,265,000 \text{ Btu/h}$$

or 18% of the fuel energy supplied. Some of this could be recovered by suitable waste heat equipment.

Figure 6-5. Heat Content of Gases Found in Flue Products, Based on Gas Volumes at 60°F (dashed line indicates dissociation)

## Heat Balance on a Boiler

Let us consider a further example, a heat balance on a boiler. A process steam boiler has the following specifications:

- Natural gas fuel with HHV = 1001.2 Btu/ft$^3$
- Gas firing rate = 2126.5 ft$^3$/min.
- Steam discharge at 150 lb/in$^2$ g saturated
- Steam capacity of 100,000 lb/h
- Condensate returned at 180°F

The heat balance on the burner is derived from measurements made *after* the burner controls had been adjusted for an optimum air/fuel ratio corresponding to 10% excess air. All values of the heat content of the fluid streams are referred to a base temperature of 60°F. Consequently the computations for each fluid stream entering or leaving the boiler are made by use of the equation below.

$$\dot{H} = \dot{m} \, (b - b_o) \qquad \qquad Formula \; (6\text{-}11)$$

Where

$\quad \dot{H} \;$ = the enthalpy rate for entrance or exit fluids

$\quad \dot{m} \;$ = mass flow rates for entrance or exit fluids

$\quad b \;$ = specific enthalpy at the fluid temperature of the fluid entering or leaving the entrance or exit

$\quad b_o \;$ = specific enthalpy of that fluid at the reference temperature $T_o = 60°F$

The first law of thermodynamics for the boiler is expressed as: Sum of all the enthalpy rates of substances entering = Sum of all enthalpy rates of substances leaving + $q$ where $q$ is the rate of heat loss to the surroundings. This can be expressed as

$$\Sigma_{in} H_i = \Sigma_{out} H_i + q \qquad \qquad Formula \; (6\text{-}12)$$

where $\Sigma$ is the summation sign, and $H_i$ is the enthalpy rate of substance $i$.

For gaseous fuels the computation for the heat content of the gases is more conveniently expressed in the form

$$H = J_o C_{pm} \, (T - T_o) \qquad \qquad Formula \; (6\text{-}13)$$

Where

$\quad J_o \;$ = the volume rate of the gas stream corrected back to 1.0 atmosphere and 60°F $(T_o = 60°F)$

$\quad C_{pm} \;$ = the specific heat given on the basis of a standard volume of gas averaged over the temperature range $(T - T_o)$ and the gas mixture components.

$$C_{pm} \; = \Sigma_i \chi_i C_{pi} \qquad \qquad Formula \; (6\text{-}14)$$

Where

$\chi_i$ = percent by volume of a component in one of the flow paths

$C_{pi}$ = average specific heat over temperature range for each component

From Formula 6-10 we derive the volume fractions of each gas component as follows:

| Component | $\chi$ |
|-----------|--------|
| $CO_2$ | 8.9 |
| $H_2O$ | 17.2 |
| $N_2$ | 72.2 |
| $O_2$ | 1.7 |
| | 100.0 |

The average specific heat is found (using Figure 6-6).

$$
\begin{aligned}
C_{pm} = \ .089 \times 0.0275 &= 0.00245 \\
0.172 \times 0.0220 &= 0.00378 \\
0.722 \times 0.0186 &= 0.01344 \\
0.017 \times 0.0195 &= 0.00033 \\
&\overline{0.02 \ \text{Btu/Scf} \cdot \text{F}}
\end{aligned}
$$

The combustion equation also tells that when the products are at standard conditions 11.88 ft$^3$ fuel and air generate 11.915 ft$^3$ of products and that for every cubic foot of gas burned, 10.880 ft$^3$ of air is introduced. Thus for a firing rate of 2126.5 ft$^3$/min. the air required is 2126.5 $\times$ 10.88 = 23,136 ft$^3$/min. or 23,136 $\times$ 60 = 1,388,179 ft$^3$/h. The total fuel and air flow rate is then almost exactly equal to 1,388,179 + 2126.5 $\times$ 60 = 1,515,769 ft$^3$/h. This corresponds to a flue gas discharge rate of 1,519,128 ft$^3$/h.

For a flue gas discharge rate of 1,519,128 ft$^3$/h and a temperature of 702°F, the total exhaust heat rate is found as

$$
\begin{aligned}
\dot{H}_{\text{EX GAS}} &= 1,519,128 \ \frac{\text{ft}^3}{\text{h}} \times 0.02 \ \frac{\text{Btu}}{\text{ft}^3 \cdot \text{F}} \\
&\times (702-60)\text{F} = 19,505,604 \ \text{Btu/h}
\end{aligned}
$$

For the steam leaving the boiler (100,000 lb/h at 150 lb/in$^2$g saturated) the energy flow rate is found using the Steam Tables in Chapter 15 and the following equation:

$$\dot{H}_{\text{Steam}} = \dot{m}\,(h-h_o)$$
$$= 100{,}000 \text{ lb/h } (1195.6-28.08) \text{ Btu/lb}$$
$$= 116{,}752{,}000 \text{ Btu/h}$$

where 1195.6 is the specific enthalpy of saturated steam at 150 lb/in²g and 28.08 is the specific enthalpy of saturated liquid water at 60°F, since we are using that temperature as our standard reference temperature for the heat balance. We have used 60°F as a reference temperature; this is not universal practice and in the boiler industry 70°F is more common, whereas in other areas 25°C is normal.

**Figure 6-6. Mean Heat Capacity of Gases Found in Flue Products from 60 to $T$°F (dashed line indicates dissociation)**

## CHEMICAL ENERGY IN FUEL

To determine the heat content of the chemical energy in fuel, find the higher heating value (HHV) for the fuel and multiply it by the volumetric flow rate for a gaseous fuel or the mass flow rate for a liquid or solid fuel. The assumed higher heating value for the natural gas used in the boiler of our example is 1001.1 Btu/ft$^3$ and the heat content rate is then

$$\dot{H} = 2126.5 \frac{ft^3}{min.} \times 60 \frac{min.}{h} \times 1001.1 \frac{Btu}{ft^3}$$

$$= 127,743,000 \frac{Btu}{h}$$

The enthalpy rates for the condensate return and make-up water are derived from data in the steam tables where the specific enthalpy of the compressed liquids is taken to be almost exactly equal to the specific enthalpy of the *saturated* liquid found at the same *temperature.*

The complete heat balance derived in the manner detailed above is presented in Figure 6-7.

**Figure 6-7. Heat Balance for a Simple Steam Generator Burning Natural Gas with 10% Excess Air**

Waste heat is available from the combustion products leaving the stack. It amounts to 19,505,604 Btu/h at a temperature of 702°F. Some energy is also available from the condensers.

## WASTE HEAT RECOVERY

The energy exhausted to the atmosphere should not be discarded. A portion of it can be recovered by using a heat exchanger. Any requirement for energy at a temperature in excess of 200°F can be satisfied. It is necessary to identify the prospective uses for the waste energy, make an economic analysis of the costs and savings involved in each of the options, and decide among those options on the basis of the economics of each. An important option in every case is that of rejecting all options if none proves economic. For the illustrative example, let us assume that the following uses for the waste heat have been identified:

- Preheating the combustion air
- Preheating the boiler feedwater
- Heating the domestic hot water supply (370 gal/hr from 50° to 170°F)
- A combination of the preceding

The first two are practices which are standard in energy-intensive high technology industries (e.g., electric companies).

Two rapid calculations give the waste heat available in the exhaust gases between 702°F and 220°F and the heat requirements for the domestic hot water supply to be 14,063,000 Btu/h and 369,900 Btu/h respectively. Since the latter constitutes only 2.6% of the available waste energy, we should reject that option. The remaining options are to preheat the combustion air and the feedwater or to divide the waste energy between those 2 options. For a small boiler we probably would not find the purchase of two separate heat exchangers an economic option, so that we shall limit ourselves to one of the other of the first two options. Without going into a detailed analysis, we may note that preheating the combustion air or the feedwater provides a double benefit. Since the preheated air or water requires less fuel to produce the same steam capacity, a direct fuel saving results. But the smaller quantity of fuel means a smaller air requirement, and this in turn means a smaller quantity of exhaust products and thus smaller stack loss at a given stack temperature.

The economic benefits can be estimated as follows: The air preheater is estimated to save 6% of the fuel. With an average boiler loading of 60% for the 8760 h in a year, this amounts to an annual fuel saving of

$$.6 \times 0.06 \times 8760 \times 127,590 = 40,237 \ \frac{ft^3}{yr}$$

which is worth, at an average rate of \$5.50/1000 $ft^3$, an annual dollar saving of

$$\frac{40,237}{1000} \times 5.50 = \$221,000/yr$$

If the costs for installing the air preheater are assumed to be

| | |
|---|---:|
| Cost of preheater | \$ 52,000 |
| Cost of installation | 57,200 |
| New burners, air piping, controls and fan | 56,600 |
| | \$165,800 |

For the feedwater heater (or economizer) the fuel savings is 9.2% or a total annual fuel saving of

$$0.092 \times 127,600 \times 8760 \times 0.60 = 61,696,000 \ \frac{ft^3}{yr}$$

and the economic benefit is

$$\frac{61,696,000 \ \dfrac{ft^3}{yr}}{1000} \times \$5.50 = \$340,000/yr$$

The cost of the economizer installed is estimated to be \$134,000. This is clearly the best option for this particular boiler, especially as there is no need in this case for modifications to the boiler and accessories beyond the heat exchanger retrofit. There are several reasons for its superiority. The first is that in the case of air preheating, we are exchanging the waste heat in the gases to the incoming air which has almost the same mass flow rate and almost the same specific heat. Thus we can expect that the final temperature of the preheater air will be almost the arithmetic mean of the ambient air temperature and the exhaust gas temperature entering the economizer. In the case of the feedwater, the mass flow rate times specific heat is over four and one-quarter times that of the combustion air. Thus we can expect to transfer more energy to the water which results in a lower flue gas temperature leaving the stack.

The second reason is that preheating the air quite often (as in this case) affects the boiler accessories which require additional modifications and thus related capital expenditures.

In the preceding example of the process steam boiler, we are analyzing an efficient process unit. The heat available in the product (process steam) constituted a large percentage of the energy introduced in the fuel. The efficiency in percentage terms is computed as

$$\eta = \frac{\text{Useful output}}{\text{Energy input}} = \frac{Q_{steam}}{Q_{fuel}} \times 100$$

$$\eta = \frac{100,000 \text{ lb/h } (1195.6-148) \times 100}{127,590 \text{ ft}^3/\text{h} \times 1001.2 \text{ Btu/ft}^3}$$

$$= \frac{1.0476 \times 10^{10}}{1.2775 \times 10^8} = 0.82, \text{ or } 82\%$$

## HEAT RECOVERY IN STEEL TUBE FURNACE

As a second and very different example, note the steel tube furnace illustrated below in Figure 6-8.

**Figure 6-8. Continuous Steel Tube Furnace**

The tubing enters the furnace from the right at a temperature of 100°F.

The specifications for the steel tube heating furnace are

Product capacity — 50 ton/h
Product specifications — 0.23% carbon steel
Final product temperature — 2000°F
Air/Fuel inlet temperature — 100°F
Air/Fuel mixture —10% excess air

Fuel — No. 5 fuel oil gravity API° 16
Fuel firing rate 48.71 gpm at 240°F (factory usage)
Utilization factor — 0.62

The useful heat leaves the furnace in the steel at 2000°F. This is called the useful furnace output and equals

$$Q_{prod} = \dot{m}_{prod} C_P (T_{out} - T_{in})$$

$$= 50 \frac{ton}{h} \times 2200 \text{ lb/ton}$$

$$\times .179 \text{ Btu/lb} \times (2000-100) \text{ F}$$

$$= 0.3741 \times 10^8 \frac{Btu}{h}$$

0.179 is used as the average specific heat of steel over the 100°F–2000°F range. The heat input to the furnace is the chemical energy in the fuel oil. The heating value for No. 5 fuel oil is found from Table 6-2.

The heat input is determined by

$$Q_{fuel} = 48.71 \text{ gpm} \times 142,300 \frac{Btu}{gal}$$

$$= 4.159 \times 10^8 \text{ Btu/h}$$

The percent efficiency of the furnace is:

$$\eta = \frac{output \times 100}{input}$$

$$= \frac{Enthalpy \text{ added to steel} \times 100}{Enthalpy \text{ entering with fuel}}$$

$$= \frac{0.3741 \times 10^8}{4.159 \times 10^8} \times 100 = 9\%$$

This means that of the 415.9 MBtu introduced per hour to the furnace, that 378.4 million are released to the atmosphere. At the present average cost of $2/gal. for No. 5 fuel oil, the heat wasted is

$$Q_{waste} = 48.71 \text{ gpm} \times (1-0.09)\frac{gal \text{ wasted}}{gal \text{ used}} \times .62$$

$$\times 60 \text{ min./h} \times \$2/gal = \$3297/h$$

$$= \$3297/h \times 8760h/yr = \$28,885,000/yr$$

Table 6-2. Physical Properties of Fuel Oil at 60°F

| Fuel oil (CS-12-48) Grade No. | Gravity, API | Sp gr | Lb per gal | Btu per lb | Net Btu per gal |
|---|---|---|---|---|---|
| 6 | 3 | 1.0520 | 8.76 | 18,190 | 152,100 |
| 6 | 4 | 1.0443 | 8.69 | 18,240 | 151,300 |
| 6 | 5 | 1.0366 | 8.63 | 18,290 | 149,400 |
| 6 | 6 | 1.0291 | 8.57 | 18,340 | 148,800 |
| 6 | 7 | 1.0217 | 8.50 | 18,390 | 148,100 |
| 6 | 8 | 1.0143 | 8.44 | 18,440 | 147,500 |
| 6 | 9 | 1.0071 | 8.39 | 18,490 | 146,900 |
| 6 | 10 | 1.0000 | 8.33 | 18,540 | 146,200 |
| 6 | 11 | 0.9930 | 8.27 | 18,590 | 145,600 |
| 6 | 12 | .9861 | 8.22 | 18,640 | 144,900 |
| 6, 5 | 14 | .9725 | 8.10 | 18,740 | 143,600 |
| 6, 5 | 16 | .9593 | 7.99 | 18,840 | 142,300 |
| 5 | 18 | .9465 | 7.89 | 18,930 | 140,900 |
| 4, 5 | 20 | .9340 | 7.78 | 19,020 | 139,600 |
| 4, 5 | 22 | .9218 | 7.68 | 19,110 | 138,300 |
| 4, 5 | 24 | .9100 | 7.58 | 19,190 | 137,100 |
| 4, 2 | 26 | .8984 | 7.49 | 19,270 | 135,800 |
| 4, 2 | 28 | .8871 | 7.39 | 19,350 | 134,600 |
| 2 | 30 | .8762 | 7.30 | 19,420 | 133,300 |
| 2 | 32 | .8654 | 7.21 | 19,490 | 132,100 |
| 2 | 34 | .8550 | 7.12 | 19,560 | 130,900 |
| 1, 2 | 36 | .8448 | 7.04 | 19,620 | 129,700 |
| 1, 2 | 38 | .8348 | 6.96 | 19,680 | 128,500 |
| 1 | 40 | .8251 | 6.87 | 19,750 | 127,300 |
| 1 | 42 | 0.8156 | 6.79 | 19,810 | 126,200 |

The relation between specific gravity and degrees API is expressed by the formula:

$$\frac{141.5}{131.5 + API} = sp\ gr\ at\ 60\ F.$$

For each 10 F above 60 F add 0.7 API.
For each 10 F below 60 F subtract 9.7 AP

In order to construct the combustion equation we must note from Table 6-3 that the carbon-hydrogen ratio is about 7.3. Thus

$$C_{6.08}H_{10} + 9.44\ O_2 + 35.49\ N_2 \rightarrow 6.08\ CO_2$$
$$+ 5\ H_2O + 0.86\ O_2 + 35.49\ N_2$$

Again referring to Table 6-2 the density of the liquid fuel is 7.99 lb/gal. Therefore, the above equation represents the reaction for 100 lb fuel/7.00 lb/gal = 12.51 gal and for 35.49 + 9.44 or 44.93

### Table 6-3. Typical Properties of Commercial Petroleum Products Sold in the East and Midwest

| | Gasolines | | | | Reduced crude, heavy gas oil or premium residuum* | Fuel Oils | | | | |
|---|---|---|---|---|---|---|---|---|---|---|
| | 12 psia Reid vapor pressure natural gasoline | Straight run gasoline | No. 1 fuel, light diesel, or stove oil | Diesel or "gas house" gas oil | | No. 2 | No. 3 | Cold No. 5 | No. 5 | Bunker C or No. 6 |
| Gravity, °API | 79 | 63 | 38–42 | 35–38 | 19–22 | 32–35 | 28–32 | 20–25 | 16-24 | 6–14 |
| Viscosity | ... | ... | 34–36† | 34–36† | 15–50‡ | 34–36† | 35–45† | <20‡ | 20–40‡ | 50–300‡ |
| Conradson carbon, wt % | None | None | Trace | Trace | 3–7 | Trace | <0.15 | 1–2 | 2–4 | 6–15 |
| Pour point, °F | <0 | <0 | <0 | <10 | ... | <10 | <20 | <15 | 15–60 | >50 |
| Sulfur, wt % | <0.1 | <0.2 | <0.1 | 0.1–0.6 | <2.0 | 0.1–0.6 | 0.2–1.0 | 0.5–2.0 | 0.5–2.0 | 1–4 |
| Water and sed., wt % | None | None | Trace | Trace | <1.0 | <0.05 | <0.10 | 0.1–0.5 | 0.5–1.0 | 0.5–2.0 |
| Distillation, °F | | | | | | | | | | |
| 10% | 110 | 150 | 390–410 | 400–440 | 550–650 | 420–440 | 450–500 | <600 | <600 | 600–700 |
| 50% | 150 | 230 | 440–460 | 490–510 | 800–900 | 490–530 | 540–560 | ... | ... | 850–950 |
| 90% | 235 | 335 | 490–520 | 580–600 | ... | 580–620 | 600–670 | <700 | >700 | ... |
| End point | 315 | 370 | 510–560 | 630–660 | ... | 630–660 | 650–700 | ... | ... | ... |
| Flash point, °F | <100 | <100 | 100–140 | 110–170 | >150 | 130–170 | >130 | >130 | >130 | >150 |
| Aniline point, °F | 145 | 125 | 140–160 | 150–170 | ... | 120–140 | 120–140 | ... | ... | ... |
| C to H ratio | 5.2 | 5.6 | 6.1–6.4 | 6.2–6.5 | 6.9–7.3 | 6.6–7.1 | 7.0–7.3 | 7.3–7.7 | 7.3–7.7 | 7.7–9.0 |
| Average MBtu recovered/gal of oil: | | | | | | | | | | |
| Carbureted water gas | 101 | 101 | 102 | 101 | 93 | 91 | 86 | 82 | 83 | 74 |
| High-Btu oil gas | 90 | 89 | 90 | 90 | 82 | 80 | 76 | 72 | 73 | 64 |
| Tar + carbon, wt % of oil | | | | | | | | | | |
| Carbureted water gas | ... | ... | 20 | 22 | 35 | 31 | 36 | 42 | 42 | 52 |
| High-Btu oil gas | ... | ... | 32 | 34 | 45 | 42 | 46 | 52 | 52 | 60 |

(Data on gasolines and data below C/H ratio not given in reports.)

* The 6 wt % Conradson carbon oil used in the Hall High Btu Oil Gas Tests (A.G.A. Gas Production Research Committee. Hall High Btu Oil Gas Process, New York, 1949.) and "New England Gas Enriching Oil" are typical examples of this group.

† Saybolt Universal seconds at 100 F.

‡ Saybolt Furol seconds at 122 F.

**Remarks:**

No. 1 Fuel Oil—A distillate oil intended for vaporizing pot-type burners. A volatile fuel.

No. 2 Fuel Oil—For general purpose domestic heating; for use in burners not requiring No. 1 oil. Moderately volatile.

No. 3 Fuel Oil—Formerly a distillate oil for use in burners requiring low viscosity oil. Now incorporated as a part of No. 2.

No. 4 Fuel Oil—For burner installations not equipped with preheating facilities.

No. 5 Fuel Oil—A residual type oil. Requires preheating to 170–220 F.

No. 6 Fuel Oil—Preheating to 220–260 F suggested. A high viscosity oil.

lb mol of air. 44.93 lb mol of air weighs 1301 lb (using the molecular weight for air found in Table 6-4). The specific volume of air at standard conditions is found from the same table as 378.5 ft³/lb mol. Therefore, the volume of air required to burn 12.51 gal of fuel is 44.93 × 378.5 = 17,000 ft³ at standard temperature and pressure. The air input is then

$$m = \frac{17,000 \text{ ft}^3}{12.51 \text{ gal}} \cdot 48.71 \text{ gpm} = 66,200 \text{ ft}^3/\text{m}$$

or

$$m_{air} = 66,200 \times 60 = 3,972,000 \text{ ft}^3/\text{h}$$
$$Q_{air} = 1.2 \text{ Btu/ft}^3 \times 3,972,000$$
$$= 4,766,000 \text{ Btu/h}$$

**Table 6-4. Gas Constants and Volume of the Pound-Mol for Certain Gases**

| | $R' = R/M$, specific gas constant, ft per F | $M$, mol wt, lb per lb-mol | $R$, universal gas constant, ft-lb per (lb-mol )( F) | $Mv$, cu ft per lb-mol * |
|---|---|---|---|---|
| Hydrogen | 767.04 | 2.016 | 1546 | 378.9 |
| Oxygen | 48.24 | 32.000 | 1544 | 378.2 |
| Nitrogen | 55.13 | 28.016 | 1545 | 378.3 |
| Nitrogen, "atmospheric" † | 54.85 | 28.161 | 1545 | 378.6 |
| Air | 53.33 | 28.966 | 1545 | 378.5 |
| Water vapor | 85.72 | 18.016 | 1544 | 378.6 |
| Carbon dioxide | 34.87 | 44.010 | 1535 | 376.2 |
| Carbon monoxide | 55.14 | 28.010 | 1544 | 378.3 |
| Hydrogen sulfide | 44.79 | 34.076 | 1526 | 374.1 |
| Sulfur dioxide | 23.56 | 64.060 | 1509 | 369.6 |
| Ammonia | 89.42 | 17.032 | 1523 | 373.5 |
| Methane | 96.18 | 16.042 | 1543 | 378.2 |
| Ethane | 50.82 | 30.068 | 1528 | 374.5 |
| Propane | 34.13 | 44.094 | 1505 | 368.7 |
| $n$-Butane | 25.57 | 58.120 | 1486 | 364.3 |
| $iso$-Butane | 25.79 | 58.120 | 1499 | 367.4 |
| Ethylene | 54.70 | 28.052 | 1534 | 376.2 |
| Propylene | 36.01 | 42.078 | 1515 | 379.1 |

   * At 60 F, 30 in. Hg, dry.
   † Includes other inert gases in trace amounts.

The volume ratio of stack gases at standard conditions to air inlet rate is found from the combustion equation as

$$\frac{6.08 + 5 + 0.86 + 35.49}{9.44 + 35.49} = \frac{47.43}{44.93} = 1.056$$

$$Q_{ex} = 1.056 \times 3,972,000$$
$$= 4,194,000 \text{ ft}^3/\text{h}$$

The volume fractions of the stack gas components are found from the combustion equation as

| | $\chi$ |
|---|---|
| $CO_2$ | 0.128 |
| $H_2O$ | 0.106 |
| $O_2$ | 0.018 |
| $N_2$ | 0.748 |
| | 1.000 |

The specific enthalpy of the stack gas components may be calculated by use of Figure 6-5 for the exhaust temperature of 2200°F leading to the average specific enthalpy as indicated on next page.

$$
\begin{array}{rrrrrr}
 & & & & & b \\
CO_2 & 0.128 & \times & 69.3 & = & 8.870 \\
H_2O & 0.106 & \times & 54.0 & = & 5.724 \\
O_2 & 0.018 & \times & 45.4 & = & 0.817 \\
N_2 & 0.748 & \times & 43.4 & = & \underline{32.463} \\
 & & & & & 47.874
\end{array}
$$

Therefore

$$Q_{ex} = 4,194,000 \text{ ft}^3/\text{h} \times 47.874 \, \frac{Btu}{ft^3}$$

$$= 200,784,000 \, \frac{Btu}{h}$$

The heat losses from the surface equal the fuel enthalpy input less the exhaust gas enthalpy and the produce enthalpy.

$$Q = 415,880,000 + 4,766,000 - 200,784,000$$

$$- 37,410,000 = 182,451,270 \, \frac{Btu}{h}$$

The complete heat balance diagram is shown in Figure 6-9. Now suppose that we install a metallic recuperator and preheat the combustion air to 750°F. The new heat balance is shown in Figure 6-10.

The economic analysis of this situation is as follows:

The waste heat saved is equal to the fuel savings times the heating value of the fuel.

$$Q_{saved} = (48.71 - 31.66) \text{ gpm} \times 142,300 \, \frac{Btu}{gal}$$

$$= 145,554,000 \text{ or } 35\% \text{ saving}$$

which equals

$$0.35 \times 48.71 \text{ gpm} \times 60 \times \$2/\text{gal} = \$2,045.82/\text{h}$$

In using Figures 6-12 and 6-13 remember that the combustion air volume (in SCF) is reduced directly proportional to the reduction in firing rate. If we maintain the same air/fuel ratio for the furnace (10% excess air), the air rate is reduced from the initial value of 3,972,000 ft³/h at a firing rate of 48.71 gpm to a rate of 31.66 gpm/ 48.71 gpm × 3,972,000 ft³/h = 2,581,800 ft³/h with the lower firing rate of 31.66 gpm when the recuperator is used.

If, as in our example, the furnace is used 0.62 × 8760 or 6531 h/ yr, then the annual saving is 5431 × \$2,045.80 = \$11.10 × 10⁶/yr.

Since a metallic recuperator can be purchased for a fraction of the savings, the savings will allow a payoff in less than a year.

**Figure 6-9.  Heat Balance for a Simple Continuous Steel Tube Furnace**

**Figure 6-10.  Heat Balance for a Continuous Steel Tube Furnace Equipped with a Recuperator Preheating Combustion Air to 750°F**

**Figure 6-11.** Fuel Savings as a Function of Flue Gas Temperature °F
Leaving Furnace

**Figure 6-12.** Heat Recovered as a Function of Air Volume through Recuperator
(SCFM)

## BOILERS

### Boiler Configurations and Components

Industrial boiler designs are influenced by fuel characteristics and firing method, steam demand, steam pressures, firing characteristics and the individual manufacturers. Industrial boilers can be classified as either firetube or watertube, indicating the relative position of the hot combustion gases with respect to the fluid being heated.

### Firetube Boilers

Firetube units pass the hot products of combustion through tubes submerged in the boiler water. Conventional units generally employ from 2 to 4 passes to increase the surface area exposed to the hot gases and thereby increase efficiency. Multiple passes, however, require greater fan power, increased boiler complexity and larger shell dimensions. (Refer to Figure 6-13.) Maximum capacity of firetube units is currently limited to 25,000 lbs of steam per hour (750 boiler hp) with an operating pressure of 250 psi due to economic factors related to material strength and thickness.

**2-pass dryback**                    **3-pass dryback**

**3-pass wetback**                    **4-pass dryback**

Figure 6-13. Typical Firetube Boiler Gas Flow Patterns

Advantages of firetube units include: (1) ability to meet wide and sudden load fluctuations with only slight pressure changes, (2) low initial costs and maintenance and (3) simple foundation and installation procedures.

**Watertube Boilers**

Watertube units circulate the boiler water inside the tubes and the flue gases outside. Water circulation is generally provided by the density variation between cold feed water and the hot water/steam mixture in the riser as illustrated in Figure 6-14. Watertube boilers may be subclassified into different groups by tube shape, by drum number and location and by capacity. Refer to Figure 6-15.

Another important determination is "field" versus "shop" erected units. Many engineers feel that shop assembled boilers can meet closer tolerance than field assembled units and therefore may be more efficient; however, this has not been fully substantiated. Watertube units range in size from as small as 1000 lbs of steam per hour to the giant utility boilers in the 1000 MW class. The largest industrial boilers are generally taken to be about 500,000 lbs of steam per hour. Important elements of a steam generator include the firing mechanism, the furnace water walls, the superheaters, convective regions, the economizer and air preheater and the associated ash and dust collectors.

## FUEL HANDLING AND FIRING SYSTEMS

**Gas Fired**

Natural gas fuel is the simplest fuel to burn in that it requires little preparation and mixes readily with the combustion air supply. Industrial boilers generally use low-pressure burners operating at a pressure of 1/8 to 4 psi. Gas is generally introduced at the burner through several orifices that generate gaseous jets that mix rapidly with the incoming combustion air supply. There are many designs in use that differ primarily in the orientation of the burner orifices and their locations in the burner housing.

Figure 6-14. Water Circulation Pattern in a Watertube Boiler

A Type                                                    D Type

O Type

**Figure 6-15. Classification of Watertube Boilers by Basic Tube Arrangement**

## Oil Fired

Oil fuels generally require some type of pretreatment prior to delivery to the burner. This may include the use of strainers to remove solid foreign material and tank and flow line preheaters to assure the proper viscosity. Oil must be atomized prior to vaporization and mixing with the combustion air supply. This generally requires the use of either air, steam or mechanical atomizers. The oil is introduced into the furnace through a gun fitted with a tip that distributes the oil into a fine spray that allows mixing between the oil droplets and the combustion air supply. Oil cups that spin the oil into a fine mist are also employed on small units. An oil burner may be equipped with diffusers that act as flame holders by inducing strong recirculation patterns near the burner. In some burners, primary air nozzles are employed.

## Pulverized Coal Fired

The pulverizer system provides four functions: pulverizing, drying, classifying to the required fineness and transporting the coal to the burner's main air stream. The furnace may be designed for dry ash removal in the hopper bottom or for molten ash removal as in a slag tap furnace. The furnace is dependent on the burning and ash characteristics of the coal as well as the firing system and type of furnace bottom. The primary objectives are to control furnace ash deposits and provide sufficient cooling of the gases leaving the furnace to reduce the buildup of slag in the convective regions. Pulverized coal fired systems are generally considered to be economical for units with capacities in excess of 200,000 lbs of steam per hour.

## Stoker Fired

Coal stoker units are characterized by bed combustion on the boiler grate with the bulk of the combustion air supplied through the grate. Several stoker firing methods currently in use on industrial sized boilers include underfed, overfed and spreader. In underfed and overfed stokers, the coal is transferred directly on to the burning bed. In a spreader stoker the coal is hurled into the furnace when it is partially burned in suspension before lighting on the grate. Several grate configurations can be used with overfed and spreader stokers including stationary, chain, traveling, dumping and vibrating grates. Each grate configuration has its own requirements as to coal fineness and ash characteristics for optimum operation. Spreader stoker units have the advantage that they can burn a wide variety of fuels including waste products. Underfed and overfed units have the disadvantage that they are relatively slow to respond to load variations. Stoker units can be designed for a wide range of capacities from 2,000 to 350,000 lbs of steam per hour. Spreader stoker units are generally equipped with overfire air jets to induce turbulence for improved mixing and combustible burnout. Stoker units are also equipped with ash reinjection systems that allow collected ash that contains a significant portion of unburned carbon to be reintroduced into the furnace for burning.

## COMBUSTION CONTROL SYSTEMS

Combustion controls have two purposes: (1) maintain constant steam conditions under varying loads by adjusting fuel flow, and (2) maintain an appropriate combustion air-to-fuel flow.

### Classification

Combustion control systems can be classified as series, parallel and series/parallel.

In series control, either the fuel or air is monitored and the other is adjusted accordingly. For parallel control systems, changes in steam conditions result in a change in both air and fuel flow. In series/parallel systems, variations in steam pressure affect the rate of fuel input, and simultaneously the combustion air flow is controlled by the steam flow.

Combustion controls can be also classified as positioning and metering controls.

Positioning controls respond to system demands by moving to a present position. In metering systems, the response is controlled by actual measurements of the fuel air flows.

### Application

The application and degree of combustion controls vary with the boiler size and is dictated by system costs. The parallel positioning jackshaft system has been extensively applied to industrial boilers based on minimum system costs. The combustion control responds to changes in steam pressure and can be controlled by a manual override. The control linkage and cam positions for the fuel and air flow are generally calibrated on start-up.

Improved control of excess air can be obtained by substituting electric or pneumatic systems for the mechanical linkages. In addition, relative position of fuel control and combustion air dampers can be modified. More advanced systems are pressure ratio control of the fuel and air pressure, direct air and fuel metering and excess air correction systems using flue gas $O_2$ monitoring. Factors that have limited the application of the most sophisticated control systems to industrial boilers include cost, reliability and maintenance.

## SAFETY

It is essential that energy engineers conducting boiler evaluation tests and tune-ups understand and be aware of boiler safety devices. Occasionally operators who don't understand safety have been known to bypass safety features in order to keep a unit operating.

Summary of requirements found in the NATIONAL BOILER SAFETY CODES:

* PREPURGE—4 to 8 air changes to insure no fuel vapors or gases remain in the boiler which could ignite or explode when the pilot light is off.

* PILOT PROVING—10 to 15 second proving period pilot must ignite and be proven before the main fuel valves open.

* MAIN FLAME TRIAL—usually 10 to 15 second trial period for natural gas and oil after main fuel valves open.

* FLAME FAILURE RESPONSE TIME—usually 3 to 4 seconds after the main flame goes out the Fuel Shut Off valves should automatically close to prevent a build up of explosive vapor concentrations within the boiler.

* INTERLOCKS—automatically shut the boiler down if certain safety features are not in the safe condition.

    Loss of atomizing means (steam/air)
    High/Low gas pressure
    Low fire for light off
    Low fuel oil temperature
    Low fuel oil pressure
    Main combustion air
    Low water level

* FUEL VALVES—Safety Shut Off Valves (SSOV). Usually two valves; closing time 1 to 5 seconds depending on fuel and size of boiler.

* FLAME DETECTORS—should be positioned and capable of detecting flame only and not sparks from the spark ignitor or hot refractory so as not to give a false flame signal.

## OPERATING GUIDES

Consult the plant engineer and the boiler manufacturers technical manuals for complete boiler operating procedures. Typical items to check before conducting efficiency checks are as follows:

## Oil Burners

Make sure the atomizer is of the proper design and size and the burner is centered with dimensions according to manufacturer's drawings.

Inspect oil-tip passages and orifices for wear (use proper size drill as a feeler gage) and remove any coke or gum deposits to assure the proper oil-spray pattern.

Verify proper oil pressure and temperature at the burner.

Verify proper atomizing-steam pressure.

Make sure that the burner diffuser (impeller) is not damaged and is properly located to the oil-gun tip.

Check to see that the oil gun is positioned properly within the burner throat and that the throat refractory is in good condition.

## Gas Burners

Inspect gas-ingestion orifices and verify that all passages are un-obstructed. Also, be sure filters and moisture traps are in place, clean and operating properly, to prevent plugging of gas orifices.

Confirm proper location and orientation of diffusers, spuds, etc. Look for any burned off or missing burner parts.

## Combustion Controls

Inspect all fuel valves to verify proper movement; clean valve internal surfaces if necessary.

Eliminate "play" in control linkages on dampers. Any play, no matter how slight, will cause a loss of efficiency since a precise tune-up will be impossible.

Make sure fuel-supply inlet pressure to pressure regulators are high enough to assure constant regulator-outlet pressures for all firing rates.

Correct any control elements that fail to respond smoothly to varying steam demand. Unnecessary hunting caused by improperly adjusted regulators or automatic master controllers can waste fuel.

Check that all gages are functioning and are calibrated.

## Furnace

Inspect boiler furnace and gas side surfaces for excessive deposits and fouling. These lead to higher stack temperatures and lower boiler efficiencies.

Inspect furnace refractory and insulation for cracks that may cause leaks and missing refractory.

Clean furnace inspection parts and make sure that burner throat, furnace walls, and leading connection passes are visible through them since flame observation is an essential part of efficient boiler operation and testing.

## BOILER EFFICIENCY IMPROVEMENT

The boiler plant should be designed and operated to produce the maximum amount of usable heat from a given amount of fuel.

Combustion is a chemical reaction of fuel and oxygen which produces heat. Oxygen is obtained from the input air which also contains nitrogen. Nitrogen is useless to the combustion process. The carbon in the fuel can combine with air to form either $CO$ or $CO_2$. Incomplete combustion can be recognized by a low $CO_2$ and high $CO$ content in the stack. Excess air causes more fuel to be burned than required. Stack losses are increased and more fuel is needed to raise ambient air to stack temperatures. On the other hand, if insufficient air is supplied, incomplete combustion occurs and the flame temperature is lowered.

### Boiler Efficiency

Boiler efficiency *(E)* is defined as

$$\%E = \frac{\text{Heat Out of Boiler}}{\text{Heat Supplied to Boiler}} \times 100 \qquad \textit{Formula (6-15)}$$

For steam-generating boilers

$$\%E = \frac{\text{Evaporation Ratio} \times \text{Heat Content of Steam}}{\text{Calorific Value of Fuel}} \times 100 \qquad \textit{Formula (6-16)}$$

For hot water boilers

$$\%E = \frac{\text{Rate of Flow from Boiler} \times \text{Heat Output of Water}}{\text{Caloric Value of Fuel} \times \text{Fuel Rate}} \times 100$$

$$\textit{Formula (6-17)}$$

The relationship between steam produced and fuel used is called the evaporation ratio.

Boilers are usually designed to operate at the maximum efficiency when running at rated output. Figure 6-16 illustrates boiler efficiency as a function of time on line.

Full boiler capacity for heating occurs only a small amount of the time. On the other hand, part loading of 60% or less occurs approximately 90% of the time.

Where the present boiler plant has deteriorated, consideration should be given to replacement with modular boilers sized to meet the heating load.

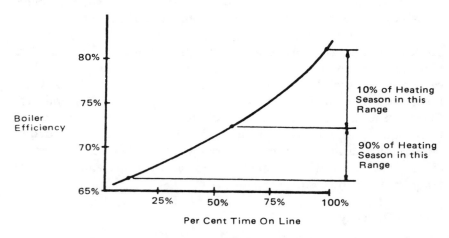

**Figure 6-16. Effect of Cycling to Meet Part Loads**

The overall thermal efficiency of the boiler and the various losses of efficiency of the system are summarized in Figure 6-17.

To calculate dry flue gas loss, Formula 6-18 is used.

$$\text{Flue gas loss} = \frac{K(T-t)}{CO_2} \qquad \textit{Formula (6-18)}$$

Where

$K$ = constant for type of fuel = 0.39 Coke
                                 = 0.37 Anthracite
                                 = 0.34 Bituminous Coal
                                 = 0.33 Coal Tar Fuel
                                 = 0.31 Fuel Oil

$T$ = temperature of flue gases in °F

$t$ = temperature of air supply to furnace in °F

$CO_2$ = percentage $CO_2$ content of flue gas measured volumetrically.

It should be noted that this formula does not apply to the combustion of any gaseous fuels such as natural gas, propane, butane, etc. Basic combustion formulas or nomograms should be used in the gaseous fuel case.

1. Overall thermal efficiency . . . . . . . . . . . . . . . . . . . . . . . . _____

2. Losses due to flue gases

    (a) Dry Flue Gas . . . . . . . . . . . . . . . . . . . . . . . . . . . . . . _____

The loss due to heat carried up the stack in dry flue gases can be determined, if the carbon dioxide ($CO_2$) content of the flue gases and the temperatures of the flue gas and air to the furnace are known.

    (b) Moisture % Hydrogen . . . . . . . . . . . . . . . . . . . . . . . _____

    (c) Incomplete combustion . . . . . . . . . . . . . . . . . . . . .

                                                       _____

3. Balance of account, including radiation and other unmeasured

   losses . . . . . . . . . . . . . . . . . . . . . . . . . . . . . . . . . . . . . . . _____

                TOTAL . . . . . . . . . . . . . . . . . . . . . . . . . . . 100%

**Figure 6-17. Thermal Efficiency of Boiler**

The savings in fuel as related to the change in efficiency is given by Formula 6-19.

$$\text{Savings in Fuel} = \frac{\text{New Efficiency} - \text{Old Efficiency}}{\text{New Efficiency}}$$
$$\times \text{Fuel Consumption} \qquad \textit{Formula (6-19)}$$

Figure 6-19 can be used to estimate the effect of flue gas composition, excess air, and stack temperature on boiler efficiency.

To estimate losses due to moisture, Figure 6-18 is used.

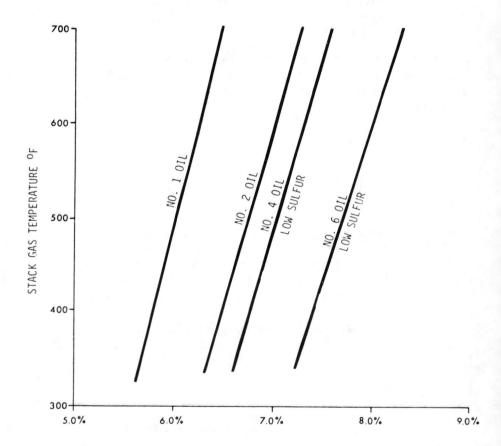

NOTE:
1. The figure gives a simple reference to heat loss in stack gases due to the formation of water in burning the hydrogen in various fuel oils.
2. The graph assumes a boiler room temperature of 80°F.

**Figure 6-18. Heat Loss Due to Burning Hydrogen in Fuel**
*(Source: Instructions For Energy Auditors, Volume 1)*

**Figure 6-19.** Effect of Flue Gas Composition and Temperature on Boiler Efficiency

*(Source: Guidelines for Saving Energy in Existing Buildings—Engineers, Architects and Operators Manual, ECM-2)*

## BOILER TUNE-UP TEST PROCEDURES

As illustrated by Figures 6-20 and 6-21, either $\% CO_2$ or $\% O_2$ can be used to determine excess air as long as the boiler is not operating on the fuel rise side of the curve.

The detailed procedures which follow illustrate how to tune-up a boiler to get the best air-to-fuel ratio. Note that for natural gas, $\% CO_2$ must also be measured, while for fuel oil, smoke spot numbers or visual smoke measurements are used.

The principal method used for improving boiler efficiency involves operating the boiler at the lowest practical excess $O_2$ level with an adequate margin for variations in fuel properties, ambient conditions and the repeatability and response characteristics of the combustion control system.

These tests should only be conducted with a through understanding of the test objectives and by following a systematic, organized series of tests.

### Cautions

Extremely low excess $O_2$ operation can result in catastrophic results.

Know at all times the impact of the modification on fuel flow, air flow and the control system.

Observe boiler instrumentation, stack and flame conditions while making any changes.

When in doubt, consult the plant engineering personnel or the boiler manufacturer.

Consult the boiler operation and maintenance manual supplied with the unit for details on the combustion control system or for methods of varying burner excess air.

### Test Description

The test series begins with baseline tests that document existing "as-found conditions" for several firing rates over the boiler's normal operating range. At each of these firing rates, variations in excess $O_2$ level from 1 to 2% above the normal operating point to the "minimum $O_2$" level are noted. Curves of combustibles as a function of excess $O_2$ level will be constructed similar to those given in Figures

# THEORETICAL AIR CURVE
## (FUEL OIL)

Figure 6-20.

Figure 6-21.

6-22 and 6-23. As illustrated in these figures, high levels of smoke or CO indicating potentially unstable operation can occur with small changes in excess $O_2$; thus small changes in excess $O_2$ should be made for conditions near the smoke or CO limit. It is important to note that the boiler may exhibit a gradual smoke or CO behavior at one firing rate and a steep behavior at another. Minimum excess $O_2$ will be that at which the boiler just starts to smoke, or the CO emissions rise above 400 ppm or the Smoke Spot Number equals the maximum value as given in Table 6-5. Once minimum excess $O_2$ levels are established, an appropriate $O_2$ margin or operating cushion ranging from 0.5 to 2.0% $O_2$ above the minimum point depending on the particular boiler control system and fuels, should be maintained.

CURVE 1 — GRADUAL SMOKE/$O_2$ CHARACTERISTIC
CURVE 2 — STEEP SMOKE/$O_2$ CHARACTERISTIC

**Figure 6-22. Typical Smoke-$O_2$ Characteristic Curves for Coal- or Oil-Fired Industrial Boilers**

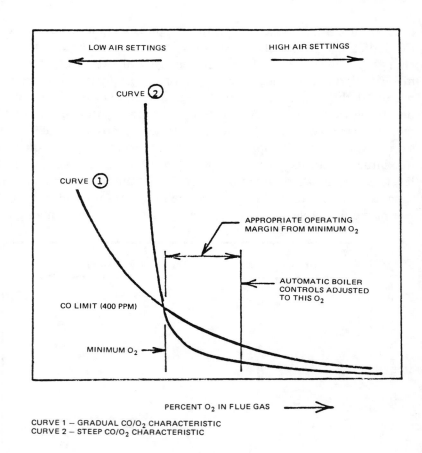

CURVE 1 — GRADUAL CO/O$_2$ CHARACTERISTIC
CURVE 2 — STEEP CO/O$_2$ CHARACTERISTIC

**Figure 6-23. Typical CO–O Characteristic Curve for Gas-Fired
Industrial Boilers**

**Table 6-5. Maximum Desirable Smoke Spot Number**

| Fuel Grade | Maximum Desirable SSN |
|---|---|
| No. 2 | less than 1 |
| No. 4 | 2 |
| No. 5 (light and heavy), and low-sulfur resid. | 3 |
| No. 6 | 4 |

Repeated tests at the same firing condition, approached from both the "high side" and "low side" (i.e., from higher and lower firing rates), can determine whether there is excessive play in the boiler controls.

Record all pertinent data for future comparisons. Readings should be made only after steady boiler conditions are reached and at normal steam operating conditions.

## Step-By-Step Boiler Adjustment Procedure for Low Excess O$_2$ Operation

1. Bring the boiler to the test firing rate and put the combustion controls on manual.
2. After stabilizing, observe flame conditions and take a complete set of readings.
3. Raise excess O$_2$ 1 to 2%, allowing time to stabilize and take readings.
4. Reduce excess O$_2$ in small steps while observing stack and flame conditions. Allow the unit to stabilize following each change and record data.
5. Continue to reduce excess air until a minimum excess O$_2$ condition is reached.
6. Plot data similar to Figures 6-22 and 6-23.
7. Compare the minimum excess O$_2$ value to the value provided by the boiler manufacturers. High excess O$_2$ levels should be investigated.
8. Establish the margin in excess O$_2$ above the minimum and reset the burner controls to maintain this level.
9. Repeat Steps 1 thru 8 for each firing rate to be considered. Some compromise in optimum O$_2$ settings may be necessary since control adjustments at one firing rate may affect conditions at other firing rates.
10. After these adjustments have been completed, verify the operation of these settings by making rapid load pick-ups and drops. If undesirable conditions are encountered, reset controls.
11. Low fire conditions may require special consideration.
12. Perform tests on any alternate fuel used. Again, some discretion will be required to resolve differences in control settings for optimum conditions on the two fuels.

## Evaluation of the New Low $O_2$ Settings

Extra attention should be given to furnace and flame patterns for the first month or two following implementation of the new operating modes. Thoroughly inspect the boiler during the next shutdown. To assure high boiler efficiency, periodically make performance evaluations and compare with the results obtained during the test program.

## Burner Adjustments

Adjustments to burner and fuel systems can also be made in addition to the low excess $O_2$ test program previously described. The approach in testing these adjustments is a "trial-and-error" procedure with sufficient organization to allow meaningful comparisons with established data.

Items that may result in lower minimum excess $O_2$ levels include changes in burner register settings, oil gun tip position, oil gun diffuser position, coal spreader position, fuel oil temperature, fuel and atomizing pressure, and coal particle size. Evaluation of each of these items involves the same general procedures, precautions and data evaluation as outlined previously. The effect of these adjustments on minimum $O_2$ is variable from boiler to boiler and difficult to predict.

## Conclusions

1.  Combustion modifications are potentially catastrophic unless caution is continuously observed.
2.  Test programs must be well thought out and planned in advance and anticipated results formulated and checked with obtained data.
3.  Any modifications at low excess $O_2$ levels must be made slowly in small steps and with continuous evaluation of the flame, furnace and stack gas conditions.
4.  All data must be recorded to allow future comparisions to be made.
5.  Practical operating excess $O_2$ levels must be given an adequate excess $O_2$ margin to allow for load variation and fuel property or ambient air condition changes.
6.  New recommended operating conditions should be moni-

tored for a sufficient length of time to gain confidence in their long-term use.

7. Periodic efficiency checks can indicate deviations from optimum performance conditions.

## AUXILIARY EQUIPMENT FOR INCREASING EFFICIENCY

The efficiency improvement potential of auxiliary equipment modifications is dependent on the existing boiler conditions.

Stack gas heat recovery equipment (air preheaters and economizers) are generally the most cost-effective auxiliary equipment additions. Addition of turbulators to firetube boilers can improve operating efficiencies and promote balanced gas flows between tube banks. Advanced combustion control systems and burners generally are less beneficial than stack gas heat recovery equipment on industrial sized units. Insulation and sootblowers, judiciously applied, can have beneficial effects on boiler efficiency. Significant energy savings potentials exist in wastewater heat recovery from blowdown water and returned condensate.

Table 6-6 summarizes the options available to improve boiler efficiency.

Air preheaters and economizers are common equipment used. A caution should be noted in that these and other options will reduce stack temperature. In order to avoid corrosion problems, exit temperatures should be as illustrated in Table 6-7. Exit gas temperature is determined by the extent of boiler convective surface or the presence of stack gas heat recovery equipment.

For boilers without heat recovery equipment, the minimum exit gas temperature is fixed by the boiler operating pressure since this determines the steam temperature. Usual design practices result in an outlet gas temperature 150°F above the saturated steam temperature.

It becomes increasingly expensive to approach boiler saturation temperatures by simply adding convective surface area. As operating pressures increase, the stack gas temperature increases to make heat recovery equipment more desirable. Economizers will permit a reduction in exit gas temperatures since the feedwater is at a lower temperature (220°F) than the steam saturation temperature. Stack gas temperatures of 300°F can be achieved with stack gas heat recovery

Table 6-6. Boiler Efficiency Improvement Equipment

| DEVICE | PRINCIPLE OF OPERATION | EFFICIENCY IMPROVEMENT POTENTIAL | SPECIAL CONSIDERATIONS |
|---|---|---|---|
| Air Preheaters | Transfer energy from stack gases to incoming combustion air. | 2.5 % for each 100 degree F decrease in stack gas temperature. | * Results in improved combustion condition.<br>* Minimum flue gas temperatures limited by corrosion characteristics of the flue gas.<br>* Application limited by space, duct orientation and maximum combustion air temperatures. |
| Economizers | Transfer energy from stack gases to incoming feedwater. | 2.5 % increase for each 100 degree F decrease in stack gas temperature (1 % increase for each 10 degree F increase in feedwater temperature. | * Minimum flue gas temperature limited by corrosion characteristics of the flue gas.<br>* Application limited on low pressure boilers.<br>* Generally preferred over air preheaters for small (50,000 lbs./hr.) units. |
| Firetube Turbulators | Increases turbulence in the secondary passes of fi-etube units thereby increasing efficiency | 2.5 % increase for each 100 degree F decrease in stack gas temperature. | * Limited to gas and oil fired units.<br>* Properly deployed, they can balance gas flows through the tubes.<br>* Increases pressure drop in the system. |
| Combustion Control Systems | Regulate the quantity of fuel and air flow | 0.25 % increase for each 1 % decrease in excess $O_2$ depending on the stack gas temperature. | * Vary in complexity from the simplest jackshaft system to cross limited oxygen correction system.<br>* Can operate either pneumatically or electrically.<br>* Retrofit applications must be compatible with existing burner hardware. |
| Instrumentation | Provide operational data | | * Provide records so that efficiency comparisions can be made. |
| Oil and Gas Burners | Promote flame conditions that result in complete combustion at lower excess air levels. | 0.25 % increase for each 1 % decrease in excess $O_2$ depending on the stack gas temperature. | * Operation with the most elaborate low excess air burners require the use of advanced combustion control systems.<br>* Flame shape and heat release rate must be compatible with furnace characteristics. |

| | | | |
|---|---|---|---|
| | | | • Flame scanners increase reliability.<br>• Advanced atomizing systems are available.<br>• Close control of oil viscosity improves atomization. |
| Insulation | Reduce external heat transfer | Dependent on surface temperature | • Mass type insulation has low thermal conductivity and release heat loss by conduction.<br>• Reflective insulation has smooth, metallic surfaces that reduce heat loss by radiation.<br>• Insulation provides several other advantages including structural strength, reduced noise and fire protection. |
| Sootblowers | Remove boiler tube deposits that retard heat transfer. | Dependent on the gas temperature | • Can use steam or air as the blowing media.<br>• Fixed position systems are used in low temperature regions whereas retractable "losses" are employed in high temperature areas.<br>• The choice of the cleaning media will depend on the characteristics of the deposits. |
| Blowdown Systems | Transfer energy from expelled blowdown liquids to incoming feedwater. | 1 - 3 % dependent on blowdown quantities and operating pressures | • Quantity of expelled blowdown water is dependent on the boiler and makeup water quality.<br>• Continuous blowdown operation not only decreases expelled liquids but also allows the incorporation of heat recovery equipment. |
| Condensate Return Systems | Reduce hot water requirements by recovering condensate. | 12 - 15 % | • Quantity of condensate returned dependent on process and contamination.<br>• Several systems available range from atmospheric (open) to fully pressurized (closed) systems. |

Table 6-7.  Minimum Exit Gas Temperatures

| | |
|---|---|
| Oil Fuel ($>$ 2.5% S) | 390°F |
| Oil Fuel ($<$ 1.0% S) | 330°F |
| Bituminous Coal ($>$ 3.5% S) | 290°F |
| Bituminous Coal ($<$ 1.5% S) | 230°F |
| Pulverized Anthracite | 220°F |
| Natural Gas (Sulfur-free) | 220°F |

equipment. Further reductions are achieved using air preheaters. Present design criteria limit the degree of cooling using stack gas heat recovery equipment to a level which will minimize condensation on heat transfer surfaces. The sulfur content of the fuel has a direct bearing on the minimum stack gas temperature since $SO_3$ combines with condensed water to form sulfuric acid and since the $SO_3$ concentration in the flue gas also determines the condensation temperature.

# 7

# Heating, Ventilation, Air Conditioning, and Building System Optimization

This chapter will review the basics of Heating, Ventilation and Air Conditioning (HVAC) and building as related to energy engineering.

## DEGREE DAYS

Degree days are the summation of the product of the difference in temperature ($\Delta T$) between the *average outdoor* and hypothetical *average indoor* temperatures (65°F) and the number of days ($t$) the outdoor temperature is below 65°F. Therefore:

$$DD = \Delta T \times t, \text{ therefore } \Delta T = DD/t \qquad \text{\textit{Formula (7-1)}}$$

Degree days divided by the total number of days on which Degree days were accumulated will yield an average $\Delta T$ for the season, based on an assumed indoor temperature of 65°F. To find the average outdoor temperature of the season, this figure must be subtracted from 65°F.

## Example Problem 7-1

If there are 6750 degree days recorded over a heating season of 270 days, what is the mean outdoor temperature for that season?

## Answer

$$\Delta T = DD/t \qquad \Delta T = \frac{6750DD}{270\,\text{days}} \qquad \Delta T = 25°F$$

The average outdoor temperature can now be found, since

$\Delta T = T$ (avg. indoor) $- T$ (avg. outdoor)

$T$ (avg. outdoor) $= T$ (avg. indoor) $- \Delta T$

$T$ (avg. outdoor) $= 65°F - 25°F$

$T$ (avg. outdoor) $= 40°F$

## RESISTANCE (R) TO HEAT FLOW AND
## CONDUCTANCE (U) AND CONDUCTIVITY (K)

The rate at which heat flows through a material depends on its characteristics. Some materials transmit heat more readily than others. This characteristic of materials which affects the flow of heat through them can be viewed either as their *resistance* to the flow of heat or as their *conductance* allowing the flow of heat.

For a section of a building, such as a wall, the conductance is expressed as the $U$-value for that wall, that is, the number of Btu's that will pass through a one-square-foot section of a building in one hour with a one-degree temperature difference between the two surfaces.

$U$ = Btu's per square foot per hour per degree Fahrenheit.

or

$U = Btu/ft^2\ h°F$

$R$-Value = Thermal Resistance = The unit time for a unit area of a particular body or assembly having defined surfaces with a unit average temperature difference established between the two surfaces per unit of Thermal Transmission.

$$\frac{hr \cdot ft^2 \cdot °F}{Btu}$$

$$R = 1/U$$

Where

$$U = \frac{1}{1/h_i + x_1/K_1 + \frac{x_2}{K_2} + \dots + 1/h_o}$$

*Formula (7-2)*

x = Respective thickness of different materials

K = Respective conductivities of layers

h = Convective heat transfer coefficients

The conductivity of a material as related to conductance and resistance is illustrated by Formula 7-3.

$$U = \frac{K}{d} = \frac{1}{R}$$

*Formula (7-3)*

where $d$ is the thickness of the material.

In buildings, Table 7-1 may be used to find acceptable conductance values for walls.

**Table 7-1. Minimum Conductance Values for Walls**

| Heating Degree Days | Min. Acceptable U- Value |
|---|---|
| 1000 | 0.40 |
| 2000 | 0.30 |
| 3000 | 0.30 |
| 4000 | 0.20 |
| 5000 | 0.20 |
| 6000 | 0.15 |
| 7000 | 0.15 |
| 8000 | 0.10 |
| 9000 | 0.10 |

## VOLUME (V) OF AIR

The volume of air within a structure is constant even though the air itself changes—new air enters; old air leaves. Total volume is equal to the volume of space in the conditioned portion of the facility. (Only the volume of conditioned space is considered since air entering and leaving the unconditioned part of the facility does not demand energy to condition it.)

To determine the volume (V) of air, multiply the height (H) of the space times the width (W) of the space times the length (L) of the space.* While this can be done for the facility as a whole, it is more accurate to calculate it for each room and then add these volumes.

## AIR CHANGES PER HOUR (AC)

The rate at which the volume of air in a structure changes per hour differs greatly from building to building. The number of air changes per hour (AC/h) has wide variation due to a number of factors such as

---

* This is only appropriate for structures with flat ceilings.

- *The number and size of openings* in the envelope—around doors and windows and in the siding itself
- *The average speed* of the wind blowing against the structure and the protection the structure has from this wind
- *The number and size* of chimneys, vents, and exhaust fans and the frequency of their use
- *The number of times* that doors and windows are opened and
- *How the structure is used.*

Currently, the general standard, or accepted value, for air changes within an average home is one air change (of entire house) per hour, unless for some reason a room contains toxic fumes, etc.

## HEATING CAPACITY OF AIR (*HC*)

Air can be heated and cooled. A certain amount of heat is necessary to change the temperature of each cubic foot (ft$^3$) of air one degree Fahrenheit (F). This amount of heat depends on the density of air which varies with temperature and pressure. This figure will generally be within the range of 0.018-0.022 Btu's/ft$^3$ °F.

## BUILDING DYNAMICS

The building experiences heat gains and heat losses depending on whether the cooling or heating system is present, as illustrated in Figures 7-1 and 7-2. Only when the total season is considered in conjunction with lighting and heating, ventilation and air conditioning (HVAC) can the energy utilization choice be decided. One way of reducing energy consumption of HVAC equipment is to reduce the overall heat gain or heat loss of a building.

## CONDUCTION HEAT LOSS

The formula used to determine the amount of heat conducted through the envelope is as follows: Degree days (*DD*) is the product of the difference in temperature ($\Delta T$) and the time (*t*) in days, providing that the days are converted to hours. This is accomplished by multiplying *DD* times 24 hours a day. This will yield the quantity of heat (*Q*) conducted through a particular section of the envelope for the entire heating season.

The formula can be written:

$$Q_{\text{(heating season)}} = U \times DD \times 24 \text{ hrs/day} \qquad \textit{Formula (7-4)}$$

or

$$Q_{\text{(heating season)}} = \frac{A \times DD \times 24 \text{ hrs/day}}{R} \qquad \textit{Formula (7-5)}$$

In general, heat flow through a flat surface is defined as

$$Q = U A \Delta T \qquad\qquad\qquad \textit{Formula (7-6)}$$

where $\Delta T$ is the temperature difference causing the heat flow.

For a composite wall, the heat flow is represented by Figure 7-3. To calculate the overall $U$ of conductance value, the resistance of each material is added in series. This is analogous to an electrical circuit.

$$R = R_1 + R_2 + R_3 + R_4 + R_5 \qquad \textit{Formula (7-7)}$$

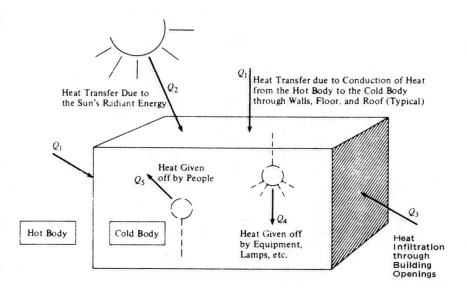

Heat Gain = $Q_1 + Q_2 + Q_3 + Q_4 + Q_5$

**Figure 7-1. Heat Gain of a Building**

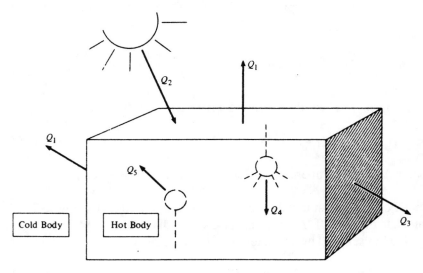

Heat Loss $= Q_1 + Q_3 - Q_2 - Q_4 - Q_5$

**Figure 7-2. Heat Loss of a Building**

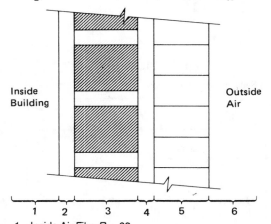

1   Inside Air Film $R = .68$
2   1/2" Plaster Board Interior Finish $R = .44$
3   8" Concrete Block, Sand & Gravel Aggregate $R = 1.1$
4   1/2" Cement Mortar $R = .1$
5   4" Brick Exterior $R = .44$
6   Outside Air Film @ 15 mph Wind $R = .17$

**Figure 7-3. Typical Wall Construction**

## FACTORS IN CONDUCTION

There are four factors which affect the conduction of heat from one area to another. They are

- *The difference in temperature (ΔT)* between the warmer area and the colder area
- *The length of time (t)* over which the transfer occurs
- *The area (A) in common* between the warmer and the colder area
- *The resistance (R) to heat flow and conduction (U)* between the warmer and the colder area.

In a typical house, the major losses due to conduction occur through the walls, the roof, and the floor.

### The Wall

— Major heat loss due to conduction, other losses are minor.
— Major concern is about the *U*-values of the walls.
— Minimum acceptable *R*-value for wall varies from 2.5 to 10.0.
— For winter the most important wall to insulate is north.
— For summer the most important wall to insulate is south.

### The Roof

—Major heat loss due to conduction, other losses are minor.
— If the funds for insulation are limited, insulate the roof first, then the north wall, then the south wall.
— Use Table 7-2 as a guide for insulating the roof.

| Heating Degree Days | Min. Acceptable *U* - Value |
|---|---|
| 1000 | 0.30 |
| 2000 | 0.20 |
| 3000 | 0.20 |
| 4000 | 0.15 |
| 5000 | 0.15 |
| 6000 | 0.10 |
| 7000 | 0.10 |
| 8000 | 0.06 |

**Table 7-2. Minimum Conductance Values for Roof**

### The Floor

— Major heat loss through conduction along the perimeter of the floor.
— Insulating floor should be the last priority compared to the walls and roof.
— Use Table 7-3 as a guide for insulating floors.

| Heating Degree Days | Min. Acceptable $U$ - Value |
|:---:|:---:|
| 1000 | 0.40 |
| 2000 | 0.35 |
| 3000 | 0.30 |
| 4000 | 0.22 |
| 5000 | 0.22 |
| 6000 | 0.18 |
| 7000 | 0.18 |
| 8000 | 0.12 |

Table 7-3. Minimum Conductance Values for Floors

## Difference in Temperature

Heat flows (much as water moves downhill) from warm areas to cold ones. The steeper the gradient between its origin and its destination, the faster it will flow. In fact, the rate at which heat is conducted is directly proportionate to the difference in temperature ($\Delta T$) between the warm area and the colder one.

## Length of Time

The longer the heat is allowed to flow across the gradient, the more heat will be conducted. The amount of heat (Btu's) is directly proportionate to the time span ($t$) of the transfer.

Btu/h is the amount of heat transferred in one hour.

## The Area ($A$) in Common

The larger the area common to the warmer and colder surfaces through which the heat flows, the greater is the rate of conducted heat. For the same material, for the same length of time, at the same $\Delta T$, the amount of heat (Btu's) transferred is directly proportionate to the area ($A$) in common.

## Example Problem 7-2

Calculate the heat loss through 20,000 ft$^2$ of building wall, as indicated by Figure 7-3. Assume a temperature differential of 17°F.

## Answer

| Description | Resistance |
|---|---|
| Outside air film at 15 mph | 0.17 |
| 4" brick | 0.44 |
| Mortar | 0.10 |
| Block | 1.11 |
| Plaster Board | 0.44 |
| Inside film | 0.68 |
| Total resistance | 2.94 |

$U = 1/R = 0.34$

$Q = U A \Delta T$

$= 0.34 \times 20{,}000 \times 17 = 115{,}600 \text{ Btu/h}$

In Figure 7-4 a surface film conductance is introduced.

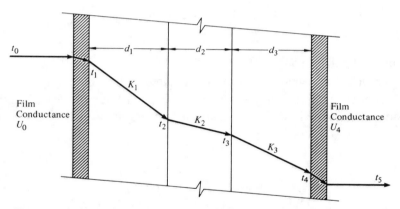

**Figure 7-4. Temperature Distribution of the Composite Wall**

The surface or film conductance is the amount of heat transferred in Btu per hour from a surface to air or from air to a surface per square foot for one degree difference in temperature. The flow of heat for the composite material can also be specified in terms of the conductivity of the material and the conductance of the air film.

$$Q = \frac{A (t_o - t_5)}{1/U_o + d_1/K_1 + d_2/K_2 + d_3/K_3 + 1/U_4}$$

*Formula (7-8)*

## LATENT HEAT AND SENSIBLE HEAT

The latent heat gain of a space means that moisture is being added to the air in the space. Moisture in the air is really in the form of superheated steam. Removing sensible heat from a space through air-conditioning equipment lowers the dry-bulb temperature of the air. On the other hand, removing latent heat from a space changes the substance state from a vapor to a liquid. The latent heat gain of a space is expressed in terms of moisture, heat units (Btu) or grains of moisture per hour (7,000 gains equals one pound). The average value for the latent heat or vaporization for superheated steam in air is 1050.

### Example Problem 7-3

2000 grains of moisture are released in a conditioned room each hour. Calculate the heat that must be removed in order to condense this moisture at the cooling coils.

**Answer**     $\dfrac{2000}{7000} \times 1050 = 299.9$ Btuh

### Example Problem 7-4

Calculate the quantity of heat ($Q$) required by infiltration in an 8,000-cubic foot ($ft^3$) home that has 1.7 air changes per hour (AC/h) when the outside temperature is 48°F and the inside temperature is 68°F ($\Delta T = 20°F$) for one day (24 hours).

### Answer

$$
\begin{aligned}
Q &= V \times AC/h \times 0.020^* \text{ Btu/ft}^3 \text{ °F} \times \Delta T \times t \\
Q &= 8,000 \text{ ft}^3 \times 1.7 \text{ } AC/h \times 0.020^* \text{ Btu/ft}^3 \text{ °F} \times 20°F \times 24 \text{ hrs} \\
Q &= 8,000 \times 1.7 \times 0.020° \times 20 \times 24 \\
Q &= 130,560 \text{ Btu's}^{**}
\end{aligned}
$$

## INFILTRATION

Leakage or infiltration of air into a building is similar to the effect of additional ventilation. Unlike ventilation, it cannot be controlled or turned

---

* This is a regional variable.
** Once again, all of the units in the formula cancel except Btu's, leaving the units for $Q$ as Btu's.

off at night. It is the result of cracks, openings around windows and doors, and access openings. Infiltration is also induced into the building to replace exhaust air unless the HVAC balances the exhaust. Wind velocity increases infiltration, and stack effects are potential problems. Air that is pushed out the window and door cracks is referred to as exfiltration.

To estimate infiltration, the Air Change Method or Crack Method is used.

## Air Change Method

The five factors which determine the amount of energy lost through infiltration can be assembled in a formula that states:

The quantity of heat ($Q$) equals the volume of air ($V$) times the number of air changes per hour ($AC/h$) times the amount of heat required to raise the temperature of air one degree Fahrenheit (0.018-0.022 Btu's) times the temperature difference ($\Delta T$) times the length of time ($t$). This is expressed as follows:

$$Q = V \times AC/h \times 0.020 \text{ Btu's} \cdot /ft^3 \text{ }^\circ F \times \Delta T \times t \qquad \qquad Formula\ (7\text{-}9)$$

The Air Change Method is considered to be a quick estimation method and is not usually accurate enough for air-conditioning design. A second method used to determine infiltration is the Crack Method.

## Crack Method

When infiltration enters a space, it adds sensible and latent heat to the room load. To calculate this gain, the following equations (7-10 and 7-11) are used.

Sensible Heat Gain $\qquad \qquad Q_S = 1.08 \text{ CFM } \Delta T \qquad \qquad Formula\ (7\text{-}10\_$

Latent Heat Gain $\qquad \qquad Q_L = .7 \text{ CFM } (HR_O - HR_i) \qquad Formula\ (7\text{-}11)$

Where
$\qquad Q_S \quad = \quad$ Sensible heat gain Btuh
$\qquad Q_L \quad = \quad$ Latent heat gain Btuh
$\qquad$ CFM $\ = \quad$ Air Flow Rate
$\qquad \Delta T \quad = \quad$ Temperature differential between outside and inside air, F
$\qquad HR_O \ = \quad$ Humidity ratio of outside air, grains per lb
$\qquad HR_i \ = \quad$ Humidity ratio of room air, grains per lb

## BODY HEAT

The human body releases sensible and latent heat depending on the degree of activity. Heat gains for typical applications are summarized in Table 7-4.

### Table 7-4. Heat Gain from Occupants

| Activity | Sensible Heat Btuh | Latent Heat Btuh | Total Heat Gain Btuh |
|---|---|---|---|
| Very Light Work – Seated (Offices, Hotels, Apartments) | 215 | 185 | 400 |
| Moderately Active Work (Offices, Hotels, Apartments) | 220 | 230 | 450 |
| Moderately Heavy Work (Manufacturing) | 330 | 670 | 1000 |
| Heavy Work (Manufacturing) | 510 | 940 | 1450 |

*Source: ASHRAE—Guide & Data Book*

## EQUIPMENT, LIGHTING AND MOTOR HEAT GAINS

It is important to include heat gains from equipment, lighting systems and motor heat gain in the overall calculations.

For a manufacturing facility, the major source of heat gains will be from the process equipment. Consideration must be given to all equipment including motors driving supply and exhaust fans.

To convert motor horsepower to heat gain in Btuh, Formula 7-12 is used.

$$Q = \frac{hp \times .746}{\eta} \times 3412$$

*Formula (7-12)*

Where

hp   is the running motor horsepower

$\eta$   is the efficiency of the motor

$Q$   is the heat gain from the motor Btuh

Similarly, the kilowatts of the lighting system can be converted to heat gain.

$$Q = (KW_F + KW_B) \times 3412$$

*Formula (7-13)*

Where

$KW_F$      is the kilowatts of the lighting fixtures

$KW_B$      if the kilowatts of the ballast

$Q$      is the heat gain from the lighting system Btuh

## RADIANT HEAT GAIN

Heat from the sun's rays greatly increases heat gain of a building. If the building energy requirements were mainly due to cooling, then this gain should be minimized. Solar energy affects a building in the following ways:

1. *Raises the surface temperature*; thus a greater temperature differential will exist at roofs than at walls.
2. A large percentage of direct solar radiation and diffuse sky radiation *passes through* transparent materials, such as glass.
3. Envelopes can be designed to have individual control loops to stop or enhance the solar radiation into the building.

## SURFACE TEMPERATURES

The temperature of a wall of roof depends upon
(a) the angle of the sun's rays
(b) the color and roughness of the surface
(c) the reflectivity of the surface
(d) the type of construction

When an engineer is specifying building materials, he should consider the above factors. A simple example is color. The darker the surface, the more solar radiation will be absorbed. Obviously, white surfaces have a lower temperature than black surfaces after the same period of solar heating. Another factor is that smooth surfaces reflect more radiant heat than do rough ones.

In order to properly take solar energy into account, the angle of the sun's rays must be known. If the latitude of the facility is known, the angle can be determined.

## SUNLIGHT AND GLASS CONSIDERATIONS

A danger in the energy conservation movement is to take steps backward. A simple example would be to exclude glass from building designs because of the poor conductance and solar heat gain factors of clear glass. The engineer needs to evaluate various alternate glass constructions and coating in order to maintain and improve the aesthetic qualities of good design while minimizing energy inefficiencies. It should be noted that the method to reduce heat gain of glass due to conductance is to provide an insulating air space.

To reduce the solar radiation that passes through glass, several techniques are available. Heat absorbing glass (tinted glass) is very popular. Reflective glass is gaining popularity, as it greatly reduces solar heat gains.

To calculate the relative heat gain through glass, a simple method is illustrated below.

$$Q = \pm U A (t_0 - t_1) + A \times S_1 \times S_2 \qquad Formula\ (7\text{-}14)$$

Where

$Q$     is the total heat gain for each glass orientation (Btuh)

$U$     is the conductance of the glass (Btu/h•ft$^2$•°F)

$A$     is the area of glass; the area used should include framing, since it will generally have a poor conductance compared with the surrounding material. (ft$^2$)

$t_0 - t_1$     is the temperature difference between the inside temperature and outside ambient. (°F)

$S_1$     is the shading coefficient; $S_1$ takes into account external shades, such as venetian blinds and draperies, and the qualities of the glass, such as tinting and reflective coatings.

$S_2$     is the solar heat gain factor. This factor takes into account direct and diffused radiation from the sun. Diffused radiation is basically caused by reflections from dust particles and moisture in the air.

## Optimizing Windows

— Use storm windows in winter.

— Use a clear plastic cover to cover windows in the winter so as to allow radiation but stop infiltration.

— Use double or triple glass windows.

— Promote thermal barriers.

— Use solar control devices.

## THE PSYCHROMETRIC CHART

Just as the steam table and the Mollier Diagram are used to relate the properties of steam, the psychrometric chart is used to illustrate the properties of moist air. The psychrometric chart is a very important tool in the design of air-conditioning systems.

## PROPERTIES OF MOIST AIR

Air expands and contracts with temperature. If pressure is held constant, then air expands or contracts at a specified rate with change in temperature as defined by Formula 7-15.

$$V_2 = V_1 \frac{T_2}{T_1}$$

$$\text{Formula (7-15)}$$

Where

$V_1$ = initial volume of air
$V_2$ = final volume of air
$T_1$ = absolute temperature
$T_2$ = final absolute temperature

$$V_2 = V_1 \frac{P_2}{P_1}$$

At constant temperature,                                    Formula (7-15a)

Where

$P_2$ = initial pressure, psia
$P_1$ = final pressure, psia

The change in the volume occupied by air at any temperature can be found by first using Formula 7-15a to calculate the change in volume with pressure and then using Formula 7-15 to calculate the change in volume with temperature.

The temperature at which the water vapor in the atmosphere begins to condense is the *Dew Point Temperature*. It should be noted that the mass of moisture per pound of dry air in a mixture of air and water vapor depends on the dew point temperature alone. If there is no condensation of moisture, the dew point temperature remains constant.

*Humidity Ratio* is defined as the mass of water vapor mixed with one pound of dry air.

*Degree of Saturation* is defined as the actual humidity ratio divided by the humidity ratio at saturation.

*Relative Humidity* is defined as the vapor pressure of air divided by the saturation pressure of pure water at the same temperature.

*Sensible Heat of an Air-Vapor Mixture* is defined as the heat which affects the dry-bulb temperature of the mixture only.

*Wet-Bulb Temperature* can be determined by covering the bulb of a thermometer with a wet wick and holding it in a stream of swiftly moving air. At first the temperature will drop quickly and then reach a stationary point referred to as the wet-bulb temperature. The wet-bulb temperature is lower than the dry-bulb temperature. The amount of water which evaporates from the wet wick into the air depends on the amount of water vapor initially in the air flowing past the wet bulb. A sling psychrometer is a convenient instrument used to measure wet-bulb temperatures.

*Total Heat* is defined as the sum of the sensible and latent heat. Sensible heat depends only upon the dry-bulb temperature, while the latent heat content depends only upon its dew point.

The *Enthalpy of an Air-Water Vapor Mixture* can be calculated by Formula 7-16.

$$h_{(mix)} = h_{(dry\ air)} + h_{(water\ vapor)}$$

or

$$h_{(mix)} = Cp \times T_{DB} + HR\ hg \qquad Formula\ (7\text{-}16)$$

Where
$h_{(mix)}$ = enthalpy of the mixture of dry air and water vapor, Btu per lb
$Cp$ = specified heat
$T_{DB}$ = dry bulb temperature
$HR$ = humidity ratio of the mixture
$hg$ = enthalpy of saturated vapor (steam) at the dew point temperature

To determine the properties of air such as the humidity ratio, relative humidity and enthalpy, the psychrometric chart is frequently used. Figure 7-5 illustrates the psychrometric chart.

### Example Problem 7-5

Given air at 70°F DB and 50% relative humidity, for the air vapor mixture find

- Wet-Bulb Temperature
- Enthalpy
- Humidity Ratio
- Dew Point Temperature
- Specific Volume
- Vapor Pressure
- Percentage Humidity

### Answer

From Figure 7-5 find the intersection of 70°F DB and 50% *HR*, point "A."

The WB temperature is found as 58.6 WB, point "B."

The Enthalpy is found as 25.5 Btu/lb, point "C."

The Humidity Ratio is found to be 56 grains of moisture per pound of dry air, point "D," and the Dew Point temperature is 53°F, point "E."

The Specific Volume is found to be 13.5 cubic feet per pound of air, point "F," with a Vapor Pressure of .38 inches of mercury, point "G."

The percentage humidity equals the actual humidity 56, point "D," divided by the humidity ratio at saturation (100% *RH*) which is found to be 110, point "H." Thus % humidity = 56/110 = .50.

### Example Problem 7-6

Given 8000 CFM of chilled air at 55°F DB and 50°F WB mixed with 3000 CFM of outside air at 90°F DB and 80°F WB, compute the properties of the mixture.

### Answer

From Figure 7-6, the intersection of 55°F DB and 50°F WB is point "A." The specific volume is then 13.1 cubic feel/lb, point "B."

Similarly, for the outside air, the specific volume is 14.3 cubic feet/lb, point "D."

The total weight and dry-bulb temperature of the mixture can be found by the following ratios:

$$\frac{8,000}{13.1} = 610.6 \text{ lb/min.}$$

$$\frac{3,000}{14.3} = \underline{209.7 \text{ lb/min.}}$$

$$\text{Total} \quad 820.3 \text{ lb/min. for total weight}$$

The dry-bulb temperature is

$$\frac{610.6}{820.3} \times 55 = 40.93°F$$

$$\frac{209.7}{820.3} \times 90 = \underline{23.0°F}$$

$$63.9°F \text{ DB}$$

The properties of the mixture, point "E," can now be determined from the chart.

*Psychrometric Chart Courtesy of Trane Company*

**Figure 7-5. Psychrometric Chart**

Figure 7-6. Psychrometric Chart for Mixture of Air

WB    = 59.8°F

  *h*    = 26.6 Btu/lb

Humidity Ratio = 70 gr of moisture/lb of dry air.

## BASICS OF FAN DISTRIBUTION SYSTEMS

In order to distribute conditioned or ventilated air, fans are the chief vehicle used. Several basic types of fans commonly used in industry are illustrated in Figure 7-7 and are listed below.

*Centrifugal, airfoil blade*—used on large heating, ventilating, and air-conditioning systems. Airfoil fans are used where clean air is handled.

*Centrifugal, backward curved blade*—used for general heating, ventilating, and air-conditioning systems. Air handling need not be as clean as above.

*Centrifugal, radial blade*—a rugged, heavy duty fan for high pressure applications. It is designed to handle sand, wood chips, etc.

*Centrifugal, forward curved blade*—ideal for low pressure applications, such as domestic furnaces or room and packaged air conditioners.

The brake horsepower of a fan is illustrated by Formula 7-17.

$$\text{Brake Horsepower} = \frac{\text{CFM x Fan PS}}{6356 \times \eta_F} \qquad \textit{Formula (7-17)}$$

Where

   CFM    is the quantity of air in CFM

    $\eta_F$    is the fan static efficiency

 Fan PS    is the fan static pressure in inches

To compute the fan static pressure

$$\text{Fan PS} = P_T(O) - P_T(i) - P_v(O) \qquad \textit{Formula (7-18)}$$

Where

  $P_T(O)$    is the total pressure at fan outlet

   $P_T(i)$    is the total pressure at fan inlet

  $P_v(O)$    is the velocity pressures at fan outlet

The excess pressure above the static pressure is known as the velocity pressure and is computed by Formula 7-19 for standard air having value of 13.33 cu ft per lb as

Figure 7-7. Fan Types: (A) Vaneaxial; (B) Backward Curved Blade; (C) Tubeaxial; (D) Radial; (E) Radial Tip Blade; (F) Airfoil Blade. *(Courtesy of Buffalo Forge Company.)*

$$P_v = \left(\frac{V}{4005}\right)^2$$

*Formula (7-19)*

where *V* is the velocity of air in FPM.

Note that the pressure of air in sheet metal ducts is so low that ordinary pressure gage (Bourdon type) cannot be used; thus a V-tube or manometer is used, which measures pressure in inches of water. A pressure of 1 psi will support a column of water 2.31 feet high or 27.7 inches.

### Fan Laws

The performance of a fan at varying speeds and air densities may be predicted by certain basic fan laws as illustrated in Table 7-5.

**Table 7-5. Fan Laws**

| | |
|---|---|
| 1. | Fan Law for variation in fan speed at constant air density with a constant system |
| | 1.1    Air volume, CFM varies as fan speed |
| | 1.2    Static velocity or total pressure varies as the square of fan speed |
| | 1.3    Power varies as cube of fan speed |
| 2. | Fan Law variation in air density at constant fan speed with a constant system |
| | 2.1    Air volume is constant |
| | 2.2    Static velocity or pressure varies as density |
| | 2.3    Power varies as density |

### Example Problem 7-7

An energy audit indicates that the ventilation requirements of a space can be reduced from 15,000 CFM to 12,000 CFM. Comment on the savings in brake horsepower if the fan pulley is changed to reduce the fan speed accordingly.

### Answer

From the Fan Laws

$$\text{hp}_1 = \text{hp}_2 \times \left(\frac{\text{CFM new}}{\text{CFM old}}\right)^3$$

$$= \text{hp}_2 \times \left(\frac{12,000^3}{15,000}\right) = \text{hp}_2 \, (.8)^3 = .512 \, \text{hp}_2$$

or a 48.8% savings.

The fan performance is affected by the density of the air that the fan is handling. All fans are rated at standard air with a density of .075 lb per cu ft and a specific volume of 13.33 cu ft per lb. When a fan is tested in a laboratory at different than standard air, the brake horsepower is corrected by using the Fan Laws.

## Fan Performance Curves

Fan performance curves are used to determine the relationship between the quantity of air that a fan will deliver and the pressure it can discharge at various air quantities. For each fan type, the manufacturer can supply fan performance curves which can be used in design and as a tool of determining the fan efficiency.

As illustrated by Ex. Prob. 7-6, one energy engineering technique to reduce fan horsepower is to reduce fan speed. An alternate way is to throttle the air flow by a damper. The fan performance curves can be used to illustrate the best choice of these options. The system characteristics can be plotted on the fan curves to show the static pressure required to overcome the friction loss in the duct system. From the Fan Laws the system friction loss varies with the square of fan speed; thus, as the air quantity increases, the friction loss will vary as illustrated by Figure 7-8.

For a detailed analysis, the fan performance curve should be used to predict how a specific fan will perform in a desired application.

**Figure 7-8. System Characteristic Curve**

## Example Problem 7-8

Given Fan Performance Curve Figure 7-9. The fan delivers 21,500 CFM at 600 rpm at a brake horsepower of 12.3. Comment on the savings in brake horsepower by reducing air flow to 14,400 CFM by each of the following methods: (a) reducing fan speed to 400 rpm, (b) throttling the air flow by a damper.

**Figure 7-9. Fan Performance Curves**

(a) From the Fan Laws the brake horsepower is reduced as follows:

$$hp = (12.3) \left(\frac{400}{600}\right)^3 = 3.64$$

Using the fan performance curve, Figure 7-10, the system characteristic curve "A" is plotted.

By reducing the rpm from 600 to 400, the system operates at point 1 and then moves to point 2. The brake horsepower is found from Figure 7-9 to be 3.7.

(b) By closing the air damper, the air flow is reduced to 14,400 CFM while still running the fan at 600 rpm. Using the fan performance curve, Figure 7-10, the system operates at point 1 and then moves to point 3. The power to operate the fan at point 3 is 7.2 hp from Figure 7-9. Thus, if the fan speed can be reduced, it is more efficient than throttling the air flow damper.

Figure 7-10. Fan Performance and System Characteristic Curves

# FLUID FLOW

Pump and piping considerations are extremely important due to the fact that energy transport losses are a part of any distribution system. Losses occur due to friction, and that lost energy must be supplied by pump horsepower.

Centrifugal pumps are commonly used in heating, ventilating and air-conditioning applications as well as utility systems. The output torque for the pump is supplied by a driver such as a motor. Liquid enters the eye of the impeller which rotates. Pressure energy builds up by the action of centrifugal force, which is a function of the impeller vane peripheral velocity.

As with fan systems, Pump Laws and curves can be used to predict system responsiveness. The affinity laws of a pump are illustrated in Table 7-6.

The horsepower required to operate a pump is illustrated by Formula 7-20.

$$hp = \frac{\Delta P \ GPM}{1715 \ \eta}$$

*Formula (7-20)*

### Table 7-6. Affinity Laws for Pumps

| Impeller Diameter | Speed | Specific Gravity (SG) | To Correct for | Multiply by |
|---|---|---|---|---|
| | | | Flow | $\left(\dfrac{\text{New Speed}}{\text{Old Speed}}\right)$ |
| Constant | Variable | Constant | Head | $\left(\dfrac{\text{New Speed}}{\text{Old Speed}}\right)^2$ |
| | | | BHP (or kW) | $\left(\dfrac{\text{New Speed}}{\text{Old Speed}}\right)^3$ |
| | | | Flow | $\left(\dfrac{\text{New Diameter}}{\text{Old Diameter}}\right)$ |
| Variable | Constant | | Head | $\left(\dfrac{\text{New Diameter}}{\text{Old Diameter}}\right)^2$ |
| | | | BHP (or kW) | $\left(\dfrac{\text{New Diameter}}{\text{Old Diameter}}\right)^3$ |
| Constant | Variable | | BHP (or kW) | $\left(\dfrac{\text{New SG}}{\text{Old SG}}\right)$ |

Where

$\Delta P$     is the differential pressure across a pump in psi

GPM    is the required flow rate in gallons per minute

$\eta$      is the pump efficiency

To convert psi to read in feet use Formula 7-21.

$$\text{Head in Feet} = \frac{\text{psi} \times 2.31}{\text{Specific Gravity of Fluid}}$$

*Formula (7-21)*

Basically, the size of discharge line piping from the pump determines the friction loss through the pipe that the pump must overcome. The greater the line loss, the more pump horsepower required. If the line is short or has a small flow, this loss may not be significant in terms of the total system head requirements. On the other hand, if the line is long and has a large flow rate, the line loss will be significant.

To calculate the pressure loss for water system piping and the corresponding velocity, Formulas 7-22 and 7-23 are used.

$$\Delta P = \frac{0.55\, CF^{1.85}}{d^{4.87}}$$

*Formula (7-22)*

$$V = \frac{.41\, F}{d^2}$$

*Formula (7-23)*

Where

   $\Delta P$    =   pressure loss per 100 feet of pipe, psi
   $V$     =   velocity of fluid, ft/sec.
   $C$     =   roughness factor
            1 for copper tubing
            1.62 for steel pipe
            .77 for plastic pipe
   $F$     =   flow rate in gallons per minute
   $d$     =   inside diameter of pipe, inches

The pressure loss due to fittings is determined by Formula 7-24.

$$\Delta P = .0067\, K V^2$$

*Formula (7-24)*

where $K$ is the loss coefficient.

## Options to Reduce Pump Horsepower

There are several alternatives that will significantly reduce pump horsepower. Summarized below are some of the options available.

1. Many pumps are oversized to very conservative design practices. If the pump is oversized, install a smaller impeller to match the load.
2. In some instances heating or cooling supply flow rates can be reduced. To save on pump horsepower, either reduce motor speed or change the size of the motor sheave.
3. Check economics of replacing corroded pipe with a large pipe diameter to reduce friction losses.
4. Consider using variable speed pumps to better match load conditions. Motor drive speed can be varied to match pump flow rate or head requirements.
5. Consider adding a smaller auxiliary pump. During part load situations a larger pump can be shut down and a smaller auxiliary pump used.

The distribution and overall load that is placed on HVAC transportation devices can additionally be optimized by using the following guidelines.

1. Reduce the resistance to flow in ducts and piping.

2. Eliminate leaks in ducts or piping.
3. Improve the efficiency of terminal devices. (Heat exchangers, diffusers, etc.)
4. Reduce or eliminate the opportunity for heat transfer during distribution.
5. Reduce the transport rate. (Flow velocity)

Remember that improvements in the envelope efficiency will automatically reduce the distribution load. This effect should always be considered and optimized by sizing distribution components accordingly.

## Common Sense Methods

Do not underestimate the effects of energy saving techniques which require only a change of habit or life style.

Some of the following guidelines can lead to significant savings with little or no initial cost.

1. Set back the thermostat. Some suggested guidelines are 68°F during the heating season and 78°F during the cooling season. When possible, reduce thermostat in unoccupied areas. This is easily done with baseboard electric heat.

2. Use or adjust thermostats to allow for a temperature deadband or range in which no action is taken by HVAC systems.

3. Regulate thermostat settings. (Hide or lock where necessary.)

4. Turn off lights in unoccupied areas and reduce overall lighting levels by illuminating the required area (task area). A good example of the potential for energy savings is illustrated on pages 310-312 of this text.

5. Reduce levels of relative humidity during the heating season. The energy required to vaporize water for humidification purposes can be significant.

6. Shut down ventilation systems during unoccupied hours and reduce levels to acceptable limit during occupied hours.

7. Check and ensure that ventilation dampers open and close as desired by HVAC system.

8. Filter indoor air to reduce outdoor air ventilation requirements. Filtering systems are available to remove cigarette smoke, cooking smoke and odors, and other common air contaminants.

9. Use storm windows or plastic to reduce infiltration as well as conduction and convection heating/cooling losses of windows.

10. Correct duct inadequacies which cause leakage, obstruction, or allow unnecessary heat transfer.

## HVAC SYSTEMS

The summary below illustrates the types of systems frequently encountered in heating and air-conditioning systems.

---

### Single Zone System

Single zone systems consist of a mixing, conditioning and fan section. The conditioning section may have heating, cooling, humidifying or a combination of capabilities. Single zone systems can be factory assembled roof-top units or built up from individual components and may or may not have distributing duct work.

---

### Terminal Reheat System

Reheat systems are modifications of single zone systems. Fixed cold temperature air is supplied by the central conditioning system and reheated in the terminal units when the space cooling load is less than maximum. The reheat is controlled by thermostats located in each conditioned space.

---

### Multizone Systems

Multizone systems condition all air at the central system and mix heated and cooled air at the unit to satisfy various zone loads as sensed by zone thermostats. These systems may be packaged roof-top units or field-fabricated systems.

## Dual Duct Systems

Dual duct systems are similar
to multizone systems
except heated and cooled
air is ducted to the conditioned spaces and mixed as
required in terminal
mixing boxes.

## Variable Air Volume Systems

A variable air volume system
delivers a varying amount of air as
required by the conditioned spaces.
The volume control may be by
fan inlet (vortex) damper,
discharge damper or fan speed
control. Terminal sections may be
single duct variable volume units
with or without reheat, controlled
by space thermostats.

## Induction Systems

Induction systems generally have units at the outside
perimeter of conditioned spaces. Conditioned primary
air is supplied to the units where it passes through
nozzles or jets and by induction draws room air
through the induction unit coil. Room temperature
control is accomplished by modulating water flow
through the unit coil.

## Fan Coil Units

A fan coil unit consists of a cabinet with heating and/or cooling coil,
motor and fan and a filter. The unit may be floor or ceiling mounted
and uses 100% return air to condition a space.

### Unit Ventilator

A unit ventilator consists of a cabinet with heating and/or cooling coil, motor and fan, a filter and a return air—outside air mixing section. The unit may be floor or ceiling mounted and uses return and outside air as required by the space.

### Unit Heater

Unit heaters have a fan and heating coil which may be electric, hot water or steam. They do not have distribution duct work but generally use adjustable air distribution vanes. Unit heaters may be mounted overhead for heating open areas or enclosed in cabinets for heating corridors and vestibules.

### Perimeter Radiation

Perimeter radiation consists of electric resistance heaters or hot water radiators usually within an enclosure but without a fan. They are generally used around the conditioned perimeter of a building in conjunction with other interior systems to overcome heat losses through walls and windows.

### Hot Water Converters

A hot water converter is a heat exchanger that uses steam or hot water to raise the temperature of heating system water. Converters consist of a shell and tubes with the water to be heated circulated through the tubes and the heating steam or hot water circulated in the shell around the tubes.

*Source: "Energy Conservation with Comfort," Honeywell*

## Reducing Energy Consumption in HVAC Systems

*Variable Air Volume (VAV) System*—A variable air volume system provides heated or cooled air at a constant temperature to all zones served. VAV boxes located in each zone or in each space adjust the quantity of air reaching each zone or space depending on its load requirements. Methods for conserving energy consumed by this system include:

1. Reduce the volume of air handled by the system to that point which is minimally satisfactory.
2. Lower hot water temperature and raise chilled water temperature in accordance with space requirements.
3. Lower air supply temperature to that point which will result in the VAV box serving the space with the most extreme load being fully open.
4. Consider installing static pressure controls for more effective regulation of pressure bypass (inlet) dampers.
5. Consider installing fan inlet damper control systems if none now exists.

*Constant Volume System*—Most constant volume systems either are part of another system—typically dual duct systems—or serve to provide precise air supply at a constant volume. Opportunities for conserving energy consumed by such systems include:

1. Determine the minimum amount of airflow which is satisfactory and reset the constant volume device accordingly.
2. Investigate the possibility of converting the system to variable (step controlled) constant volume operation by adding the necessary controls.

*Induction System*—Induction systems comprise an air handling unit which supplies heated or cooled primary air at high pressure to induction units located on the outside walls of each space served. The high pressure primary air is discharged within the unit through nozzles inducing room air through a cooling or heating coil in the unit. The resultant mixture of primary air and induced air is discharged to the room at a temperature dependent upon the cooling and heating load of the space. Methods for conserving energy consumed by this system include:

1. Set primary air volume to original design values when adjusting and balancing work is performed.

2. Inspect nozzles. If metal nozzles, common on most older models, are installed, determine if the orifices have become enlarged from years of cleaning. If so, chances are that the volume/pressure relationship of the system has been altered. As a result, the present volume of primary air and the appropriate nozzle pressure required must be determined. Once done, rebalance the primary air system to the new nozzle pressures and adjust individual induction units to maintain airflow temperature. Also, inspect nozzles for cleanliness. Clogged nozzles provide higher resistance to air flow, thus wasting energy.
3. Set induction heating and cooling schedules to minimally acceptable levels.
4. Reduce secondary water temperatures during the heating season.
5. Reduce secondary water flow during maximum heating and cooling periods by pump throttling or, for dual-pump systems, by operating one pump only.
6. Consider manual setting of primary air temperature for heating, instead of automatic reset by outdoor or solar controllers.

*Dual-Duct System*—The central unit of a dual-duct system provides both heated and cooled air, each at a constant temperature. Each space is served by two ducts, one carrying hot air, the other carrying cold air. The ducts feed into a mixing box in each space which, by means of dampers, mixes the hot and cold air to achieve that air temperature required to meet load conditions in the space or zone involved. Methods for improving the energy consumption characteristics of this system include:
1. Lower hot deck temperature and raise cold deck temperature.
2. Reduce air flow to all boxes to minimally acceptable level.
3. When no cooling loads are present, close off cold ducts and shut down the cooling system. Reset hot deck according to heating loads and operate as a single duct system. When no heating loads are present, follow the same procedure for heating ducts and hot deck. It should be noted that operating a dual-duct system as a single-duct system reduces air flow, resulting in increased energy savings through lowered fan speed requirements.

*Single Zone System*—A zone is an area or group of areas in a building which experiences similar amounts of heat gain and heat loss. A single zone system is one which provides heating and cooling to one zone controlled by the zone thermostat. The unit may be installed within or remote from the space it serves, either with or without air distribution ductwork.

1.  In some systems air volume may be reduced to the minimum required, therefore reducing fan power input requirements. Fan brake horsepower varies directly with the cube of air volume. Thus, for example, a 10% reduction in air volume will permit a reduction in fan power input by about 27% of original. This modification will limit the degree to which the zone serviced can be heated or cooled as compared to current capabilities.

2.  Raise supply air temperatures during the cooling season and reduce them during the heating season. This procedure reduces the amount of heating and cooling which a system must provide, but, as with air volume reduction, limits heating and cooling capabilities.

3.  Use the cooling coil for both heating and cooling by modifying the piping. This will enable removal of the heating coil, which provides energy savings in two ways. First, air flow resistance of the entire system is reduced so that air volume requirements can be met by low-ered fan speeds. Second, system heat losses are reduced because the surface area of cooling coils is much larger than that of heating coils, thus enabling lower water temperature requirements. Heating coil removal is not recommended if humidity control is critical in the zone serviced and alternative humidity control measures will not suffice.

*Multizone System*—A multizone system heats and cools several zones—each with different load requirements—from a single, central unit. A thermostat in each zone controls dampers at the unit which mix the hot and cold air to meet the varying load requirements of the zone involved. Steps which can be taken to improve energy efficiency of multizone systems include:

1.  Reduce hot deck temperatures and increase cold deck temperatures. While this will lower energy consumption, it also will reduce the system's heating and cooling capabilities as compared to current capabilities.

2.  Consider installing demand reset controls which will regulate hot and cold deck temperatures according to demand. When properly installed, and with all hot deck or cold deck dampers partially closed, the control

will reduce hot and raise cold deck temperature progressively until one or more zone dampers is fully open.

3. Consider converting systems serving interior zones to variable volume. Conversion is performed by blocking off the hot deck, removing or disconnecting mixing dampers, and adding low pressure variable volume terminals and pressure bypass.

*Terminal Reheat System*—The terminal reheat system essentially is a modification of a single zone system which provides a high degree of temperature and humidity control. The central heating/cooling unit provides air at a given temperature to all zones served by the system. Secondary terminal heaters then reheat air to a temperature compatible with the load requirements of the specific space involved. Obviously, the high degree of control provided by this system requires an excessive amount of energy. Several methods for making the system more efficient include:

1. Reduce air volume of single zone units.
2. If close temperature and humidity control must be maintained for equipment purposes, lower water temperature and reduce flow to reheat coils. This still will permit control but will limit the system's heating capabilities somewhat.
3. If close temperature and humidity control is not required, convert the system to variable volume by adding variable volume valves and eliminating terminal heaters.

## THE ECONOMIZER CYCLE

The basic concept of the economizer cycle is to use outside air as the cooling source when it is cold enough. There are several parameters which should be evaluated in order to determine if an economizer cycle is justified. These include:

- Weather
- Building occupancy
- The zoning of the building
- The compatibility of the economizer with other systems
- The cost of the economizer

### What Are the Costs of Using
### The Economizer Cycle

Outside air cooling is accomplished usually at the expense of an additional return air fan, economizer control equipment, and an additional burden on the humidification equipment. Therefore, economizer cycles must be carefully evaluated based on the specific details of the application.

Using outside air to cool a building can result in lower mechanical refrigeration cost whenever outdoor air has a lower total heat content (enthalpy) than the return air. This can be accomplished by an "integrated economizer" or enthalpy control. See Figure 7-11 for a comparison of controls.

### Dry-Bulb Economizer

Operation of the "integrated economizer" can be made automatic by providing (1) dampers capable of providing 100% outdoor air, and (2) local controls that sequence the chilled water or DX (direct expansion) coil and dampers so that during economizer operation, on a rise in discharge (or space) temperature, the outdoor damper opens first; then on a further rise, the cooling coil is turned "on."

Economizer operation is activated by outside air temperature, for example, 72°F DB[*]. If outside air is below 72°F, the above described economizer sequence occurs. Above 72°F, outside air cooling is not economical, and the outdoor air damper closes to its minimum position to satisfy ventilation requirements only.

### Enthalpy Control

If an economizer system is equipped with enthalpy control, savings will accrue due to a more accurate changeover point. The load on a cooling coil for an air handling system is a function of the *total heat* of air entering the coil. Total heat is a function of *two* measurements, dry-bulb (DB) and relative humidity (RD) or dew point (DP). The enthalpy control measures both conditions (DB and RH) in the return air duct and outdoors. It then computes which air source would impose the lowest load on the cooling system. If outside air is the smallest load, the controller enables the economizer cycle. (Dry-bulb economizer control savings will be less than those shown for enthalpy control, except in dry climates.)

### Enthalpy Control Savings Calculations

Savings are based on the assumption that the system previously had either (a) no 100% O.A. damper or (b) a fixed minimum outdoor air

---

[*] This varies according to location.

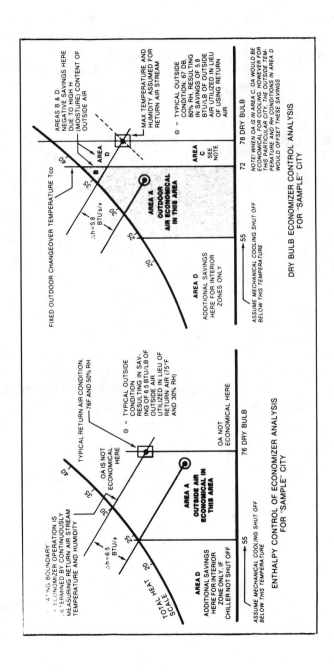

Figure 7-11. Comparative Economizer Controls

setting (whenever the fan operated) sufficient for ventilation purposes; it is also assumed that (c) minimum outdoor air has already been reduced, and the minimum damper opening will be at the *new value*.

  Step 1.  Determine minimum CFM of outdoor air to be used during occupied hours.

  Step 2.  Calculate annual savings.

$$A \frac{\text{ft}^3}{\text{min.}} \left(1 - \frac{B\%}{100}\right) \times K \frac{10^6 \text{ Btu}}{\text{yr } 1000 \text{ CFM}} \times \frac{\text{operating hrs/wk}}{50} \times J \frac{\$}{10^6 \text{ Btu}}$$

$$= \$ \text{ SAVED PER YEAR} \qquad\qquad\qquad \textit{Formula (7-25)}$$

Where

  $A$ = air handling capacity $\left(\dfrac{\text{ft}^3}{\text{min.}}\right)$

  $B$ = present ventilation air (%)

  $J$ = cost of cooling $\left(\dfrac{\$}{10^6 \text{ Btu}}\right)$

  $K$ = seasonal cooling savings $\left(\dfrac{10^6 \text{ Btu}}{\text{yr } 1000 \text{ CFM}}\right)$

  Formula 7-25 is used to calculate the savings resulting from enthalpy control of outdoor air. The calculated savings generally will be greater than the savings resulting from a dry-bulb economizer. To estimate dry-bulb economizer savings, multiply the enthalpy savings by .93.

# 8

# HVAC Equipment

## PERFORMANCE RATIOS

### Measuring Efficiency by Using the Coefficient of Performance

The coefficient of performance *(COP)* is the basic parameter used to compare the performance of refrigeration and heating systems. *COP* for cooling and heating applications is defined as follows:

$$COP \text{ (Cooling)} = \frac{\text{Rate of Net Heat Removal}}{\text{Total Energy Input}} \qquad \textit{Formula (8-1)}$$

$$COP \text{ (Heating, Heat Pump*)} = \frac{\text{Rate of Useful Heat Delivered*}}{\text{Total Energy Input}}$$
$$\textit{Formula (8-2)}$$

### Measuring System Efficiency Using the Energy Efficiency Ratio

The energy efficiency ratio *(EER)* is used primarily for air-conditioning systems and is defined by Formula 8-3.

$$EER = \frac{COP \ (3412)}{1000} \text{ Btu/watt-hr} \qquad \textit{Formula (8-3)}$$

---

*For Heat Pump Applications, exclude supplemental heating.

## THE HEAT PUMP

The heat pump has gained wide attention due to its high potential *COP*. The heat pump in its simplest form can be thought of as a window air conditioner. During the summer, the air on the room side is cooled while air is heated on the outside-air side. If the window air conditioner is turned around in the winter, some heat will be pumped into the room. Instead of switching the air conditioner around, a cycle reversing valve is used to switch functions. This valve switches the function of the evaporator and condenser, and refrigeration flow is reversed through the device. Thus, *the heat pump is heat recovery through a refrigeration cycle.* Heat is removed from one space and placed in another. In Chapter 7, it was seen that the direction of heat flow is from hot to cold. Basically, energy or pumping power is needed to make heat flow "up hill." The mechanical refrigeration compressor "pumps" absorbed heat to a higher level for heat rejection. The refrigerant gas is compressed to a higher temperature level so that the heat absorbed by it, during the evaporation or cooling process, is rejected in the condensing or heating process. Thus, the heat pump provides cooling in the summer and heating in the winter. The source of heat for the heat pump can be from one of three elements: air, water or the ground.

### Air to Air Heat Pumps

Heat exists in air down to 460°F below zero. Using outside air as a heat source has its limitations, since the efficiency of a heat pump drops off as the outside air level drops below 55°F. This is because the heat is more dispersed at lower temperatures, or more difficult to capture. Thus, heat pumps are generally sized on cooling load capacities. Supplemental heat is added to compensate for declining capacity of the heat pump. This approach allows for a realistic first cost and an economical operating cost.

An average of two to three times as much heat can be moved for each kW input compared to that produced by use of straight resistance heating. Heat pumps can have a COP of much more than 2 or 3 in industrial processes, depending on the temperature involved. Commercially available heat pumps range in size from 2 to 3 tons for residences up to 40 tons for commercial and industrial users. Figure 8-1 illustrates a simple scheme for determining the supplemental heat required when using an air-to-air heat pump.

Figure 8-1. Determining Balance Point of Air to Air Heat Pump

## Hydronic Heat Pump

The hydronic heat pump is similar to the air to air unit, except the heat exchange is between water and refrigerant instead of air to refrigerant, as illustrated in Figure 8-2. Depending on the position of the reversing valve, the air heat exchanger either cools or heats room air. In the case of cooling, heat is rejected through the water-cooled condenser to the building water. In the case of heating, the reversing valve causes the water to refrigerant heat exchanger to become an evaporator. Heat is then absorbed from the water and discharged to the room air.

Imagine several hydronic heat pumps connected to the same building water supply. In this arrangement, it is conceivable that while one unit is providing cool air to one zone, another is providing hot air to another zone; the first heat pump is providing the heat source for the second unit, which is heating the room. This illustrates the principle of energy conservation. In practice, the heat rejected by the cooling units does not equal the heat absorbed. An additional evaporative cooler is added to the system to help balance the loads. A better heat source would be the water from wells, lakes or rivers which is thought of as a constant heat source. Care should be taken to insure that a heat pump connected to such a heat source does not violate ecological interests.

**Figure 8-2. Hydronic Heat Pump**

### Liquid Chiller

A liquid chilling unit (mechanical refrigeration compressor) cools water, brine, or any other refrigeration liquid, for air-conditioning or refrigeration purposes. The basic components include a compressor, liquid cooler, condenser, the compressor drive, and auxiliary components. A simple liquid chiller is illustrated in Figure 8-3. The type of chiller usually depends on the capacity required. For example, small units below 80 tons are usually reciprocating, while units above 350 tons are usually centrifugal.

Factors which affect the power usage of liquid chillers are the percent load and the temperature of the condensing water. *A reduced condenser water temperature saves energy.* In Figure 8-4, it can be

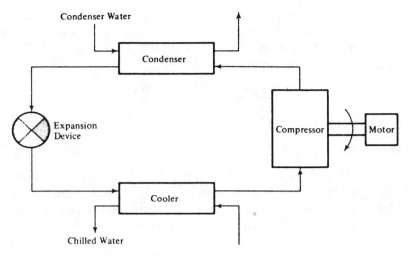

Figure 8-3. Liquid Chiller

seen that by reducing the original condenser water temperature by 10 degrees, the power consumption of the chiller is reduced. Likewise, a chiller operating under part load consumes less power. The

Figure 8-4. Typical Power Consumption Curve for Centrifugal Liquid Chiller

"ideal" coefficient of performance *(COP)* is used to relate the measure of cooling effectiveness. Approximately 0.6 kW is required per ton of refrigeration (0.6 kW is power consumption at full load, based on typical manufacturer's data).

$$COP = \frac{1 \text{ ton} \times 12,000 \text{ Btu/ton}}{0.6 \text{ kW} \times 3412 \text{ Btu/kW}} = 5.8$$

### Chillers in Series and in Parallel

Multiple chillers are used to improve reliability, offer standby capacity, reduce inrush currents and decrease power costs at partial loads. Figure 8-5 shows two common arrangements for chiller staging namely, chillers in parallel and chillers in series.

**Figure 8-5. Multiple Chiller Arrangements**

In the parallel chiller arrangement, liquid to be chilled is divided among the liquid chillers and the streams are combined after chilling. Under part load conditions, one unit must provide colder than designed chilled liquid so that when the streams combine, including the one from the off chiller, the supply temperature is provided. The parallel chillers have a lower first cost than the series chillers counterparts but usually consume more power.

In the series arrangement, a constant volume of flow of chilled water passes through the machines, producing better temperature control and better efficiency under part load operation; thus, the upstream chiller requires less kW input per ton output. The waste of energy during the mixing aspect of the parallel chiller operation is avoided. The series chillers, in general, require higher pumping costs. The energy conservation engineer should evaluate the best arrangement, based on load required and the partial loading conditions.

### The Absorption Refrigeration Unit

Any refrigeration system uses external energy to "pump" heat from a low temperature level to a high temperature. Mechanical refrigeration compressors pump absorbed heat to a higher temperature level for heat rejection. Similarly, absorption refrigeration changes the energy level of the refrigerant (water) by using lithium bromide to alternately absorb it at a low temperature level and reject it at a high level by means of a concentration-dilution cycle.

The single-stage absorption refrigeration unit uses 10 to 12 psig steam as the driving force. Whenever users can be found for low pressure steam, energy savings will be realized. A second aspect for using absorption chillers is that they are compatible for use with solar collector systems. Several manufacturers offer absorption refrigeration equipment which uses high temperature water (160°–200°F) as the driving force.

A typical schematic for a single-stage absorption unit is illustrated in Figure 8-6. The basic components of the system are the evaporator, absorber, concentrator and condenser. These components can be grouped in a single or double shell. Figure 8-6 represents a single-stage arrangement.

*Evaporator*—Refrigerant is sprayed over the top of the tube bundle to provide for a high rate of transfer between water in the tubes and the refrigerant on the outside of the tubes.

*Absorber*—The refrigerant vapor produced in the evaporator migrates to the bottom half of the shell where it is absorbed by a

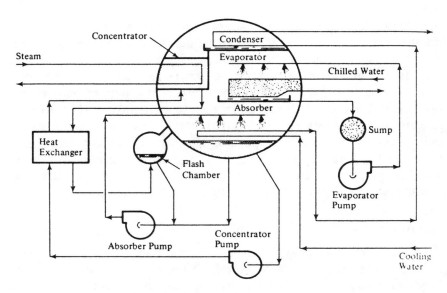

**Figure 8-6. One-Shell Lithium Bromide Cycle Water Chiller**
*(Source: Trane Air Conditioning Manual)*

lithium bromide solution. Lithium bromide is basically a salt solution which exerts a strong attractive force on the molecules of refrigerant (water) vapor. The lithium bromide is sprayed into the absorber to speed up the condensing process. The mixture of lithium bromide and the refrigerant vapor collects in the bottom of the shell; this mixture is referred to as the dilute solution.

*Concentrator*—The dilute solution is then pumped through a heat exchanger where it is preheated by a hot solution leaving the concentrator. The heat exchanger improves the efficiency of the cycle by reducing the amount of steam or hot water required to heat the dilute solution in the concentrator. The dilute solution enters the upper shell containing the concentrator. Steam coils supply heat to boil away the refrigerant from the solution. The absorbent left in the bottom of the concentrator has a higher percentage of absorbent than it does refrigerant; thus it is referred to as concentrated.

*Condenser*—The refrigerant vapor boiling from the solution in the concentrator flows upward to the condenser and is condensed. The condensed refrigerant vapor drops to the bottom of the condenser and from there flows to the evaporator through a regulating orifice. This completes the refrigerant cycle.

The single-stage absorption unit consumes approximately 18.7 pounds of steam per ton of capacity (steam consumption at full load based on typical manufacturer's data). For a single-state absorption unit

$$COP = \frac{1 \text{ ton} \times 12{,}000 \text{ Btu/ton}}{18.7 \text{ lb} \times 955 \text{ Btu/lb}} = 0.67$$

The single-stage absorption unit is not as efficient as the mechanical chiller. It is usually justified based on availability of low pressure steam, equipment considerations or use with solar collector systems.

### Example Problem 8-1

Compute the energy wasted when 15 psig steam is condensed prior to its return to the power plant. Comment on using the 15 psig steam directly for refrigeration.

### Answer

From Steam Table 15-14 for 30 psia steam, hfg is 945 Btu per pound of steam; thus, 945 Btu per pound of steam is wasted. In this case where *excess low pressure* steam cannot be used, absorption units should be considered in place of their electrical-mechanical refrigeration counterparts.

### Example Problem 8-2

2000 lb/hr of 15 psig steam is being wasted. Calculate the yearly (8000 hr/yr) energy savings if a portion of the centrifugal refrigeration system is replaced with single-stage absorption. Assume 20 kW additional energy is required for the pumping and cooling tower cost associated with the single-stage absorption unit. Energy rate is $.09 kWh and the absorption unit consumes 18.7 lb of steam per ton of capacity.

The centrifugal chiller system consumes 0.8 kWh per ton of refrigeration.

Answer

Tons of mechanical chiller capacity replaced = 2000/18.7 = 106.95 tons. Yearly energy savings = 2000/18.7 × 8000 × 0.8 × $.09 = $61,602.

### Two-Stage Absorption Unit

The two-stage absorption refrigeration unit (Figure 8-7) uses steam at 125 to 150 psig as the driving force. In situations where excess medium pressure steam exists, this unit is extremely desirable. The unit is similar to the single-stage absorption unit. The two-stage absorption unit operates as follows:

Medium pressure steam is introduced into the first-stage concentrator. This provides the heat required to boil out refrigerant from the dilute solution of water and lithium bromide salt. The liberated refrigerant vapor passes into the tubes of the second-stage concentrator, where its temperature is utilized to again boil a lithium bromide solution, which in turn further concentrates the solution and liberates additional refrigerant. In effect, the concentrator frees an increased amount of refrigerant from solution with each unit of input energy.

The condensing refrigerant in the second-stage concentrator is piped directly into the condenser section. The effect of this is to reduce the cooling water load. A reduced cooling water load decreases the size of the cooling tower which is used to cool the water. The remaining portions of the system are basically the same as the single-stage unit.

The two-stage absorption unit consumes approximately 12.2 pounds of steam per ton of capacity; thus, it is more efficient than its single-stage counterpart. The associated *COP* is

$$COP = \frac{1 \text{ ton} \times 12{,}000 \text{ Btu/ton}}{12.2 \text{ lb} \times 860 \text{ Btu/lb}} = 1.14$$

Either type of absorption unit can be used in conjunction with centrifugal chillers when it is desirable to reduce the peak electrical demand of the plant or to provide for a solar collector addition at a later date.

**Figure 8-7. Two-Stage Absorption Unit**

# 9

# Cogeneration: Theory and Practice

Because of its enormous potential, it is important to understand and apply cogeneration theory. In the overall context of Energy Management Theory, cogeneration is just another form of the conservation process. However, because of its potential for practical application to new or existing systems, it has carved a niche that may be second to no other conservation technology.

This chapter is dedicated to development of a sound basis of current theory and practice of cogeneration technology. It is the blend of theory and practice, or praxis of cogeneration, that will form the basis of the most workable conservation technology during the remainder of the 20th century.

## DEFINITION OF "COGENERATION"

Cogeneration is the sequential production of thermal and electric energy from a single fuel source. In the cogeneration process, heat is recovered that would normally be lost in the production of one form of energy. That heat is then used to generate the second form of energy. For example, take a situation in which an engine drives a generator that produces electricity: With cogeneration, heat would be recovered from the engine exhaust and/or coolant, and that heat would be used to produce, say, hot water.

Making use of waste heat is what differentiates cogeneration facilities from central station electric power generation. The overall fuel utilization efficiency of cogeneration plants is typically 70-80% versus 35-40% for utility power plants.

This means that in cogeneration systems, rather than using energy in the fuel for a single function, as typically occurs, the available energy is cascaded through at least two *useful* cycles.

To put it in simpler terms: Cogeneration is a very efficient method of making use of *all* the available energy expended during any process generating electricity (or shaft horsepower) and then utilizing the waste heat.

A more subjective definition of cogeneration calls upon current practical applications of power generation and process needs. Nowhere more than in the United States is an overall system efficiency of only 30% tolerated as "standard design." In the name of limited *initial* capital expenditure, all of the waste heat from most processes is rejected to the atmosphere.

In short, present design practices dictate that of the useful energy in one gallon of fuel, only 30% of that fuel is put to useful work. The remaining 70% is rejected randomly.

If one gallon of fuel goes into a process, the designer may ask, "How much of that raw energy can I make use of within the constraints of the overall process?"

In this way, cogeneration may be taken as a way to use a maximum amount of available energy from any raw fuel process. Thus, cogeneration may be thought of as *just good design.*

## COMPONENTS OF A
## COGENERATION SYSTEM

The basic components of any cogeneration plant are

- A prime mover
- A generator
- A waste heat recovery system
- Operating control systems

The prime mover is an engine or turbine which, through combustion, produces mechanical energy. The generator converts the mechanical energy to electrical energy. The waste heat recovery system is one or more heat exchangers that capture exhaust heat or engine coolant heat and convert that heat to a useful form. The operating control systems insure that the individual system components function together.

The prime mover is the heart of the cogeneration system. The three basic types are steam turbines, combustion gas turbines and internal combustion engines. Each has advantages and disadvantages, as explained below.

## Steam Turbines

Steam turbine systems consist of a boiler and turbine. The boiler can be fired by a variety of fuels such as oil, natural gas, coal and wood. In many installations, industrial by-products or municipal wastes are used as fuel. Steam turbine cogeneration plants produce high-pressure steam that is expanded through a turbine to produce mechanical energy which, in turn, drives a device such as an electric generator. Thermal energy is recovered in several different ways, which are discussed in Chapter 5. Steam turbine systems generally have a high fuel utilization efficiency.

## Combustion Gas Turbines

Combustion gas turbine systems are made up of one or more gas turbines and a waste heat recovery unit. These systems are fueled by natural gas or light petroleum products. The products of combustion drive a turbine which generates mechanical energy. The mechanical energy can be used directly or converted to electricity with a generator. The hot exhaust gases of the gas turbine can be used directly in process heating applications, or they can be used indirectly, with a heat exchanger, to produce process steam or hot water.

A variation on the combustion gas turbine system is one that uses high-pressure steam to drive a steam turbine in conjunction with the cogeneration process. This is referred to as a combined cycle.

## Internal Combustion Engines

Internal combustion engine systems utilize one or more reciprocating engines together with a waste heat recovery device. These are fueled by natural gas or distillate oils. Electric power is produced by a generator which is driven by the engine shaft. Thermal energy can be recovered from either exhaust gases or engine coolant. The engine exhaust gases can be used for process heating or to generate low-pressure steam. Waste heat is recovered from the engine cooling jacket in the form of hot water.

Cogeneration plants that use the internal combustion engine generate the greatest amounts of electricity for the amount of heat produced. Of the three types of prime movers, however, the fuel utilization efficiency is the lowest, and the maximum steam pressure that can be produced is limited.

## AN OVERVIEW OF
## COGENERATION THEORY

As discussed in the introduction, and as may be seen from Figure 9-1, standard design practices make use of, at best, 30% of available energy from the raw fuel source (gas, oil, coal).

Of the remaining 70% of the available energy, approximately 30% of the heat is rejected to the atmosphere through a condenser (or similar) process. An additional 30% of the energy is lost directly to the atmosphere through the stack, and finally, approximately 7% of the available energy is radiated to the atmosphere because of the high relative temperature of the process system.

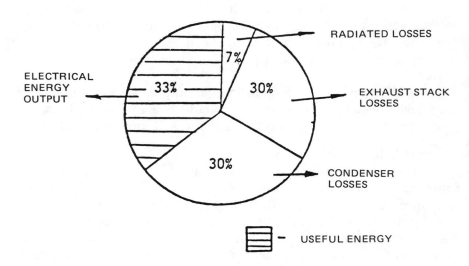

Figure 9-1. Energy Balance Without Heat Recovery

With heat recovery, however, potential useful application of available energy more than doubles. Although in a "low quality" form, *all* of the condenser-related heat may be used, and 40% of the stack heat may be recovered. This optimized process is depicted (in theory) as Figure 9-2.

Thus, it may be seen that effective use of all available energy may more than double the "worth" of the raw fuel. System efficiency is increased from 30% to 75%.

This higher efficiency allows the designer to use low grade energy for various cogeneration cycles.

**Example Problem 9-1**

A cogeneration system vendor recommends the installation of a 20-megawatt cogeneration system for a college campus. Determine the approximate *range* of useful *thermal* energy. Use the energy balance of Figure 9-2.

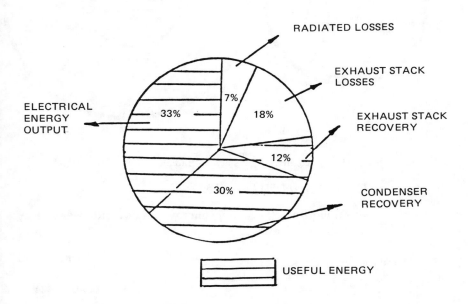

Figure 9-2. Energy Balance With Heat Recovery

**Analysis**

Step 1: "Range" is defined by the best and worst operating *times* of the installed system:

At best, system will operate 365 days per year, 24 hours per day.

At worst, system will operate 5 days per week, 10 hours per day.

Step 2: Perform heat balance. See diagram.

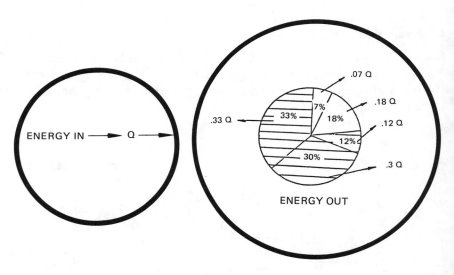

$$Q = E_{\text{out}} = E_{\text{electricity}} + E_{\text{condenser}} + E_{\text{stack}} + E_{\text{radiated}} \qquad \textit{Formula (9-1)}$$

Step 3: Calculate *available thermal* energy.

From Figure 9-2, available energy equals condenser energy and 40% stack energy.

$$E_{\text{available}} = E_{\text{condenser}} + E_{\text{stack}} \qquad \textit{Formula (9-2)}$$

$$E_{\text{condenser}} = .3Q \qquad \textit{Formula (9-3)}$$

$$E_{\text{stack}} = 0.12Q$$

$$E_{available} = 0.30Q + 0.12Q \qquad \qquad \textit{Formula (9-4)}$$

$$E_{electrical} = .33Q \qquad \qquad \textit{Formula (9-5)}$$

$$Q = \frac{E_{electrical}}{.33} \qquad \qquad \textit{Formula (9-6)}$$

By substitution $\quad E_{available} = 0.3\dfrac{E_{electrical}}{0.33} + 0.12\dfrac{E_{electrical}}{0.33}$

$$E_{available} = 1.272 \text{ Electrical} \qquad \qquad \textit{Formula (9-7)}$$

Calculate available energy

$$Q = Q_{1 \times t} = K_1 \, E_{available} \times K_2 \times t$$

$K_1$ = 3413 Btu/kWh
$K_2$ = 1000 Kilowatts per megawatt
$t$ = equipment hours of operation

$$Q_1 = 3413 \text{ Btu/kWh} \times (1.272 \times 20 \text{ megawatt}) \times 1000\,\frac{kW}{MWatt}$$

$$Q_1 = 86.82 \times 10^6 \text{ Btu/hr}$$

Worst Case: System operates
5 days/wk $\times$ 52 wks/yr $\times$ 10 hrs/day = 2600 hrs/yr

Best Case: System operates
365 days/yr $\times$ 24 hrs/day = 8670 hrs/yr

Thus range is

$Q =$ $\quad$ 86.82 $\times 10^6 \times 2600 = 225.7 \times 10^9$ Btu/yr
$Q =$ $\quad$ 86.82 $\times 10^6 \times 8760 = 760.5 \times 10^9$ Btu/yr

## APPLICATION OF THE COGENERATION CONSTANT

The *cogeneration constant* may be used as a fast check on any proposed cogeneration installation. Notice from the sample problem which follows, the ease with which a thermal vs. electrical comparison of end needs may be made.

### Example Problem 9-2

A cogeneration system vendor recommends a 20 megawatt installation. Determine the approximate rate of *useful* thermal energy.

### Analysis

$$E_{available} = E_{electrical} \times K_c \qquad \textit{Formula (9-8)}$$

$E$   is the cogeneration system electrical rated capacity
$K_c$   is the cogeneration constant

$$E_{available} = 20 \, MW \times 1.272$$
$$= 25.4 \, MW \text{ of } \textit{useful} \text{ heat}$$

or

$$25.4 \, MW \times 1000 \frac{kW}{MW} \times \frac{3413 \, Btu}{kWh} \times \frac{Therm}{100,000 \, Btu}$$

$$E_{available} = 866.9 \text{ therms/hour}$$

## APPLICABLE SYSTEMS

To ease the complication of matching power generation to load, and because of newly established laws, it is most advantageous that the generator operate in parallel with the utility grid which thereby "absorbs" all generated electricity. The requirement for "qualify facility (QF) status," and the consequent utility rate advantages which are available when "paralleling the grid," is that a significant portion of the thermal energy produced in the generation process must be recovered. Specifically, Formula 9-10 must be satisfied:

$$\frac{\text{Power Output} + \frac{1}{2} \text{ Useful Thermal Output}}{\text{Energy Input}} \geqslant 42.5\% \text{ (for any calendar year)}$$

$$\textit{Formula (9-9)}$$

Note that careful application of the *cogeneration constant* will generally assure that the qualifying facility status is met.

## BASIC THERMODYNAMIC CYCLES

### Bottoming and Topping Cycles

Cogeneration systems can be divided into "bottoming cycles" and "topping cycles."

#### Bottoming Cycles

In a bottoming cycle system, thermal energy is produced directly from the combustion of fuel. This energy usually takes the form of steam that supplies process heating loads. Waste heat from the process is recovered and used as an energy source to produce electric or mechanical power.

Bottoming cycle cogeneration systems are most commonly found in industrial plants that have equipment with high-temperature heat requirements such as steel reheat furnaces, clay and glass kilns and aluminum remelt furnaces. Some bottoming cycle plants operating in Georgia are Georgia Kraft Company, Brunswick Pulp and Paper Company and Burlington Industries.

#### Topping Cycles

Topping cycle cogeneration systems reverse the order of bottoming cycle systems: Electricity or mechanical power is produced first; then heat is recovered to meet the thermal loads of the facility. Topping cycle systems are generally found in facilities which do not have extremely high process temperature requirements.

Figure 9-3 on the following page shows schematic examples of these two cogeneration operating cycles.

A sound understanding of basic cogeneration principles dictates that the energy manager should be familiar with two standard thermodynamic cycles. These cycles are

1. Brayton Cycle
2. Rankine Cycle

The Brayton Cycle is the basic thermodynamic cycle for the simple gas turbine power plant. The Rankine Cycle is the basic cycle for a vapor-liquid system typical of steam power plants. An excellent theoretical discussion of these two cycles appears in Reference 10.

Figure 9-3. Cogeneration Operating Cycles

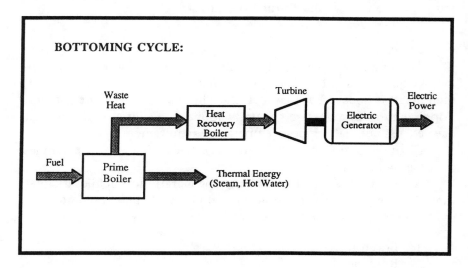

## The Brayton Cycle

In the open Brayton Cycle plant, energy input comes from the fuel that is injected into the combustion chamber. The gas/air mixture drives the turbine with high-temperature waste gases exiting to the atmosphere. (See Figure 9-4.)

**Figure 9-4. Open Gas-Turbine Power Plant (Brayton Cycle)**

The basic Brayton Cycle, as applied to cogeneration, consists of a gas turbine, waste heat boiler and a process or "district" heating load. This cycle is a *full-load* cycle. At part loads, the efficiency of the gas turbine goes down dramatically. A simple process diagram is illustrated in Figure 9-5.

**Figure 9-5. Process Diagram — Brayton Cycle**

Because the heat rate of this arrangement is superior to that of all other arrangements at full load, this simple, standard Brayton Cycle merits consideration under all circumstances. Note that to complete the loop in an efficient manner, a deaerator and feedwater pump are added.

### The Rankine Cycle

The Rankine Cycle is illustrated in the simplified process diagram, Figure 9-6. Note that this is the standard boiler/steam turbine arrangement found in many power plants and central facility plants throughout the world.

The Rankine Cycle, or steam turbine, provides a real-world outlet for waste heat recovered from any process or generation situation. Hence, it is the steam turbine which is generally referred to as the topping cycle.

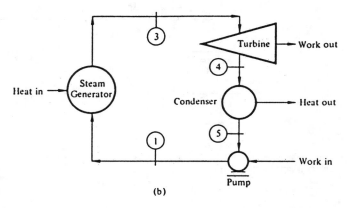

**Figure 9-6. Simple Rankine Cycle**

### Combined Cycles

Of major interest and importance for the serious central plant designer is the Combined Cycle. This cycle forms a hybrid which includes the Brayton Cycle on the "bottoming" portion and a standard Rankine Cycle on the "topping" portion of the combination. A process diagram with standard components is illustrated in Figure 9-7.

The Combined Cycle, then, greatly approximates the cogeneration Brayton Cycle but makes use of a knowledge of the plant requirements and an understanding of Rankine Cycle theory. Note also that the ideal mix of power delivered from the Brayton and Rankine portions of the Combined Cycle is 70% and 30%, respectively.

Even within the seemingly limited set of situations defined as "Combined Cycle," many variations and options become available. These options are as much dependent on any local plant requirements and conditions as they are on available equipment.

Some examples of Combined Cycle variations are:

1. Gas turbine exhaust used to produce 15 psi steam for Rankine Cycle turbine with no additional fuel burned. This situation is shown in Figure 9-7.

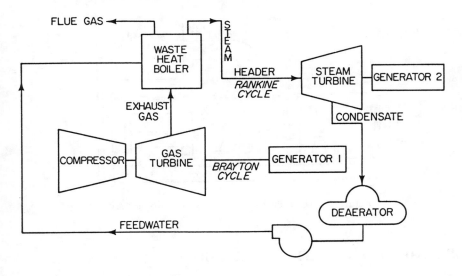

**Figure 9-7. Combined Cycle Operation**

2. Gas turbine exhaust fired in the duct with additional fuel. This provides a much greater amount of produced power with a correspondingly greater amount of fuel consumption. This situation generally occurs with a steam turbine pressure range of 900–1259 psig.

3. Gas turbine exhaust fired directly and used directly as combustion air for a conventional power boiler. Note here that

the boiler pressure range may vary between 200 and 2600 psig. (See Figure 9-8.)

One other note to keep in mind is that in any Combined Cycle case, the primary or secondary turbine may supply direct mechanical energy to a refrigerant compressor. As discussed, the variations are endless. However, a thorough understanding of the end process generally will result in a final, and best, cogeneration system selection.

**Figure 9-8. Gas Turbine Used As Combustion Air**

## DETAILED FEASIBILITY EVALUATION[1]

This section will introduce the parameters affecting the evaluation, selection, sizing and operation of a cogeneration plant and shows a means of evaluating those parameters where it counts—on the bottom line.

This section will enable you to answer two basic questions:

1. Is cogeneration technically feasible for us, given our situation?

2. If it is technically feasible, is it also economically feasible,

---

[1]*Georgia Cogeneration Handbook,* Governor's Office of Energy Resources, August 1988.

considering estimated costs, energy savings, current utility rates and regulatory conditions in Georgia?

Keep in mind that this chapter is not designed to take the place of a full-scale feasibility study. It is designed to help you and your engineering staff decide—in-house and at minimal expense—whether such a study is warranted. You will reach one of the following conclusions:

- No, cogeneration is not feasibile in our case.

- Yes, cogeneration looks promising and a feasibility study is warranted.

- Maybe, but I have some questions.

If your conclusion is "no," you will have saved yourself the expense of a feasibility study. If your conclusion is "yes," you will be assured that the expense of employing outside consultants to perform the study is justified by the potential benefits. If your conclusion is "maybe," you may wish to discuss your situation with qualified engineers before deciding whether to proceed.

## Utility Data Analysis

The first step in the feasibility evaluation is the gathering and analysis of utility data. These data are necessary to determine the limits of the thermal and electric energy consumed at your facility. They are also necessary for determining the timing of maximum use of each of these energy sources. From this information, you will also determine the thermal and electric load factors for your facility.

### Load Factor Defined

A load factor is defined as the average energy consumption rate for a facility divided by the peak energy consumption rate over a given period of time, and it is an important simplification of energy use data. These values are usually expressed as either a decimal number or a percent. To calculate a load factor you need two pieces of data: the total energy consumption for a given time period and the maximum energy demand observed during that time period.

In the electric power industry, load factors serve as a measure of the utilization of generation equipment. A utility load factor of 0.90, or 90%, indicates very good utilization of power generation facilities, while a load factor of 0.30 or 30% indicates poor utilization. The lower the load factor, the larger and more expensive your generating equipment must be merely to meet peak power demands for short time periods.

The thermal and electric load factors of your facility are of great importance in sizing a cogeneration plant. In fact, they are even more important to a cogeneration plant than to an electric utility because the cogeneration plant does not have the advantage of an electric power grid to diversify the variations in energy use.

A high thermal or electrical load factor generally indicates that a cogeneration plant would be utilized a major portion of the time and therefore would provide a favorable return on your investment. An ideal situation exists when both load factors are high, indicating that a properly sized cogeneration plant would efficiently utilize most of its output. If both the thermal and electrical load factors are small, you may not be a practical candidate for cogeneration.

In the analysis of your electric and thermal energy consumption data, you will calculate the load factors for your facility. In this evaluation, we will concentrate on annual load factors. It is important to note, however, that monthly and daily load factors may be important in your final analysis.

*Electric Energy Consumption Analysis*

In order to obtain the necessary energy-use information, you must refer to your monthly electric utility bills for one complete year. From those bills, calculate the following information:

- Determine your Annual kWh consumption by tabulating and adding the monthly kWh consumption.

- Tabulate the actual kW demand metered for each month. Be sure to use the actual metered demand and not the billed demand, as the billed demand may be based on time of year and a percent ratchet of the previous year's peak demand.

- Identify the maximum monthly kW demand value as the Annual Peak kW.

- Determine the Annual Average kW Demand by dividing the Annual kWh consumption by 8,760 hours per year.

$$\text{Annual Average kW} = \frac{\text{Annual kWh}}{8,760 \text{ hours/year}}$$

- Determine the Annual Electric Load Factor by dividing the Annual Average kW demand by the Annual Peak kW demand.

$$\text{Annual Electric Load Factor} = \frac{\text{Annual Average kW}}{\text{Annual Peak kW}}$$

**Thermal Energy Consumption Analysis**

If your boiler plant has an accurate steam or Btu metering system, the following data can be gathered directly from your metered output data. If you do not have this degree of metering, your boiler plant fuel bills for one complete year will be required to obtain the necessary data. Monthly natural gas bills can be analyzed with the same method used for analyzing electrical consumption. If you use fuel oil, propane or coal, make sure that accurate measurements of the reserve supply were taken at the beginning and end of the year.

- Determine your annual fuel consumption. From Figure 9-9, select the Btu value per unit of fuel for the type of fuel you use. To obtain the Annual Fuel Btu Input, multiply the annual fuel consumption by this value.

- Determine your boiler's Fuel-to-Steam Efficiency. An estimated fuel-to-steam efficiency of 78% may be used for a well-maintained boiler plant that operates only enough boilers to keep them well loaded. A poorly maintained boiler plant with some leaks, missing insulation, and oversized boilers that cycle frequently may have a Fuel-to-Steam Efficiency of 60% or lower.

- Determine your Annual Btu Output by multiplying your Annual Fuel Btu Input by your Fuel-to-Steam Efficiency.

$$\text{Annual Btu Output} = \frac{\text{Annual Fuel}}{\text{Btu Input}} \times \frac{\text{Fuel-to-Steam}}{\text{Efficiency}}$$

- Determine the Annual Average Btu/Hour Demand by dividing the Annual Btu Output by 8,760 hours per year.

$$\text{Annual Average Btu/Hour Demand} = \frac{\text{Annual Btu Output}}{8{,}760 \text{ hours/year}}$$

**Figure 9-9. Typical Fuel Caloric Values (Btu/CF)**

| Fuel Type | HHV<br>Higher Heating Value<br>(Approximate) | LHV/HHV<br>Lower/Higher Heating Value<br>(Approximate) |
|---|---|---|
| Natural Gas (Dry) | 1,000 | 0.90 |
| Butane | 3,200 | 0.92 |
| Propane | 2,500 | 0.92 |
| Sewage Gas | 300-600 | 0.90 |
| Landfill Gas | 300-600 | 0.90 |
| No. 2 Oil | 139,000 Btu/Gal | 0.93 |
| No. 6 Oil | 154,000 Btu/Gal | 0.96 |
| Coal, Bituminous | 14,100 Btu/lb | |

If you have metered steam production data, select the maximum value for Btu/hour output that occurred last year as the Annual Peak Btu/hour Demand. If these data are not available, estimate your Annual Peak Btu/hour Demand as a percentage of the maximum possible output of your boiler plant.

NOTE: Boilers that operate at approximately 150 psig and below use the terms "lbs per hour" and "mbh output" interchangeably to represent 1,000 Btu/hour. We will use the unit "mbh" to simplify notation when referring to 1,000 Btu/hour units.

- Determine the Annual Thermal Load Factor by dividing the Annual Average Btu/hour Demand by the Annual Peak Btu/hour Demand.

$$\frac{\text{Annual Thermal}}{\text{Load Factor}} = \frac{\text{Annual Average Btu/hour Demand}}{\text{Annual Peak Btu/hour Demand}}$$

If your Annual Minimum Btu/Hour Demand is extremely low or your boiler plant actually shuts down for several months during the summer, your Annual Thermal Load Factor is not an accurate indicator of your cogeneration potential. It is probably too large, and a closer examination of the number of hours at the Annual Minimum Btu/hour Demand may be required to assess feasibility. This can be done using monthly and daily load profiles.

If your Annual Minimum Btu/hour Demand is zero for a large number of hours per year (for example, if your boiler plant shuts down during the summer months), you may not be a good candidate for cogeneration.

### Thermal/Electric Load Ratio

For cogeneration to be feasible, the demands for thermal and electric energy must overlap much of the time. Therefore, once the thermal and electric demands of your facility are known, you must determine the ratio of heat demand to electric demand that may be expected to occur together. This is done by using the Thermal/Electric (T/E) Load Ratio. The T/E Load Ratio is defined as the quantity of heat energy that is coincident with a quantity of electrical energy. In making these calculations, you will attempt to find an optimum match between your facility's T/E Load where

$$\text{Thermal/Electric Load Ratio} = \frac{\text{Thermal Demand}}{\text{kW Demand}}$$

If the thermal and electric load factors calculated earlier are high, there is a good possibility that your facility's demand for thermal energy occurs at about the same time as your demand for electric energy. In that case, your facility's Annual Average Thermal/Electric Load Ratio is approximately equal to the Annual Average Btu/hour Demand divided by the Annual Average kW Demand. For convenience, we will use the term "mbh" to represent 1,000 Btu/hr.

$$\frac{\text{Annual Average Thermal/}}{\text{Electric Load Ratio}} = \frac{\text{Annual Average mbh Demand}}{\text{Annual Average kW Demand}}$$

If either the thermal or the electric load factor is small, then a worst-case assumption should be made. This assumption is called the Minimum Demand Thermal/Electric Load Ratio and is calculated as follows:

$$\frac{\text{Mimimum Demand Thermal/}}{\text{Electric Load Ratio}} = \frac{\text{Annual Minimum mbh Demand}}{\text{Annual Minimum kW Demand}}$$

The use of either of the Thermal/Electric Load Ratios above must be tempered with the knowledge that these are only approximations of the load correlations and are useful only for a preliminary evaluation. Most facilities will require a more detailed examination of thermal and electric load profiles to obtain the number of hours per year that loads overlap and can be served with a cogeneration plant.

Generally, cogeneration opportunities are good for facilities with T/E Load Ratios above 5 and are best for ratios above 10, with Average Annual Btu/hour Demand above 10,000,000 Btu/hour, or 10,000 mbh.

If your T/E Load Ratio is less than two, it is reasonable to assume that you are not a good candidate for cogeneration. This would certainly be the case if your thermal demand is extremely small during the summer months and is only significant during a few winter months. However, if your Annual Electric Load Factor is small, you may consider using electric power generation only, in a strategy known as peak shaving. This is discussed in more detail under Operating Strategies in this chapter.

### Equipment Sizing Considerations

For efficient cogeneration, the T/E Load Ratio in your facility must correspond to the T/E Output Ratio of the cogeneration systems. T/E Output Ratios vary for different prime movers and cogeneration plant configurations.

In an analysis of cogeneration options at your facility, you must decide whether to size the cogeneration plant to match your peak electric load, which will produce some waste heat and lower overall plant efficiency, or to size the cogeneration plant to match your heat load and supplement your electric needs with more expensive purchased power. If your load factor is small for either thermal or electric demand, you should size toward the minimum value of that demand.

### Operating Strategies

#### Electric Dispatch

One method of operating a cogeneration plant is to supply your facility's total electrical requirements as a first priority and generate steam as a second priority. This mode of operation is referred to as "electric dispatch."

Under the electric dispatch mode, the cogeneration plant is sized for Annual Peak kW Demand, with some additional capacity for growth. A cogeneration plant sized in this manner can operate totally independently of the electrical utility company.

Standby electric service from your local utility is required for scheduled and unscheduled equipment shutdown. The cost of this standby power can be very expensive, especially if demand is high or if the utility company must provide additional generation or transmission equipment to serve your facility. Total on-site backup using standby emergency power generating equipment is rarely economical.

Electric dispatch requires that you operate your cogeneration plant to meet your electric load demand requirement. Operation of the major electric loads of your facility should be coordinated with the cogeneration plant to prevent sudden load spikes or irregular, repetitive load spikes that could exceed the capacity of your generating plant.

Also, if your annual electric load factor is low, your cogeneration equipment will be oversized for your facility load most of the time. This will reduce efficiency, increase maintenance costs and lower the return on investment.

### Peak Shaving

Peak shaving is the practice of selectively dropping electric loads or generating on-site electricity during periods of peak electric demand. This procedure is not cogeneration; normally, no heat recovery equipment is installed and no heat is recovered from the generating process. Peak shaving is most commonly used to reduce annual electric utility costs. Peak shaving with an electric generator has a lower initial cost than a true cogeneration system since no heat exchangers or associated recovery equipment is installed.

### Thermal Dispatch

Thermal dispatch operation is the complement to electric dispatch. In this mode of operation, the cogeneration plant is sized to meet your facility's Annual Peak Btu/hour (or mbh) Demand. You will then purchase a part of your electric power from the local utility and cogenerate the rest. For this mode of operation to be successful, thermal load requirements should parallel each other in the same way required for electric dispatch.

For maximum savings on electric energy, electric generation should be near its maximum capacity during the summer peak electric demand period. This means that the ideal candidate for thermal dispatch operation would have maximum thermal loads occurring during summertime electric peak periods. If this is not the case, you will save less on electric energy.

*Hybrid Strategy*

A "hybrid strategy" utilizes the best features of the electrical and thermal dispatch strategies. As the name suggests, this operating strategy is a hybrid of the electric dispatch and thermal dispatch operating strategies periodically adjusted to minimize operating costs and maximize return on investment.

With the hybrid strategy, the cogeneration plant is sized smaller than for electrical or thermal dispatch. The hybrid strategy calls for the plant to be operated to achieve maximum savings during electric peak demand periods. At other times, it would be operated in a thermal load following mode. To satisfy the total thermal load requirements, it may be necessary to maintain existing boilers or install additional ones. As in the thermal dispatch strategy, your facility still must purchase a portion of its electric energy from the utility.

## Prime Mover Selection

Choosing the most appropriate prime mover for a cogeneration project involves evaluation of many different criteria, including

- Hours of operation
- Maintenance requirements
- Fuel requirements
- Capacity limits

Following are some general guidelines for evaluating each of the different prime movers with respect to these criteria.

*Hours of Operation*

Most cogeneration plants are designed for continuous operation, with the exception of some small reciprocating engine packages. Large plants are not economical if they cycle on and off on an hourly or even daily basis.

Because cogeneration equipment is relatively expensive as compared with boilers of similar thermal capacity, the equipment must be operated as much as possible to achieve an acceptable return on investment.

If an analysis of your facility shows a low electric load factor, indicating relatively few hours of high electric demand, cogeneration is probably not a cost-effective option. Instead, you may wish to examine the potential of peak shaving with reciprocating engine generators.

### Maintenance Requirements

All cogeneration equipment requires some type of periodic maintenance. The frequency and amount of maintenance varies considerably between prime movers.

**Reciprocating Engine Generators** have the highest maintenance requirements. Like automotive engines, diesel engines require routine maintenance such as oil and filter changes as often as once a week. However, natural gas fired reciprocating engines require more overhauls of the head and block, and they have a shorter life expectancy.

**Gas Turbine Engines** are mechanically simpler and require less frequent maintenance. They can operate for longer periods between major maintenance intervals, and the major maintenance is usually simpler, such as bearing inspection and replacement.

**Steam Turbines** require even less mechanical maintenance than gas turbines since they have no combustion equipment attached. Occasional bearing inspection and replacement is generally the extent of steam turbine maintenance. The cost is a function of the size and number of turbines.

### Fuel Requirements

**Reciprocating Engines** are not flexible with respect to their fuel requirements. Generally, you buy either a diesel engine or a natural gas fired engine. Special design adaptations can produce engines that burn other fuels, such as low Btu gas or heavier oils. Occasionally, natural gas engines can be adapted to switch between natural gas and propane as an alternate fuel. Diesel engines are more efficient in their fuel-to-energy conversion. However, air pollution regulations may impose greater constraints on these engines than gas turbines, usually requiring catalytic converters and carburetion limits.

**Gas Turbines** can be switched from natural gas to diesel fuel to take advantage of price fluctuations and backup fuel requirements. In addition, they can be adjusted for other fuel oils, low Btu gas, and, occasionally, organic by-products of industrial processes such as "black liquor" from pulp and paper mills.

**Steam Turbines** are limited only by the fuel for their steam source. In addition to the above fuels, coal, wood, waste, peanut shells, any suitable biomass, and incinerated municipal waste can be used to generate steam for steam turbines.

### Capacity Limits

The three prime movers discussed here have separate and distinct capacity ranges. Reciprocating engine generators range in size from about 40 kW to over 3,000 kW. Generally, small electric demand plants with still smaller heat requirements can be satisfied with reciprocating engines. The quality of the heat recovered from these engines can be a limitation. Only about 35% of the recoverable heat is available as 125 psig steam; the rest of it is available only as 180°F hot water.

Almost all of the gas turbine heat is recoverable as 125 psig steam. The lower limit of gas turbine equipment size is about 480 kW, and the upper limit is over 30,000 kW. Combined-cycle gas and steam turbine plants can produce over 100,000 kW.

Steam turbines are the most limited with respect to power generation. Their practical lower limit is about 1,000 kW. Their conversion efficiency for power generation is below 15% if practical upper limits of 200 psig of superheated inlet steam and 100 psig of saturated outlet steam are maintained. This efficiency can be improved by increasing the inlet pressure and temperature, which dramatically increases the cost of the steam production plant. Another alternative for improving electric generating efficiency is to use a condensing turbine, but this lowers the temperature of the outlet steam to the point that it cannot be used for much more than low-temperature hot water production.

Figure 9-10 on the following page shows the thermal and electric energy output typically available from several different cogeneration system configurations.

## Figure 9-10. Energy Production in Cogeneration Systems

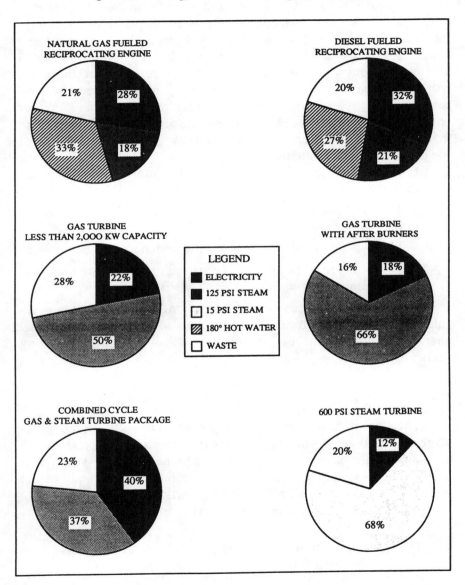

**Energy Savings Analysis**

To arrive at an estimate of the energy savings potential of cogeneration at your facility, you will need to make two simplifying assumptions. For this analysis we will assume that the proposed cogeneration system is sized such that 100 percent of its thermal and electric output is utilized. We will also assume that the plant will operate at or near full output capacity at all times. These assumptions represent a "best case" operating scenario for your cogeneration system. In reality, your plant probably will not operate in this manner; however, these assumptions are necessary here for the purpose of this evaluation.

In addition to the above assumptions, you will need to obtain the following information on your proposed cogeneration plant:

- The thermal and electric output capacities of the cogeneration equipment

- The fuel consumption rate of the cogeneration equipment

- The number of hours per year that you plan to operate the cogeneration system

This information can be obtained from equipment data available from the manufacturer. You will need to determine the annual operating hours based on knowledge of your facility operating characteristics.

*Energy Production*

The thermal and electric energy produced by your cogeneration plant will be used at your facility to offset thermal energy produced by your boiler plant and electricity purchased from your local utility. The next step is to determine the annual thermal and electric energy production of the proposed cogeneration equipment.

The annual electric energy production is calculated as follows:

$$\frac{\text{Annual Electric}}{\text{Energy Production}} = \frac{\text{Electric Output}}{\text{Capacity}} \times \frac{\text{Annual}}{\text{Operating Hours}}$$

The annual thermal energy production is calculated as follows:

$$\frac{\text{Annual Thermal}}{\text{Energy Production}} = \frac{\text{Thermal Output}}{\text{Capacity}} \times \frac{\text{Annual}}{\text{Operating Hours}}$$

### Fuel Consumption

To determine the fuel use requirements of your proposed cogeneration plant, you will need to know the rate of fuel consumption for the equipment being evaluated here. This information can be obtained from the equipment manufacturer. Your cogeneration plant fuel consumption is calculated as follows:

$$\frac{\text{Cogeneration Fuel}}{\text{Consumption}} = \frac{\text{Equipment Fuel}}{\text{Consumption Rate}} \times \frac{\text{Annual}}{\text{Operating Hours}}$$

Since you have assumed that all of the output will be used by your facility, you will now calculate the value of the cogenerated energy based on your current utility costs.

### Current Utility Costs

To determine your annual average cost per kWh, first tabulate electric energy costs from the bills that you used for gathering the electric demand and consumption data. (Generally, a more detailed analysis of your electric rate structure is needed to develop an accurate unit cost of the purchased power you will offset with cogeneration. For this evaluation, we will assume the unit cost of electricity is the same as the annual average unit cost of electricity.) Next, divide your annual electric cost by your annual electric consumption.

$$\text{Annual Average Cost/kWh} = \frac{\text{Annual Electric Cost}}{\text{Annual Electric Consumption}}$$

You can use the following method to determine your thermal energy cost:

- Tabulate monthly fuel costs from the bills used to obtain the fuel consumption data earlier.

- Determine your average Fuel Unit Cost by dividing the Annual Fuel Cost by your Annual Fuel Consumption.

$$\text{Fuel Unit Cost} = \frac{\text{Annual Fuel Cost}}{\text{Annual Fuel Consumption}}$$

- Determine your current cost per thermal MMBtu of steam or hot water. To do this, first multiply the Fuel Unit Cost by the Btu/unit value from Figure 9-9, then divide by your boiler's Fuel-to-Steam Efficiency. Divide the result by 1,000,000 to convert to MMBtu units.

$$\frac{\text{Current Cost}}{\text{Thermal MMBtu}} = \frac{\text{Fuel Unit Cost X Btu/Fuel Unit}}{(\text{Fuel-to-Steam Efficiency}) (1,000,000)}$$

*Annual Savings*

The annual energy savings of your cogeneration plant is calculated as the savings from electric and thermal energy production less the cost of operating the plant. The annual electric energy savings can be determined from the following equation:

$$\frac{\text{Annual Electric}}{\text{Energy Savings}} = \frac{\text{Annual Electric}}{\text{Energy Production}} \, X \, \frac{\text{Annual Average}}{\text{Cost/kWh}}$$

The annual thermal energy savings is given by the following:

$$\frac{\text{Annual Thermal}}{\text{Energy Savings}} = \frac{\text{Annual Thermal}}{\text{Energy Production}} \, X \, \frac{\text{Current Cost/}}{\text{Thermal MMBtu}}$$

*Operating and Maintenance Costs*

Operating and maintenance costs are, to a large degree, dependent on plant operating hours and, therefore, proportional to fuel consumption. For the purpose of this evaluation, maintenance costs can be approximated as 15% of fuel costs for reciprocating engines and 7% of fuel costs for gas or steam turbines. Use a lower figure for steam turbine maintenance costs if some boiler maintenance is already included in your operating expenses.

The cost of the fuel to operate your cogeneration plant can be determined using the following equation:

$$\frac{\text{Cogeneration}}{\text{Fuel Cost}} = \frac{\text{Cogeneration Fuel}}{\text{Consumption}} \, X \, \frac{\text{Fuel}}{\text{Unit Cost}}$$

You now have all the information necessary to determine your annual savings from the operation of a cogeneration plant. This is calculated as follows:

$$\frac{\text{Annual}}{\text{Savings}} = \frac{\text{Annual Electric}}{\text{Energy Savings}} + \frac{\text{Annual Thermal}}{\text{Energy Savings}} -$$

$$\frac{\text{Cogeneration}}{\text{Fuel Cost}} - \frac{\text{Operating and}}{\text{Maintenance Cost}}$$

At this point, it is important to note that the annual savings derived from this analysis is only a preliminary estimate of the savings potential at your facility and is based on simplifying assumptions you have made.

## Economic Analysis

### Initial Cost
To develop the total initial cost of the cogeneration plant you are evaluating, first refer to Figure 9-11 below for approximate equipment cost ranges per kW of installed cogeneration electric capacity.

**Figure 9-11.**
**Approximate Cost Per kW of Cogeneration Plant Capacity**

---

*Reciprocating Engine Packages (High Speed)*

---

Large 900 to 3,000 kW packages . . . . . . . .$600 to $400/kW
Medium 400 to 800 kW packages . . . . . . . .$500 to $700/kW
Small 45 to 300 kW packages. . . . . . . . . $700 to $1200/kW

---

*Gas Fired Turbine Engine Packages*

---

Large 4,000 to 10,000 kW. . . . . . . . .$800 to $1,000/kW
Medium 500 to 4,000 kW . . . . . . . . .$1,200 to $1,800/kW

---

*Steam Powered Turbine Engines\**

---

Less than 125 psig inlet turbines. . . . . . . $100 to $130/kW
Less than 250 psig inlet turbines. . . . . . . .$90 to $120/kW

\*Note initial costs for steam powered turbines are highly dependent on the entering and leaving pressures of the turbine. The best efficiencies are for 1,000 to 3,000 psi superheated steam. The cost of small-scale steam boilers in this range is prohibitive, and the cost of condensing steam turbines is not listed since considerable extra equipment is required. The costs as given do not include the cost of heat recovery equipment.

The cost figures in Figure 9-11 do not include the cost of new buildings to house the equipment, nor do they include allowances for the electrical wiring necessary to connect your facility to your utilities. If additional space must be constructed or major work is required to connect the utilities, these expenses must be added to the equipment cost estimates to determine your total initial cost. Also, these costs were assembled at an earlier time and may need to be revalidated for your current analysis.

**Investment Analysis**

The final decision to build a cogeneration plant is usually based on investment analysis. Broadly defined, this is an evaluation of costs versus savings. The costs for cogeneration include the initial capital cost for the equipment or the cost of operating and maintaining that equipment, costs; fuel costs, finance charges, tax liabilities, and other system costs, that may be specific to your application. Savings include offset electrical power costs; offset fuel costs; revenues from excess power sales, if any; tax benefits; and any other applicable savings that are offset by the cogeneration plant, such as planned replacement of equipment.

A number of economic analysis techniques are available for comparing investment alternatives. The Simple Payback Period is the most commonly used and the least complex, and is usually adequate for a go, no-go decision at this stage of project development. It is the period of time required to recover the initial investment cost through savings associated with the project. Simple Payback Period is calculated as follows:

$$\text{Simple Payback Period} = \frac{\text{Initial Cost, \$}}{\text{Annual Savings, \$/Yr}}$$

The Simple Payback Period will give you an indication of the attractiveness of cogeneration at your facility and will help you decide whether a full-scale feasibility study is warranted.

Figure 9-12 is a summary of the information used in this evaluation of cogeneration at your facility.

It is important to note that many of the variables used in this evaluation are assumed values based on your knowledge of the operating characteristics of your facility and the simplifying assumptions made for this preliminary analysis. In a full-scale cogeneration feasibility study, you will want to evaluate several equipment types and sizes in conjunction with different operating strategies for each system configuration. This type of evaluation should be performed for you by an experienced consulting engineer.

**Figure 9-12.**
**Cogeneration Feasibility Evaluation Summary**

**UTILITY DATA ANALYSIS**

| | | |
|---|---|---|
| Electric Energy Consumption | _____ | KWH/year |
| Annual Peak Electric Demand | _____ | KW |
| Electric Energy Cost | $_____ | /year |
| Thermal Energy Consumption | _____ | BTU/year |
| Annual Peak Thermal Demand | _____ | BTU/hour |
| Annual Fuel Cost | $_____ | /year |
| Annual Electric Load Factor | _____ | % |
| Annual Thermal Load Factor | _____ | % |
| Thermal/Electric Load Ratio | _____ | |

**ENERGY ANALYSIS**

| | | |
|---|---|---|
| Cogeneration Plant Capacity | | |
|     Electric Output | _____ | KW |
|     Thermal Output | _____ | BTU/hour |
| Annual Operating Hours | _____ | hours/year |
| Annual Electric Energy Production | _____ | KWH/year |
| Annual Thermal Energy Production | _____ | BTU/year |
| Cogeneration Plant Fuel Consumption | _____ | BTU/year |
| Current Utility Costs | | |
|     Average Annual Electric Cost | $_____ | /KWH |
|     Average Thermal Energy Cost | $_____ | /MMBTU |
| Annual Savings | | |
|     Electric Energy Savings | $_____ | /year |
|     Thermal Energy Savings | $_____ | /year |
| Cogeneration Fuel Cost | $_____ | /year |
| Operating and Maintenance Costs | $_____ | /year |

**ECONOMIC ANALYSIS**

| | | |
|---|---|---|
| Annual Savings | $_____ | /year |
| Initial Cogeneration Equipment Cost | $_____ | |
| Simple Payback Period | _____ | years |

# 10

# Control Systems

This chapter will focus on the following subjects:

## INTRODUCTION

Energy engineers have been and will continue to make significant contributions for the accomplishment of energy-efficient processes, buildings and transportation systems. In the hands of the energy engineer, control systems are an important tool to make these systems energy efficient; the majority of the presently existing systems were designed primarily as labor- and materials-saving (low first costs) rather than as energy-saving (low operating costs) equipment.

For example, the practice in the design and selection of heating, ventilating and air-conditioning (HVAC) systems has been based on maximum load requirements. Usually, HVAC systems operate at less than design load conditions and require controllers to regulate operations during partial load conditions. Design optimization of the configuration and selection of HVAC systems and controllers is essential to meet the required environmental conditions with minimum energy consumption. Energy-related characteristics of the HVAC equipment and processes have been discussed in Chapters 7 and 8. The purpose of the present chapter is to describe the controls needed for an energy-efficient design and operation of building systems.

## BASIC CONCEPTS

A *control system* is a means by which some quantity of interest in a machine, mechanism or other equipment is maintained or altered, in accordance with a desired manner. A control system consists essentially of five parts:

1. Process (system)
2. Measuring means (sensor)
3. Error detector
4. Controller
5. Final control element

These five elements can be interconnected either in an *open-loop* format or in a *closed-loop* format.

An open-loop system is a system in which the output has no effect upon the input signal. A clothes dryer may be considered as an open-loop control system in which input is the status of the electric switch (on/off) and the output is the dryness (moisture content) of the clothes. When the dryer is turned on, it runs for a pre-set time and then automatically turns off. The actual moisture content of the clothes has no effect on the on-time.

The dryer could be put in a closed-loop system by installing a moisture-sensing probe which will continuously measure the moisture content of the clothes being dried and compare this value with a desired value. The difference between the two values could be used by a controller to control the on/off switch to the dryer.

Similarly, if a home thermostat were installed outside the house, the output (furnace heat) would have no effect on the input to the thermostat (in this case, outdoor temperature). If we closed this loop by bringing the thermostat indoors, the output (furnace heat) would now affect the input (indoor temperature). This closed-loop thermostat would continue to operate until the indoor temperature reached the desired value (setpoint). In this closed-loop control system, the output is fed back and compared with the desired value, and the difference between the two is used as an actuating signal to actuate the control device (gas valve) for a gas furnace.

Thus a closed-loop (feedback) control system is a system in which the output has an effect upon the input quantity to maintain the desired output value. A block diagram of a feedback control system showing all basic elements is shown in Figure 10-1.

If Figure 10-1 were to represent a residential heating system using a natural gas furnace, different elements would be as shown in Table 10-1.

## CONTROL MODES

Figure 10-1 may be condensed to Figure 10-2. In this figure, for the residential heating system, the manipulated variable $m(t)$, gas flow rate at any instant in time, could have one of the two possible values, namely zero or one, depending upon the thermostat status $e(t)$ which could be on or off. Such a control action is called on/off or two-position control mode.

Thus, the two-position control is a type of control action in which the manipulated variable is quickly changed to either a maximum or minimum value depending upon whether the controlled variable is greater or less than the setpoint. The minimum value of the manipulated variable is usually zero (off).

The equations for two-position control are

$m = M_1$       when $e > 0$            *Formula (10-1)*

$m = M_0$       when $e < 0$

Figure 10-1. Block Diagram of a Feedback Control System
Containing All Basic Elements

**Table 10-1. Basic Elements of a Residential Heating System**

| Element | Example in a residential heating system |
|---|---|
| Command v | Desire to set the indoor temperature |
| Reference input element $G_v$ | Thermostat knob |
| Reference input r | Position of thermostatic switch |
| Primary feedback b | Position of bimetallic strip |
| Actuating signal e | On/off status of switch |
| Control elements $G_1$ | Relay-controlled gas valve |
| Manipulated variable m | Gas flow rate (on or off) |
| Controlled system $G_2$ | Furnace burner, ducts, house, etc. |
| Controlled variable c | Indoor air dry-bulb temperature |
| Disturbance d | Heat gains, losses, infiltration, etc. |
| Feedback element H | Bimetallic strip in the thermostat |
| Indirectly controlled system z | Occupant |
| Indirectly controlled variable q | Thermal comfort level |
| Idealized system $G_i$ | Imaginary system which would result in an indoor environment where level of occupant's satisfaction is 100 percent. |
| System error | Thermal discomfort level |

Where

$M_1$ = maximum value of manipulated variable.
$M_0$ = minimum value of manipulated variable.

A differential or dead-band in two-position control causes the manipulated variable to maintain its previous value until the controlled variable has moved beyond the setpoint by a predetermined amount. In actual operation, this action may be compared to hysteresis as shown in Figure 10-3.

A differential may be intentional, as is common in domestic thermostats when employed for the purpose of preventing rapid operation of switches and solenoid valves and to enhance the life of the system.

Two-position control is simple and inexpensive, but it suffers from inherent drifts. Rapid changes of the controlled variable are

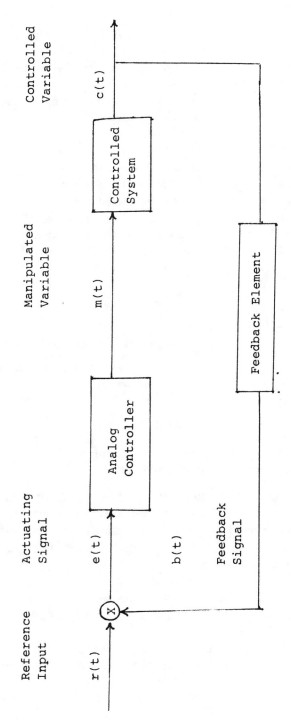

Figure 10-2. Generalized Block Diagram for a Closed-Loop Control System

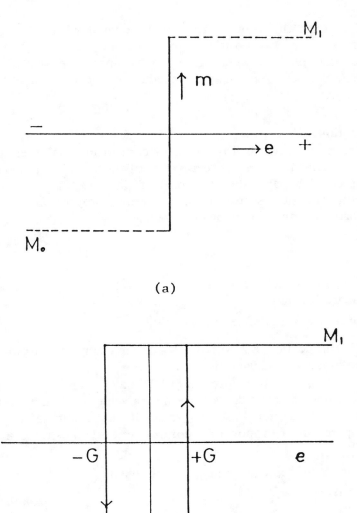

(a)

(b)

**Figure 10-3. (a) Two-Position Control Action;
(b) Two-Position Control Action with Differential Gap**

possible with this type of control. Compared with other types of control actions, two-position control can be more energy-intensive. Some examples of applications of two-position control include residential heating/cooling systems and rooftop units in commercial buildings.

### Proportional Control

Proportional control is a type of control in which there is a continuous linear relation between values of the actuating signal and the manipulated variable. For purposes of flexibility, an adjustment of the control action is provided and is termed proportional sensitivity. Proportional control may be described as

$$m = K_c e + M \qquad\qquad\qquad \textit{Formula (10-2)}$$

Where

$K_c$ = proportional sensitivity
$M$ = a constant

and other terms as defined previously.

The proportional sensitivity, $K_c$, is the change of output variable caused by a unit change of input variable.

The constant M in Formula (10-2) may be termed as the calibration constant because the selection of a value for M determines the normal (zero actuating signal) value of the manipulated variable. The operation of proportional control action is illustrated in Figure 10-4.

For a unit step change in actuating signal

$$
\begin{aligned}
e &= 0 \qquad t < 0 \\
e &= E \qquad t \geqslant 0
\end{aligned}
\qquad\qquad \textit{Formula (10-3)}
$$

where E is a constant; substituting in (10-2)

$$m - M = K_c E \qquad\qquad\qquad \textit{Formula (10-4)}$$

The change in manipulated variable corresponds exactly to the change in deviation with a degree of amplification depending upon the setting of proportional sensitivity $K_c$. Thus, a proportional controller is simply an amplifier with adjustable gain.

**Figure 10-4. Proportional Control Action**

Proportional control is more sensitive to the error signals than is two-position control. It is the least expensive of continuous-type controls but is more expensive compared to the two-position control. Calibration procedures are more difficult for proportional control than for the two-position control. Proportional control is used in air distribution systems. hydronic systems and in central systems for commercial buildings.

**Integral Control**

Integral action is a type of control action in which the value of the manipulated variable m is changed at a rate proportional to the actuating signal. Thus, if the actuating signal is doubled over a previous value, the final control element is moved twice as fast. When the controlled variable is at the set point (zero actuating signal), the final control element remains stationary.

Mathematically, integral control may be expressed as

$$\dot{m} = \frac{1}{t_{int}} e \qquad\qquad\qquad \text{Formula (10-5)}$$

or, in integrated form

$$m = \frac{1}{t_{int}} \int e\, dt + M \qquad\qquad \text{Formula (10-6)}$$

Where

$M$ = constant of integration
$t_{int}$ = integral time (defined as the time of change of manipulated variable caused by a unit step change of e)

The operational form of the equation is

$$m(s) = \frac{1}{t_{int}\, s} e(s) \qquad\qquad \text{Formula (10-7)}$$

and is shown in Figure 10-5.

For a step change of actuating signal

$$\begin{array}{ll} e = 0 & t < 0 \\ e = E & t \geqslant 0 \end{array} \qquad\qquad \text{Formula (10-8)}$$

where E is a constant. Substituting in Formula (10-6) and integrating

$$m - M = \frac{1}{t_{int}} Et \qquad\qquad \text{Formula (10-9)}$$

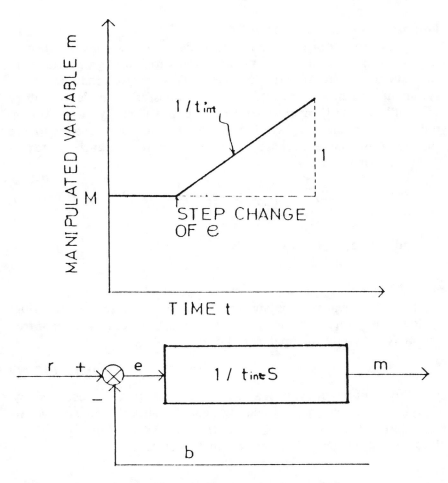

**Figure 10-5. Operation of Integral Control Action**

Thus, the manipulated variable changes linearly with time and "integrates" the area under the actuating signal function. For a unit step change, if actuating signal (E = 1.0), the slope of the line is inverse of integral time (Figure 10-5).

Integral control has the advantage over proportional control in that it tends to zero the offset, but it requires more expensive calibration procedures. Its maintenance is more difficult. Some applications of integral control include control of boilers, solar storage systems and meat-processing plants.

**Proportional-Derivative Control**

Derivative control action may be defined as a control action in which the magnitude of the manipulated variable is proportional to the rate of change of actuating signal. This control mode has many synonyms, such as "pre-set," "rate," "booster" and "anticipatory control" action. Derivative control response is always used in conjunction with the proportional mode. It is not satisfactory to use this response alone because of its inability to recognize a steady-state actuating signal.

Mathematically, a proportional-derivative (PD) control action is defined by

$$m = K_c e \ | \ + K_c t_d \dot{e} \ | \ + \quad M \qquad \qquad Formula\ (10\text{-}10)$$
$$\quad [\text{proportional}] \quad [\text{derivative}]$$

where $t_d$ = derivative time and other variables as described previously. PD is the simple addition of proportional control and rate control action as shown by the operation Formula

$$m(s) = K_c (1 + t_d\ s)\ e(s) \qquad \qquad Formula\ (10\text{-}11)$$

Proportional-derivative action is not adequately described by applying a step change of actuating signal because the time derivative of a step change is infinite at the time of change. Consequently, a linear (ramp) change of actuating signal must be used:

$$e = Et \qquad \qquad\qquad\qquad Formula\ (10\text{-}12)$$

Where

$E$ = a constant

$t$ = time

Substituting Formula (10-12) and its first time derivative into Formula (10-10)

$$m - M = K_c E\ (t + t_d) \qquad \qquad Formula\ (10\text{-}13)$$

The actuating signal is defined at time t, whereas the manipulated variable is defined at $(t + t_d)$. The net effect is to shift the manipulated variable ahead by time $t_d$, the derivative time. As shown in Figure 10-6, the controller response leads the time change of actuating signal. Derivative time is defined as the amount of lead, expressed in units of time, that the control action is given. In other words, derivative time is the time interval by which the rate action advances the effect of proportional control action.

Figure 10-6. Proportional-Derivative Control Action

Proportional-derivative control has the advantage of a rapid response to the magnitude and to the rate of change in loads. However, PD control can become unstable easily because it has no zeroing capability. This control is useful for controlling environments in buildings with large variations in occupancy.

**Proportional-Integral Control**

Integral control action is often combined additively with proportional control action. The combination is termed proportional-integral or proportional plus reset control and is used to obtain advantages of both control actions.

Proportional-integral control action is defined by the following differential formula:

$$\frac{dm}{dt} = \frac{K_c}{t_{int}} e \Big|_{[integral]} + K_c \frac{de}{dt} \Big|_{[proportional]} \qquad \textit{Formula (10-14)}$$

or, in integrated form

$$m = [\frac{K_c}{t_{int}} \int e \ dt]_{[integral]} + [K_c \ e + M]_{[proportional]} \qquad \textit{Formula (10-15)}$$

where terms are as previously described. These formulas illustrate the simple addition of proportional and integral control actions. In operational form:

$$m(s) = K_c (\frac{1}{t_{int}s} + 1) \ e(s) \qquad \textit{Formula (10-16)}$$

where the system function $K_c/(t_{int}s)$ identifies the integral action and the system function $K_c$ identifies the proportional action.

Proportional-integral (PI) control action has two adjustment parameters, the proportional sensitivity $K_c$ and integral time $t_{int}$. The propotional sensitivity is defined the same way as for the proportional control action. With the integral response turned off ($t_{int} \rightarrow \infty$), the proportional sensitivity is the number of units change in manipulated variable in per-unit change of actuating signal e. As clear from Formula (10-16), the proportional sensitivity $K_c$ affects both the proportional and integral parts of the action.

The integral action adjustment is achieved through integral time. For a step change of actuating signal e, the integral time, $t_{int}$, is the time required to add an increment of response equal to original step change of response as shown in Figure 10-7. Another term used with this type of control is reset rate, defined as the number of times per minute that the proportional part of response is replicated. Reset rate is therefore called "repeats per minute" and is the inverse of integral time.

**Figure 10-7. Operation of Proportional-Integral Control**

For a step change of deviation

$e = 0 \qquad t < 0$

$e = E \qquad t > 0$

*Formula (10-17)*

Substituting in Formula (10-14),

$$m - M = K_c E \left( \frac{t}{t_{int}} + 1 \right) \qquad \textit{Formula (10-18)}$$

This is the formula for a straight line. The first term, $t/t_{int}$, is the integral response, and the second term is the proportional response. The latter is indicated by the dotted line of Figure 10-7.

PI control has the advantage of compensating for changes in input in addition to compensating the deviations in the controlled variable. High cost of maintenance is a disadvantage with this type of control.

### Proportional-Integral-Derivative Control

The additive combination of proportional action, integral action and derivative action is termed proportional-integral-derivative action. It is described by the differential formula

$$\dot{m} = \underbrace{\frac{K_c \, e}{t_{int}}}_{\text{[integral]}} \qquad \underbrace{+ K_c \, \dot{e}}_{\text{[proportional]}} \qquad \underbrace{+ K_c t_d \dot{e}}_{\text{[derivative]}} \qquad \textit{Formula (10-19)}$$

or

$$m = \underbrace{\frac{K_c}{t_{int}} \int e \, dt}_{\text{[integral]}} + \underbrace{K_c \, e + M}_{\text{[proportional]}} \qquad \underbrace{+ K_c t_d \dot{e}}_{\text{[derivative]}} \qquad \textit{Formula (10-20)}$$

$$m(s) = K_c \left( \frac{1}{st_{int}} + 1 + t_d s \right) e(s) \qquad \textit{Formula (10-21)}$$

Proportional-integral-derivative (PID) control action is illustrated in Figure 10-8 in which the change of manipulated variable is shown for a ramp function of the actuating signal.

$$e = Et \qquad \textit{Formula (10-22)}$$

Substituting this ramp function and its time derivative into Formula (10-20):

$$m - M = K_c E \left( \frac{1}{t_{int}} \int t \, dt + t + t_d \right) \qquad \textit{Formula (10-23)}$$

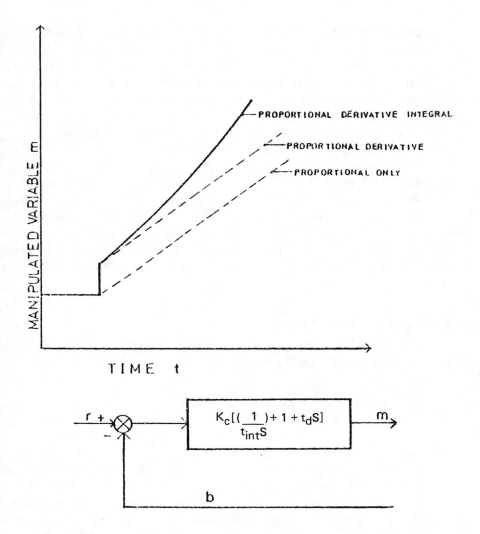

**Figure 10-8. Operation of Proportional-Integral-Derivative Control**

Integrating the first term

$$m - M = K_c E \left( \frac{t^2}{2t_{int}} + t + t_d \right)$$

*Formula (10-24)*

The proportional part of the control action repeats the change of actuating signal (lower straight line) in Figure 10-8. The derivative part of the control action adds an increment of manipulated variable proportional to the area under the deviation line and, as Figure 10-8 shows, the increment increases because the area increases at an increasing rate. The combination of proportional, integral and derivative actions may be made in any sequence, because these actions are described by linear differential equations.

PID control action has many advantages. It can compensate for magnitude and rate of change in input. It zeroes the deviation in controlled variable and is the most energy-efficient. However, it is the most expensive of all the controls and is difficult to calibrate. It is very hard to keep PID control stable; it requires frequent calibration checks. PID control is applied in environmental control chambers for scientific research and in variable air volume HVAC systems.

### TYPES OF CONTROLLERS

Controllers for an energy (conversion and/or utilization) system may be one of the following types:

    1. Electric or electronic controllers
    2. Pneumatic controllers
    3. Computer-based controllers

### Electric or Electronic Controllers

Electric switches operated manually, or with relays and timers, are an important class of controllers for implementing several control strategies like start-stop of equipment, duty cycling, night setback, control of lighting, etc. Single-pole, double-throw (SPDT) switching circuits are used to control three-wire uni-directional motors. SPDT circuits are also used for heating-cooling applications.

Manually controlled electric switches used to "turn it off when it's not needed" offer great opportunities for energy savings. For example, a manufacturing company happened to hire an energy engineer whose first assignment was to identify and implement two energy-saving strategies within the first month of his starting the job. The condition was that the first cost of these projects should be zero and the savings should at least be equal to his salary. Manually operated electric switches were the answer for the engineer. He identified that the company had a policy of leaving all its office lights on for the janitor crew at night. He implemented a policy that all the employees were required to turn off their office lights manually, while leaving at 5:00 p.m. Further, the janitors would turn on the lights in an office only when they are cleaning that office. It turned out that 10,000 fluorescent tubes (40 watts each) could be turned off for 4 hours/day—5 days/week—52 weeks/year. The company was paying $0.07/kWh and $8.00/kW. Thus, savings from this strategy alone were

$$\frac{10,000 \times 40 \times 4}{1000} \text{ (kWh/day) } \times \text{ 5 days/week } \times$$
$$52 \text{ weeks/year } \times \$0.07/\text{kWh} = \$29,120/\text{year.}$$

Secondly, he identified that an exhaust fan in the grinding shop was kept running all the time, even though the shop operated for 16 hours/day and 5 days/week. He implemented a policy that the workers at the end of the second shift would *manually turn off* this fan and the workers at the start of the first shift would manually start this fan. The specifications of the fan and heating equipment were as follows:

Horsepower of the fan = 10 hp
Air exhausted = 20,000 ft$^3$/minute
Cost of electricity = $0.07/kWh
Efficiency of the gas heater = 0.70
Cost of gas = $0.60/therm
Heating season = 24 weeks/year
Setpoint = 70°F
Avg. outdoor temperature = 35°F

Calculations:
a. *Gas savings*

$$\frac{(20,000 \text{ ft}^3)}{\text{minute}} \times \frac{(60 \text{ minutes})}{\text{hour}} \times \left[ \frac{(5 \text{ days})}{\text{week}} \frac{(8 \text{ hours})}{\text{day}} + \frac{(2 \text{ days})}{\text{week}} \frac{(24 \text{ hours})}{\text{day}} \right] \times$$

$$\frac{(24 \text{ hours})]}{\text{day}} \times \frac{24 \text{ weeks}}{\text{year}} \times \frac{(0.077 \text{ lbm})}{\text{ft}^3} \frac{(0.248 \text{ Btu})}{\text{lbm}^\circ \text{F}} \times$$

$$(70^\circ \text{F} - 35^\circ \text{F}) \times \frac{(1 \text{ Therm})}{10^5 \text{ Btu}} \left(\frac{1}{0.70}\right) \left(\frac{\$0.60}{\text{Therm}}\right) = \$14,518/\text{year.}$$

b. *Electricity savings*

$$10 \text{ hp} \times \frac{0.746 \text{ kW}}{\text{hp}} \left[ \frac{5 \text{ days}}{\text{week}} \times \frac{8 \text{ hours}}{\text{day}} + \frac{2 \text{ days}}{\text{week}} \times \frac{24 \text{ hours}}{\text{day}} \right] \times$$

$$\frac{(52 \text{ weeks})}{\text{year}} \times \frac{\$0.07}{\text{kWh}} = \$2,389/\text{year.}$$

Total savings from manually operated switches were $46,027/ year, which covered the starting salary of the energy engineer, which was $40,000/year.

The example illustrates the usefulness of manually controlled switches.


**TIMERS**

Either single-pole, single-throw (SPST) or SPDT circuits can be interfaced to timers. These timers can range from very simple clocks to sophisticated central-time clocks with multiple-channel capability for controlling equipment on different time schedules.

Timers can be used to implement control strategies like night setback, duty cycling and automatic start-stop of equipment.

**Night Setback**

Energy expended to heat unoccupied buildings up to comfort conditions is wasted, and most buildings are indeed unoccupied at night-time. Energy is saved by setting back the temperature levels at these times.

Night setback offers opportunities for significant energy savings with little or no capital expenditure in most of the residential, institutional, commercial and industrial buildings. However, one should be careful that the night setback does not:

(i)Effect process conditions, e.g., tolerances on the manufactured goods

(ii) Cause the air conditioners to turn on to meet the setback temperatures

**Duty Cycling**

In duty cycling, loads are cycled on and off by a timing device. Duty cycling has been used to duty cycle the operation of exhaust fans and space heating/cooling equipment with some possible savings. Savings are only a possibility because some experts believe that duty cycling is not an economical way of achieving savings in the long run. The idea that duty cycling can cause detrimental damage to equipment must be considered. If, for example, a company can save $10,000 per year on its electricity bill but has to spend about the same amount in maintenance and repair, then the use of duty cycling is not feasible.

In some instances, the breakdown of equipment is not acceptable at all. This is the case in a hospital. If an unforeseen problem occurs in the operating room, no financial savings could pay for the problems that might result.

Duty cycling has been with us for years. Ever since the energy crisis, duty cycling has become a broader area. Since air-conditioning systems are usually based on the most extreme operating conditions, the basic theory behind duty cycling is that the mechanical equipment is oversized during a large part of the operating life.

The standard thermostat has been used to cycle air conditioners for decades. Thermostats essentially cycle the air-conditioning equipment on and off to maintain comfortable environmental conditions. Consequently, mechanical equipment controlled by thermostats is designed to tolerate cycling for normal operation. Problems may be caused by the addition of the cycle-timing devices in addition to the thermostat controls. Therefore, the problem of accelerated deterioration of mechanical equipment has been attributed to duty cyclers.

The idea that the age of the equipment may be altered by duty cycling has to be considered first. There have been ideas on both sides. Some feel that the life will be extended and some feel otherwise.

In one case, it was reported that duty cycling control towers caused the breaking up of gear boxes. When discussed at a forum on duty cycling, it was found that, when the fan changed speed from high to low, the gears acted as a brake for slowing the fan. The added stress on the gears was not taken into consideration when the towers were designed. This problem was easily fixed by a) shutting off the fan for approximately 45 seconds so the fan speed decreased enough to avoid any more damage to the gears or b) a more sophisticated way to use decelerating relays.

A serviceman mentioned trouble he had had with air-handling equipment in a television studio. The cycling pattern was 27 minutes off and 21 minutes on. Within one year, 25% of the fan drives had failed. The problem was either a motor-drive end bearing had failed or the fan-drive end bearing had failed. The solution to this problem was simple. There was over-tension in the belts. This was initially done to keep the air-conditioning equipment from squeaking. After the belts were loosened to the proper specification, there were no problems reported within the next year.

The same serviceman also duty cycled air-handling equipment at a hospital. The off time was never less than 5 minutes. There were no problems reported there.

Another concern has been the decrease in compressor or starter motor life. It is estimated that the starter motor life would be reduced by one half for every 10-degree Celsius increase in winding temperature. For example, a motor with a 15-year normal life that is cycled once a day with an increase in winding temperature of 20°C will last only about 4 years. In other words, if the same motor with a 30°C increase in temperature were cycled four to five times a day, it would last only 2 years.

In addition, the bearings in the compressor and starter motor may get damaged by duty cycling. Moreover, the contacts and control equipment could also have a shorter life. This decrease in life would be dependent upon the number of cycles.

The problem with the contacts' wearing out prematurely occurs because the original equipment is not made for the excessive use that occurs during duty cycling. All that has to be done is to put in heavy-duty contacts.

When it comes to the problem with bearings, an HVAC dealer in Peoria stated that, if the roller bearings are greased properly, there will be no problem. However, he did not recommend duty cycling sleeve bearings. Heavy-duty or over-specified sleeve bearings must be used. For collar-mounted bearings (either screw-locking or eccentric), the collars may work loose if duty cycled; so these must be watched very carefully.

As for the life of the compressor and compressor motor, this is a more complex problem. Some motors cannot be duty cycled. This will depend upon several parameters. Problems could occur because the motor was not properly maintained during its years of operation. Such a motor should not be cycled. Even if the older motor was properly maintained, it should not be aggressively cycled (cycled too often). Improper cycling can inhibit oil return to the compressor and cause it to cease. Most older motors were not manufactured for duty cycling; that is why special care should be taken.

Currently, larger motors are not used as often. A 100-hp motor will consist of four 25-hp motors. The amperage of the in-rush at start is much less for these smaller motors, and the chance of damage actually occurring is significantly less. Duty cycling of these smaller motors is much more efficient also. They are either on or off and run at optimum efficiency at all times.

Generally, heat is the worst enemy of motors. If a motor is duty cycled, it runs cooler. Even though the highest temperature occurs during start-up and there may be very many start-ups when duty cycling is used, the manufacturer takes this into consideration. This is done when the minimum on and minimum off times are specified. The average temperatures at which the starter motors and compressors run are determined. The average duration without duty cycling will not be exceeded, when duty cycling, if the minimum on and off times are followed. Thus, if a compressor is cycled off 10% of the time, the number of run hours will not be decreased. In this case, the number of days or years of the compressor life will be increased by 10%.

Publicity concerning the problems with duty cyclers led mechanical equipment manufacturers to issue stipulations in the warranties of air-conditioning equipment in the past. One such stipulation was that, if the air conditioner was duty cycled at all, the warranty was voided. This was true on some units until approximately 3 years ago.

Now, the warranties are straightforward. For example, the warranty for a compressor is for 5 years. This is straight across the board (no stipulations against duty cycling). For a top-of-the-line compressor, the warranty could be for 15 years because of the additional safety equipment that is included. The only quidelines specify that the minimum on and off times stated by the manufacturer must be followed to protect the equipment against short cycling.

While some people consider duty cycling is bad for the mechanical equipment, many manufacturers have introduced duty-cycle timers to their lines and will give the buyer information on cycling on and off times. Why so? Can duty cycling actually reduce energy usage without the added expense of premature equipment failure? The answer is yes. Duty cycling can reduce energy costs and, if the cycle rates are not excessive, duty cycling will not cause the premature failure of well-designed mechanical equipment.

### Determining On Time and Off Time

The National Electrical Manufacturers Association (NEMA) has published a standard (NEMA Standard MG 10) in which the problems of duty cycling of electrical motors are examined. The standard was prepared as a guide for selecting the optimal on and off times of a given electrical motor, taking into account motor type, horsepower rating, speed, starting frequency, restrictions on in-rush current, power demand charges and the extra winding stress imposed by repeated accelerations. A portion of the table outlining NEMA recommendations is shown in Table 10-2.

**Table 10-2.\* Allowable Number of Starts and Minimum Time
Between Starts for Design A and Design B Motors**

|      | 2-Pole |      |      | 4-Pole |      |      | 6-Pole |      |      |
|------|--------|------|------|--------|------|------|--------|------|------|
| HP   | A      | B    | C    | A      | B    | C    | A      | B    | C    |
| 2.0  | 11.5   | 2.4  | 77   | 23     | 11   | 39   | 26.1   | 30   | 35   |
| 5.0  | 8.1    | 5.7  | 83   | 16.3   | 27   | 42   | 18.4   | 71   | 37   |
| 7.5  | 7.0    | 8.3  | 88   | 13.9   | 39   | 44   | 15.8   | 104  | 39   |
| 15.0 | 5.4    | 16   | 100  | 10.7   | 75   | 50   | 12.1   | 200  | 44   |
| 20.0 | 4.8    | 21   | 110  | 9.6    | 99   | 55   | 10.9   | 262  | 48   |
| 25.0 | 4.4    | 26   | 115  | 8.8    | 122  | 58   | 10.0   | 324  | 51   |
| 30.0 | 4.1    | 31   | 120  | 8.2    | 144  | 60   | 9.3    | 382  | 53   |
| 40.0 | 3.7    | 40   | 130  | 7.4    | 189  | 65   | 8.4    | 503  | 57   |
| 50.0 | 3.4    | 49   | 145  | 6.8    | 232  | 72   | 7.7    | 620  | 64   |

\*Adapted from NEMA Standard MG10.

The parameters above represent the following:

hp = NEMA horsepower rating of the motor
A = Maximum number of starts per hour
B = Maximum product of A times load inertia (lb x ft$^2$)
C = Minimum reset time in seconds (convert to minutes)

The table lists the minimum off time and maximum number of starts permitted per hour. What is actually needed is the information for determining the duty period (on time plus off time). Using the numbers in the table, Item A (maximum starts per hour), the shortest duty period permissible for a 2-pole, 25-hp motor, taking into consideration that the motor and load are sized correctly, can be calculated as follows:

$$\frac{60 \text{ min/hr}}{4.4 \text{ starts/hr}} = 60 \text{ min/4.4 starts} = 13.53 \text{ min/start}$$

Therefore, the duty period = 14 minutes.
The minimum off time is 115 seconds (approx. 2 minutes).

On Time + Off Time = Duty Period
On Time + 2 Min. = 14 Minutes
On Time = 12 Minutes.

To verify the above data, exchange the 14-minute duty period for the 4.4 starts per hour in the above equation. The number of starts per hour is 4.28.

Since 4.28 is less than 4.4, which was specified as the maximum number of required starts per hour, the duty period of 12 minutes on and 2 minutes off will meet the requirements of the NEMA Standard MG 10. Increasing either the off time and/or the on time from the minimums listed above will decrease the number of starts per hour and save wear and tear on the mechanical equipment.

The NEMA standards recommend the minimum on and off time for cycling motor loads. It is also imperative to contact the compressor manufacturer for their recommendation as a minimum off time to allow the compressor to pump down and not start against a high-pressure head. Remember that duty cycling must be done properly in order to get the maximum savings.

**Energy-Saving Cycle Strategy**

Now that the minimum on and off times have been calculated to protect the equipment from short cycling and premature failures, an energy-saving duty cycling strategy must be determined. The measure of the effectiveness of a duty cycle program is to determine that the loads are cycled in a manner to save electrical demand (to be discussed in programmable controllers) as well as consumption (kWh). Straight-time duty cycling is accomplished by starting the duty cycler at one point in time (for example, at 8:00 a.m.). Monday through Friday for a cycle of 15 minutes on time and 15 minutes off time, then stopping the duty cycle (say at 5:00 p.m.).

Whenever more than one load is to be duty cycled, there are two methods to provide the cycling program. The first method is to start all the loads at the same time and have them cycle on and off at the same time. This method is called parallel duty cycling. The second method to provide duty cycling to more than one load is to stagger the on and off times to avoid the in-rush current while evenly spreading the distribution time to each load. The decision now is which method to choose and what times to use.

The demand period is the time interval that the utility company checks the power consumption of a user to determine the maximum number of kW used during this interval. In most cases this period is 15 minutes.

The duty period is equal to the on time plus the off time. So, if the duty period is less than the demand period interval, the parallel duty cycling can be used. During the demand interval, the utility can detect if all of the loads are running at the same time. If the duty cycle period exceeds the demand time, then the staggered duty cycling is necessary if savings in kW demand are to be achieved.

Mechanical equipment manufacturers have specified that minimum duty cycle periods must be, in most cases, greater than the demand periods to ensure their equipment's life span. So, staggering load starts are usually a must to save demand charges. The mechanical equipment manufacturers are primarily concerned about short-cycle protection for the air-conditioner compressors. This allows the compressor to pump down before trying to start against a high-pressure head and thus burn out the motor.

## Stagger Cycling and Savings

To show an example of the stagger method of duty cycling, assume that the manufacturer was contacted and he recommends a duty cycle of 20 minutes on and 10 minutes off as the minimum duty cycle. Anything greater in the on or off time would be acceptable. The duty period is 30 minutes. Say three compressors, each rated at 10 kW (approx. 3 tons), have to be cycled and the demand interval is 15 minutes.

$$\underline{\text{ON PERIOD} + \text{OFF PERIOD}} = \underline{(20 + 10) \text{ min.}}$$

number of loads = 3 loads

The stagger rate equals 10 minutes per load.

Figure 10-9 shows the relationship between the stagger rate of 10 minutes per load, demand interval and the kW savings possible using the stagger function. Figure 10-10 has been generated from Figure 10-9. The demand for the three loads would be the highest demand recorded over any 15-minute interval. Since the highest demand recorded was 19.99 kW and the maximum potential demand would have been 30 kW, the savings are:

$$\frac{30 \text{ kW} - 19.90 \text{ kW}}{30 \text{ kW}} = 0.2933 = 2933\%$$

If duty cycling were done using the parallel method, there would be no savings of demand because all three loads would have been on for 20 minutes. This peak would show up during the first demand period and the second demand period.

Regardless of the cycling method, the actual energy consumption would be the same because the on time would, of course, be the same.

Percent consumption savings:
OFF TIME X 100/DUTY PERIOD
10 MIN OFF X 100/30 MIN = 33.33%

Duty cycling can save electrical expenses when tied to air-conditioning equipment, provided that the proper installation is done and proper maintenance followed. (For example, the mechanical equipment is not short-sycled—the correct on and off times are calculated for the equipment). Also, maintenance routines including proper lubrication, tightened belts, and clean filters should be regularly performed.

Figure 10-9. Stagger Rate

Figure 10-10. Load Stagger Duty Cycle

## Thermostats Considered

The duty cycle example discussed did not take into account the cycle rate of the thermostat in the estimates of potential savings. To calculate the actual savings, the thermostatic cycle rate must be known for a given condition of cooling requirements. For example, if the thermostat cycle rate were 30%, the savings in both demand and consumption would be divided by .30 to calculate the additional savings caused by cycling the load in addition to having the thermostat wired in series with the duty cycler.

The thermostat is a temperature-compensating duty cycler. It is essential in many applications where environmental temperature conditions along with energy savings are paramount.

Designers using duty cycles traditionally must balance energy savings against space comfort. With the newer compensated duty cycling, savings plus comfort can be attained.

Duty cycling has been done for years, but a duty cycler was not accomplishing the duty cycling—a thermostat was cycling the load on and off to maintain the set temperature conditions. Duty cyclers and compensated duty cycling programs based on time rather than set specific temperature have replaced the standard thermostat. The time element must be taken into account, particularly since the utility company is charging for electricity based on time.

## Pneumatic Controllers

Pneumatic controllers are usually combined with sensing elements with a force or position output to obtain a variable air pressure output. The control mode is usually proportional, but other modes such as proportional-integral can be used. These controllers are generally classified as non-relay, relay, direct- or reverse-acting type.

The *non-relay* pneumatic controller uses a restrictor in the air supply and a bleed nozzle. The sensing element positions an air exhaust flapper that varies the nozzle opening, causing a variable air-pressure output applied to the controlled device.

A *relay-type* pneumatic controller, directly or indirectly through a restrictor, nozzle and flapper, actuates a relay device that amplifies the air volume available for control.

*Direct-acting* controllers increase the output signal as the controlled variable increases. For example, a direct-acting pneumatic thermostat increases output pressure when the sensing element detects a temperature rise.

*Reverse-acting* controllers increase the output signal as the controlled variable decreases. A reverse-acting pneumatic thermostat increases output pressure when the temperature drops.

As pointed out earlier, pneumatic controllers can be operated in either proportional or proportional-integral mode. System characteristics in this control mode having influence on energy efficiency are proportional band, reset time, over-capacity of the heating/cooling equipment and part-load operation. Proportional band is the ratio of the percent change in the actuating signal to the percent change in the manipulated variable. Reset time is defined as the time required by the reset to repeat the proportional action.

**Programmable Controllers**

A programmable controller is a control device that has logic potential but is not powerful enough to be called a computer-interfaced controller. This type of control fills a need for applications requiring more than a timer but not a computer. A programmable controller can do all that timers can do and much more but at a cost considerably less than that of a computer.

Because of the logic capability, programmable controllers are very useful in implementing control strategies like

a) Demand control
b) Economizer control
c) Enthalpy control
d) Combustion air control
e) Temperature reset

## Demand Control

Demand charge is that portion of the electric bill which reflects the utility's capital requirements in generating plant, transmission lines, transformers, etc., to provide power to a facility. The demand load is the maximum concurrent load to occur during any 15- or 30-minute interval during the billing month, and it is measured in kilowatts. It is in the interest of the customer and the utility to keep the peak demand as low as possible. While this does not directly save energy, it definitely saves money. The following methods can be used to reduce peak demand:

(i)   Off-peak schedule
(ii)  Thermal storage (see Chapter 12)
(iii) Demand-limiting controls
(iv) Programmable controllers

In demand control by a programmable controller, when the controller senses that the electrical demand is approaching a critical (preset) level, it shuts off equipment on a programmed priority basis to restrict the demand from passing the critical level. Since demand charge generally amounts to approximately 25%-35% of the total electric bill, savings from controlling demand can be significant.

## Economizer Cycle

An economizer cycle is the adaptation to the fresh-air intake which permits the use of outside air for cooling when temperatures are sufficiently low. When nighttime outdoor temperatures are below indoor setpoint by 5°F or more, the controller shuts off the refrigeration system and the return air dampers and opens the outdoor air dampers fully. Using outdoor air for night cooling will save energy in most areas.

## Enthalpy Control

During occupied periods, the opportunities to use outdoor air for cooling will depend on the outdoor wet-bulb (WB) temperature as well as dry-bulb (DB) temperature. If the outdoor air is brought into the building above 60°F WB, the cooling load is increased. The wet-bulb temperature is a measure of the total energy content (enthalpy) of the air.

In enthalpy control, the controller is interfaced to DB sensors and to WB sensors for both the outdoor as well as the indoor air.

The controller is programmed to calculate the enthalpies; if the enthalpy of the outdoor air is less than the enthalpy of the indoor air, 100 percent of the outdoor air is used for cooling, resulting in energy savings. (See Chapter 7 for Enthalpy Control Savings Calculations.)

### Combustion Air Control

The programmable controllers can be used to control the excess air in a combustion process. By sensing $CO_2$ or $O_2$ or $CO$ levels in exhaust, the controller can adjust the combustion air intake for optimal combustion efficiency. Continuous control through the use of a programmable controller allows the air intake to be adjusted in accordance with the demand on the unit.

### Temperature Reset

The logic capability of the programmable controllers can be used to reset the setpoints on the chillers and the boilers. The controller can be interfaced to DB, WB, wind velocity and solar radiation sensors. The arithmetic logic of the controller can be used to calculate the optimal setpoints and reset accomplished which can result in significant energy savings and can enhance the comfort levels in the buildings.

## COMPUTER CONTROL

The control of physical systems with a digital computer is becoming more and more common. Aircraft autopilots, mass transit vehicles, oil refineries, paper-making machines and countless electromechanical servomechanisms are among the many existing examples. Furthermore, many new digital control applications are being simulated by microprocessor technology, including on-line computer controllers in automobiles and household appliances. Among the advantages of digital logic for control are the increased flexibility of the control programs and the decision-making or logic capability of digital systems.

One desirable characteristic of a control system is to have a satisfactory response. "Satisfactory response" means that the output, $c(t)$, is to be forced to follow or track the reference input, $r(t)$, (See Figure 10-2) despite the presence of disturbance inputs to the plant (system) and despite errors on the sensor. It is also essential that the tracking succeeds even if the dynamics of the plant should change somewhat during the operation. The process of holding c(t) to r(t) is called *regulation*. A system which has good regulation in the face of changes in the system dynamics is said to have low *sensitivity* to system's parameters. A system which has both good disturbance rejection and low sensitivity is called *robust*. Use of digital computers as controllers can help create robust systems.

Figure 10-11 shows the use of a digital computer as a control element. Note that the notation on this figure has been changed compared with Figure 10-2 to incorporate digital computers as controllers.

In Figure 10-11, consider first the action of the analog-to-digital (A/D) converter on a signal. This device acts on a physical variable, most commonly an electrical voltage, and converts it into a stream of numbers (pulses). In Figure 10-11, the A/D converter acts on the indicated error signal, ê, and applies the numbers to the digital computer. It is also common for the sensor output, y, to be

Notation:

r   = reference or command inputs
u   = control or actuator input signal
y   = controlled or output signal
ŷ   = instrument or sensor output, usually an approximation to or estimate of y. (For any variable, say θ, the notation θ̂ is now commonly taken from statistics to mean an estimate of θ).
ê   = r−ŷ = indicated error
e   = r−y = system error
w   = disturbance input to the plant
v   = disturbance or noise in the sensor
A/D = analog-to-digital converter
D/A = digital-to-analog converter

**Figure 10-11. Digital Control System Block**

sampled and have the error formed in the computer. We need to know the times at which these numbers arrive if we are to analyze the dynamics of the system.

We may assume that all the numbers (pulses) arrive with the same fixed period T, called the *sampling period*. In practice, digital control systems sometimes have varying sample periods and/or different periods in different feedback paths. Typically, there is a clock as part of the computer logic which supplies a pulse every T seconds (See Figure 10-12), and the A/D converter sends a pulse (number) to the computer each time the pulse from the clock arrives. Thus, in Figure 10-11 we identify the sequence of numbers as m(kT). We conclude from the periodic sampling action of the A/D converter that some of the signals in the digital control system, like m(kT), are variable only at discrete times. We call these variables *discrete signals* to distinguish them from variables like ê and y which change continuously in time. A system containing only discrete variables is called a *discrete-time* system. A system having both discrete and continuous signals is called a *sampled-data* system.

In addition to generating a discrete signal, however, the A/D converter also provides a *quantized* signal. By this we mean that the output of the A/D converter must be stored in digital logic composed of a finite number of digits. Most commonly, of course, the logic is based on binary digits (i.e., bits) composed of 0's and 1's, but the essential feature is that the representation has a finite number of digits. A common situation is that the conversion of ê to m is done so that m may be thought of as a number with a fixed number of places of accuracy. If we plot ê versus m, we may get a plot as shown in Figure 10-13.

In Figure 10-13, we would say that m has been truncated to one decimal place, or that m is *quantized* with q of 0.1 since m changes only in fixed quanta of, in this case, 0.1 units. A signal which is both discrete and quantized is called a *digital signal.*

In a real sense, the problems of analysis and design of *digital controls* are concerned with accounting for the effects of the sampling period T and the quantization size q. If both T and q are small, digital signals are nearly continuous, and continuous methods of analysis can be used.

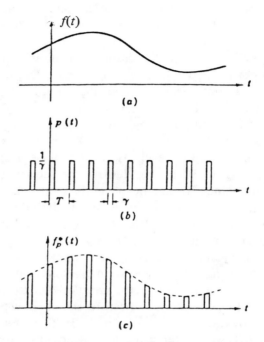

Figure 10-12. (*a*) Continuous Function $f(t)$:
(*b*) Sampling Pulse Train $p(t)$; (*c*) Sampled Function $f_p^*(t)$.

Figure 10-13. Plot of Output vs Input Characteristics of A/D Device

**Design of Digital Systems**

As mentioned previously, we can place the systems of interest in three categories according to the nature of the signals present. These are:

1. Discrete systems
2. Sampled-data systems
3. Digital systems

In *discrete systems,* all signals vary at discrete times only. For the analysis of discrete systems, one must learn z-transforms of discrete signals and "pulse"-transfer functions for linear constant discrete systems. Of special interest is the characterization of the dynamic response of discrete systems. As a special case of discrete systems, we consider discrete equivalents of continuous systems, which is one aspect of the current field of *digital filters* (Kalman filter).

If quantization effects are ignored, digital filters are discrete systems which are designed to process discrete signals in such a fashion that the digital device (a digital computer program, for example), can be used to replace a continuous filter.

A *sampled-data* system has both discrete and continuous signals. In these systems, we are concerned with the question of data extrapolation to convert discrete signals (as they emerge from a digital computer) into the continuous signals necessary for providing the input to one of the plants. This action typically occurs in conjunction with the D/A conversion.

Digital controllers for the building systems may be one of the following types:

a. Direct digital control (DDC) or distributed control or remote control or on-site control
b. Supervising control or central control
c. Hierarchy control (DDC plus supervisory control)

A central control system provides the energy engineer with a means of constant surveillance of the building and helps in making efficient and effective use of physical plant systems and personnel. Proprietary central computerized control systems are marketed by each of the major temperature control manufacturers. These systems have common features and can accomplish a similar range of tasks, but each manufacturer uses coding and computer languages which are unique to the system.

Each manufacturer's system is made up of standard hardware, but the application is always tailored to the specific project. Basically, any system is composed of the following four major components:

- Interface panels which are located at strategic points throughout the building.

- The transmission system between the central console and all interface panels. This system can be single-core cable for digital transmission or multi-core cable for multiple transmissions.

- A central control console and associated computer hardware located in a control room. The console, computer and associated hardware form the point at which the operator enters all instructions and retrieves all data.

- Software programs. Programs for common applications such as start-stop action, load shedding, reset, alarm, optimization, etc. are available from the major control system suppliers.

Experiences with central control systems reported thus far in literature have not been very positive. Some of the reasons cited for ineffective use of the central control systems are inadequate training facilities for the operators, inadequate buyer commitment, poor vendor assessment and ineffective interface with the building system. In spite of these difficulties, if the energy engineer is involved with larger, more complex buildings (or groups of buildings), possible use of central computer-based control systems should be fully explored.

**References**
1. D'Azzo, J.J.; and Houpis, C.H. *Feedback Control Systems: Analysis and Synthesis*. New York. McGraw Hill Book Co., 1983.
2. Mehta, D. Paul. Dynamic Thermal Responses of Buildings and Systems. Ph.D. Dissertation. Iowa State University, Ames, IA, 1979.
3. _____ *HVAC Systems and Applications Handbook*. ASHRAE, Atlanta, Ga., 1987.
4. Smith, C.B. *Energy Management Principles*. Pergamon Press, 1981.
5. Kennedy W.J.; and Turner, W.C. *Energy Management*. Prentice-Hall Inc.; Englewood Cliffs, N.J., 1984.
6. Dubin, F.S.; and Long, C.G. *Energy Conservation Standards*. New York. McGraw Hill Book Co., 1978.

# 11

# Computer Applications

This chapter will focus on the following subjects:

Introduction
Basic Hardware
Basic Software
Classification of Computer Programs
 Heating and/or Cooling Load Calculation Programs
 Energy Analysis Programs
 Duct Design Programs
 Piping Design Programs
 Equipment Selection Programs
 Solar Programs
 CAD for HVAC Systems
Selection of Computer Programs
 General Information
 Plant Capacity Estimated by the Program
 Plant Capacity Estimated by Another Program
 Weather Data
 Load Profile Calculations
 Building Structure
 Radiation Transmission Through Glass and Blinds
 Gains Through Opaque Walls and Roofs
 Simulation of Building Services Systems
 Consumed Energy Profiles
Future Trends

## INTRODUCTION

Digital computers became available in the 1940's. In the following four decades, the computational power of computers has increased more than a million times, while the cost, power requirements and programming difficulties have reduced dramatically. This has led to a rapid increase in the use of computers by engineers in general and energy engineers in particular during the last decade. Several areas can be identified where computers have been incorporated by the energy engineers in their practice:

a) Information and communication
b) Computation
c) Design
d) Instrumentation
e) Control

The details of the use of computers as control elements were discussed in Chapter 10. The objective of this chapter is to describe the application of a computer in other areas, especially as it impacts the engineering analysis and design. For analysis, computer applications basically involve the use of computers to do calculations. Computers find their greatest advantage in making large quantities of simple calculations, a task where energy engineers typically find them very helpful.

## BASIC HARDWARE

The main components of a computer are shown in Figure 11-1. All computers, no matter how simple or complex they are, contain these components.

The heart of a computer is the CPU (Central Processing Unit), which is simply a group of electronic devices which can exist in one of the two formats or states. These states could be interpreted as "yes" or "no"; "1" or "0"; "on" or "off," etc. The computer is sensitive to the status of these states and makes calculations using binary arithmetic. The operation is simple but sophisticated, incorporating speed and accuracy with which these calculations can be made. The main functions of the various components are described as follows:

Figure 11-1. Computer Components

1. *The Central Processing Unit (CPU)*
   a. fetches and decodes each instruction
   b. performs arithmetic and logical operations on data as required by each instruction
   c. determines internal status (flags) resulting from operations with data
   d. can transfer data to and from the memory and I/O sections of the computer
   e. generates and receives handshake control signals from the memory and I/O sections
   f. provides master system timing

2. *The Address Bus*
   a. generally under CPU control
   b. uni-directional
   c. provides a unique numerical identifying code for selecting a particular memory or I/O element

3. *The Data Bus*
   a. generally under CPU control
   b. bi-directional (data travel in both directions)
   c. provides route for data transfer among computer parts

4. *The Memory Section*
   a. stores the instructions for CPU operation
   b. stores data to be operated upon by the CPU
   c. may be physically divided into program/data sections
   d. generally performs under CPU direction

5. *The Input/Output Section (I/O)*
   a. provides the interface between the environment and the rest of the computer
   b. transmits data and/or commands to the computer peripherals
   c. receives data from the peripherals and makes it available to the CPU
   d. can alert the CPU when data is available
   e. generally performs under CPU direction

6. *The Control Lines*
   a. provide timing signals to synchronize the various computer sections

    b.   provide path for handshake signals

    c.   provide a means of interrupting CPU operations

    d.   individual lines generally uni-directional

## BASIC SOFTWARE

As a result of the rapid increase in the use of computers by the energy engineers, a large number of computer programs have appeared. These programs, in most cases, are available to users other than the developer. Generally, these programs are available in one of the following ways:

a. The program is not proprietary and the source code can be purchased for in-house implementation.

b. The program is considered proprietary and can be used on a time-sharing basis or by leased-object language codes.

c. The program is proprietary and can be used only by submit-timing input data to the program developer.

d. The program is developed to meet the specific needs of a specific customer. This could be accomplished either by using the outside services or by developing in-house facilities.

## CLASSIFICATION OF COMPUTER PROGRAMS

The computer programs which are available to an energy engineer for computation and design could be divided into the following classes:

### 1. *Heating and/or Cooling Load Calculation Programs*

The programs within this class are meant to calculate heating and/or cooling loads for a building or zone. Load calculations are well suited to computer analysis due to the large number of relatively simple calculations involved, the amount of data manipulation needed and the fact that these calculations are routinely performed on nearly all building projects. It was, therefore, one of the earliest applications of the computer to building design. A large number of programs currently exist which vary in the methods used. In general, these programs can be classified as either design condition load calculations (usually a single design condition or design day, e.g.,

'Bless'*) or as hour-by-hour load calculations, which are usually more suited for energy analysis programs, as is the case with the program "NBSLD."

2. *Energy Analysis Programs*

Programs meant to estimate the annual energy consumption of buildings are contained in this class. Energy analysis programs can be classified according to the method by which they treat the annual weather data necessary to perform the estimation. The degree-day method proportions a design heating load over a heating season based on the number of degree days in the heating season. The bin method is based on the number of hours per year that the outside temperature falls within specified temperature ranges. Hour-by-hour calculations consider each hour of a typical year's weather data. They generally consist of a load subroutine which calculates the heating and cooling loads and a system simulation portion, as is the case with the program "AXCESS." Several programs, e.g., "RUND1620" and "ECUBE," also contain an economic analysis section which calculates the cost of the energy required and makes various economic calculations.

3. *Duct Design Programs*

Duct design programs are used to assist in the areas of duct sizing, cost estimation and sound analysis. Examples of such programs are DUCT, HVDUCT, DUCTAN, EXHAUST, DUCTS, DADDS and so on. Input to these programs normally consists of required CFM for each terminal unit and a nodal network which is developed based on a line diagram of the ductwork system. Some programs are designed to also calculate heat losses or gains from ductwork and to recalculate CFM's accordingly. The analysis of fan noise attenuation is part of several of these programs.

4. *Piping Design Programs*

The piping design programs can generally be classified as: (1) pipe-sizing programs, which calculate the size of pipe required based on specified GPM and a description of the piping system, e.g., "MOIP" or (2) pipe-flexibility programs, which are utilized to estimate the stress and deflections of piping systems, as is the case with the program "AUTOFLEX." The sizing programs will normally size

---

*See Table 11-1 for details on these programs.

piping and estimate pump head for systems based on velocity and pressure drop limits, and some will calculate heat gain or loss from piping sections. A number of the programs produce bills of material or cost estimates for the piping system. The piping flexibility programs are intended to assist in a stress analysis of piping systems. These programs can generally handle thermal, uniform or constant loadings and will calculate deflections, stresses, forces and moments.

### 5. *Equipment Selection Programs*

The programs contained in this class have to do with the selection of HVACR equipment, as is the case in "MFOURTEEN," "RH" and "CONDX." Areas covered include coil selection (hot water, chilled water, steam or direction expansion), refrigeration units, air-handling units and ceiling diffusers.

### 6. *Solar Programs*

The utilization of solar energy for the heating and cooling of buildings has been receiving considerable attention, and a number of programs for simulating the performance of these systems have appeared, e.g., "SUN," "TRANSYS" and "SIMSHAC." Solar programs can be considered to be specialized energy analysis programs. These programs will take into account collector area and performance characteristics, storage capacity, building load and weather data to predict the performance of the solar heating and/or cooling system.

### 7. *CAD for HVAC Systems*

Computer-aided design and drafting programs (CAD) have been developed that not only provide designers with computerized methods for the basic drafting of HVAC systems and shop drawings but that also simplify HVAC analytical and design tasks. Further, CAD programs enhance the interdisciplinary design of buildings, their mechanical systems and other systems and ease the task of modifying designs. Communications among the various design groups increase through the use of CAD programs.

All systems display a drawing or part of a drawing on a screen and provide a means of constructing straight lines, circles, polygons, arcs, section crosshatching, components, dimensioning and lettering. Pre-defined elements can be added at any scale and in any orientation. Elements can be added or deleted on the screen. Close-up views of areas of the drawing are available on the screen. Options such as a background grid, forcing of orthogonality of lines, snap-to-grid, snap-to-tangent and filleting can speed up the drafter's work.

CAD programs usually have layering capabilities, so different parts of a drawing can be displayed or modified separately or together. For example, in a duct drawing, supply ducts may be put on one layer and return ducts on another so either or both can be seen, changed or removed. Layering can also help an HVAC designer coordinate designs with other disciplines. For example, an HVAC designer working on a duct drawing can have the CAD system display the building's architecture, structure, piping and electrical systems in the background to check for space conflicts. Errors and obstructions are easily detected and corrected with the use of CAD programs.

Scale drawings express not only physical size but also how different parts are connected and where they are located within an area. As an example, an architectural designer can draw a building's walls, partitions and windows and have the computer automatically extract the number and size of windows, the areas and lengths of walls and the areas and volumes of rooms and zones.

Some CAD programs can link the graphic elements to the non-graphic characteristics of those elements. With these programs, an operator can store information from a drawing for later use in reports, schedules, design procedures or notes. A pump, for example, can be displayed accurately on a drawing, and its flow rate, voltage, weight, manufacturer, model number, cost and other data can be stored in a file or database. The link between graphics and attributes makes it possible to review the characteristics of an item simply by pointing to the graphics on the drawing or to generate schedules of items located in a certain area on a drawing. Conversely, the graphics-to-data links allow the designer to see the graphic items on a drawing that have a certain characteristic by searching for those characteristics in the data files and then making the associated graphics stand out as brighter, flashing, bolder or different-color displays.

Computer-aided design also makes building design drawings and information more useful than manually generated drawings. CAD HVAC drawings and their associated data can be used by building owners and maintenance personnel for ongoing facilities' management, strategic planning and maintenance. They are useful for computer-aided manufacturing such as duct construction, where duct design drawings serve as the basis for flat patterning, nesting and automatic cutting of ductwork and fittings. Two-dimensional CAD drawings can be used to create a three-dimensional model of the building and its components. This 3-D model can then help to visualize all or part of the building; to check for interferences between building parts; and to perform lighting, shading, daylighting, acoustic or energy analysis of the building and its spaces. The 3-D building model can also aid in developing sections and details for building drawings.

Further, CAD drawings and data can aid in specification preparation by allowing automated cross-referencing of drawings, drawing notation and building documents. CAD drawings of HVAC systems also serve as a basis for extracting quantity and size information of HVAC components that can be used for drawing equipment schedules and cost estimating.

## SELECTION OF COMPUTER PROGRAMS

Selection of a suitable computer program for a specific application by an energy engineer requires a careful approach for evaluating the available programs. Generally, answers to the following questions are needed:

a) Are there any programs available which would satisfy the specified need?

b) Which of the available programs would be best for the situation?

c) How can those programs be obtained?

ASHRAE-sponsored research has produced a document titled *"A Bibliography of Available Computer Programs in the General Area of Heating, Refrigerating, Air Conditioning and Ventilating."* This document is one of the most comprehensive listings of HVAC-related software available today. The reader is urged to refer to this document to find answers to the above-listed questions.

Once a group of programs has been identified in the general area of specified application, each should be subjected to detailed scrutiny. Searching questions should be asked to make sure that a program of proper size is selected. This phase of program selection is very important. CIBS (Chartered Institute of Building Services) has stressed this point very well by developing a questionnaire to aid in the selection of a program for estimating the building energy use. CIBS suggests that the following information should be available:

### 1. *General Information*

The program name, giving a brief description of the program, and the name of the company or organization owning the program; also the company or organization responsible for the development of the program should be stated if different from owner. If the program can be made available for general use, can this be on another company's computer equipment or on a bureau basis? On what computer system has the program run successfully and in which modes is the program used? Can changes be made to individual items of input data, without repeating the entire input? In what computer language is the program written? Is there available documentation for the program user such as user manuals, pro forma data sheets, details of theoretical bases of programs, detailed specifications and flowcharts and program listings? Finally, what input data verification is performed by the program, and does the program output include a printout of the input data in a checkable form?

### 2. *Plant Capacity Estimated by Program*

If heating and/or cooling plants are sized by this program, is it necessary to specify the means by which calculations are made (for example, by peak demand in a given year)? Is it further necessary to specify how the start-up demand (after the weekend, say) is controlled in such circumstances or by any other peak design criteria? What assumptions are made for partial load efficiencies of plant chosen automatically?

### 3. *Plant Capacity Estimated by Another Program*

If plant capacity is calculated using another program, give name and details. Is a minimum information list available for this program?

### 4. *Weather Data*

What standard weather data is available to the engineer and how is it constructed (for example, by hourly values all year, or by mean weekly or monthly)? If the weather data is based on a design condition, the source of this data should be given, e.g., the CIBS Guide, Average of Meteorological Office data naming the year, etc. If weather tapes are used, details of source and year should be given, e.g., Meteorological Office, 1967. If the user wishes to supply his own weather data, what details are required of solar radiation, either direct or diffuse, temperature, humidity, wind speed and direction for the approximate time interval (days, weeks or months)?

### 5. Load Profile Calculations

a) Calculation of Time Interval. The frequency at which calculations are performed should be given (for example, hour-by-hour calculations of every day in the year, representative days from every week or representative days from every month). Other combinations available should be stated.

b) Internal Load Profiles. How are load profiles for commodity use, hot water, equipment, lighting and occupancy considered?

c) Internal Comfort. What criteria are used for maintaining internal comfort in terms of temperature, humidity, etc. for both heating and cooling seasons?

d) Infiltration. State on what basis infiltration and air change are handled during heating season (infiltration) and cooling season (natural ventilation).

e) External Heat Gains/Losses. Are calculations performed identically to an existing A/C load program? If so, the name and availability of information lists for that program should be stated.

### 6. Building Structure

Consideration should be given to building thermal storage, the method of defining internal environment, comfort conditions and gains.

### 7. Radiation Transmission Through Glass and Blinds

The availability of the program to cope with:
a. Varying glass thickness
b. Double- or triple-glazing
c. Heat-absorbing/heat-reflecting glass

    d.   Internal/external blinds

    e.   Combinations of special glazing and blinds

    f.   Facade shading

    g.   Range and intervals or orientation possible

    h.   Air conduction through glazing

    i.   Solar blind operation times

    8.   *Gains Through Opaque Walls and Roofs*

Consideration should be given to the method by which solar radiation on external opaque surfaces is considered and the conduction calculation procedure for walls (lag and decrement).

    a.   Plant Operation. Is the plant operational 7 days a week or 5 days a week, and can the plant running hours be varied for intermittent operation?

    b.   Use of Load Profiles. These load profiles are space loads, i.e., the heat which must be added to or removed from any given space to maintain the specified design condition. Summation of these loads over the year can provide a means of comparison of performance of various building envelopes, although such summations are by no means a measure of energy used by heating or cooling equipment. Power requirements for lights, etc., cannot meaningfully be added at this stage. Does the program thus compare space load requirements and, if this is so, then are heating and cooling requirements shown separately? By use of COP/boiler efficiency, can these loads be converted to provide equipment loads, and, further, is the conversion done hourly by using variable COP/boiler efficiency or alternatively using constant COP/boiler efficiency? Finally, are the power loads from lighting, lifts, etc., then added?

    9.   *Simulation of Building Services Systems*

    a.   Heating Systems. Are heating controls possible by using time clock, or clocks, night setback, optimizers or by any other method? Can frost and condensation protection be included? Is heating plant response time taken into account, and, if so, in what way? Is heating plant efficiency considered as constant?

    b.   Air-Conditioning Systems. Can only heating and cooling periods be specified? Is heating by A/C equipment or by direct heating? Is it possible for both to operate if required? Can each zone have different systems? Which individual systems can be modeled (for example, single duct, dual duct, V.A.V., 2-pipe induction, 4-pipe induction or others)? Are the individual performance and part-load efficiencies of plant items considered (for example, fans, pumps, chillers, cooling towers/air-cooled condensers, cooling coils, heat recovery, controls—proportional band, etc., and thermostats)? Can overall controls be investigated, such as economizer and enthalpy control? Also, is plant response time taken into account and, if this is so, how?

c. Operating Conditions. Can variations in building operation be considered, such as weekend, holidays, plant operating continuously or intermittently? Can a characteristic part-load efficiency curve be input for multiple unit installations, such as multiple boilers and multiple refrigeration units?

10. *Consumed Energy Profiles*
How are profiles presented: by overall totals, by type of energy or by type of use? Give a brief description of output, specifying options where appropriate.

## FUTURE TRENDS

Future applications of computers are likely to develop "intelligent buildings" based on the use of "expert systems." Expert systems are the systems which make decisions based on "knowledge base" accumulated from one or more human experts. Some powerful expert systems have already been developed for medical diagnosis, chemical structure predictions, aircraft design and computer system configuration. Buildings seem to be the next potential candidate for using expert systems in their design and operation.

**References**
1. Howell, R.H., and Sauer, H.J., *Bibliography on Available Computer Programs in the General Area of Heating, Refrigerating, Air Conditioning and Ventilating.* ASHRAE, Atlanta, GA, 1981
2. Spielvogel, L.G., *Computer Energy Analysis for Existing Buildings.* ASHRAE Journal, August, 1975.
3. Smith, C.B., *Energy Management Principles.* Pergamon Press, 1981.
4. _____ *HVAC Systems and Applications Handbook.* ASHRAE, Atlanta, GA, 1987.
5. _____ *CIBS Computer Applications Panel/Task Group Program Evaluation/Building Energy Use Estimation Programs.* BSE, Vol. 45, Feb. 1978.

Table 11-1. Details of Commercial Programs

| Program | Availability | Purpose | Input Data Needed |
|---------|--------------|---------|-------------------|
| BLESS | V.C. Thomas; Daverman Associates; 200 N.W. Monroe; Grand Rapids, MI 49502 616/451-3525 | To calculate heating/cooling loads. | Summer and winter design conditions; building design parameters; room exposure data; glass, wall, and roof information. |
| NBSLD | James P. Barnett, Thermal Engr-Sec; Center for Building Technology, NBS Washington, DC 20234 | To calculate heating/cooling loads and to evaluate room temperature fluctuations. | Building parameters like wall, ceiling and floor characteristics; window area and location, operating schedules. Weather data from U.S. Weather Bureau tapes. |
| AXCESS | E.S. Douglas, Edison Electric Institute 90 Park Ave. New York, NY 10016 | To estimate the energy requirements of various alternatives. | Weather data and control information; base load profiles; base design heating/cooling load; space type data; zone data; HVAC system details. |
| RUN01620 | Duke Power Co. 422 S. Church St. Charlotte, NC 28242 704/373-4304 | To estimate building energy usage comparing input fuel requirements. | Base load items and durations of operation, heating/cooling loads; details of HVAC system. |
| ECUBE | American Gas Assoc. 1515 Wilson Blvd. Arlington, VA 22209 703/524-2000 | To calculate annual energy requirements in buildings. | Details of internal heat gains; electrical load and process load; solar load; hourly weather data; part load equipment performance data; fuel type, economic input data. |

*(Continued)*

| Program | Availability | Purpose | Input Data Needed |
|---------|-------------|---------|-------------------|
| DUCT | APEC Executive Office<br>Grant Deneau Tower<br>Suite M-15<br>Dayton, OH 45402<br>513/228-2606 | To design, size and analyze ducts. | Duct specifications; design criteria; insulation; cost data. |
| HVDUCT | ACTS Computing Corp.<br>29200 Southfield Rd.<br>Southfield, MI 48076<br>313/557-6800 | To design high velocity air duct systems. | Pressure losses in different ducts; line lengths; fitting types; duct diameters. |
| DUCTAN | Deere and Company<br>John Deere Road<br>Moline, IL 61265 | To size and design ducts. | Duct dimensions; angle in degrees of grille deflection; static regain efficiency factor; air flow. |
| EXHAUST | Deere and Company<br>John Deere Road<br>Moline, IL 61265 | To design industrial dust collection and air systems. | Hood entry loss factors; duct geometry and dimensions; hood air flow rates. |
| DUCTS | General Electric Co.<br>401 N. Washington St.<br>Rockville, MD 20850<br>301/340-4000 | To size and design ducts. | Duct size, length, air flow and fittings code. |
| DADDS | Daverman Associates<br>200 Monroe, N.W.<br>Grand Rapids, MI 49502<br>616/451-3525 | To design ducts. | Length and fitting size; outlet air quantities; fan and noise criteria. |

*(Continued)*

| Program | Availability | Purpose | Input Data Needed |
|---------|-------------|---------|-------------------|
| MOIP | APEC Executive Office Grant-Deneau Towers Suite M-15 Dayton, OH 45402 513/228-2602 | To size piping. | Pipe specifications, design criteria, material and labor costs; Nodal description of piping. |
| AUTOFLEX | Auton Computing Corp. 10 Columbus Circle New York, NY 10019 212/581-5250 | To determine piping flexibility and to analyze stresses. | Piping code; system description; geometry; external forces, moments, restraints, and anchors. |
| MFOURTEEN | Giffels Assoc. Inc. Marquette Building 243 W. Congress Detroit, MI 48226 313/961-2084 | To calculate hot, cold water demand, size supply piping and hot water heater. | Fixture count; continuous demands; and water heater information. |
| RH | Hankins & Anderson, Inc. 2117 N. Hamilton St. P.O. Box 6872 Richmond, VA 23230 804/353-1221 | To select reheat coils for variable volume and induction reheat boxes. | Heating load; box size; water and air temperatures. |
| CONDX | ACTS Computing Corp. 29200 Southfield Rd. Southfield, MI 48076 313/557-6800 | To match air cooled condensing units and direct expansion cooling coils. | Cooling load; entering wet-bulb and dry-temperatures; CFM of air; face velocity. |

*(Continued)*

| Program | Availability | Purpose | Input Data Needed |
|---------|-------------|---------|-------------------|
| SUN | Berkeley Solar Group 1815 Francisco Street Berkeley, CA 94703 | To estimate the performance of solar systems. | Average weather data; system parameters; loads. |
| TRANSYS | Solar Energy Lab University of Wisconsin Madison, WI 53706 | To simulate solar systems for dynamic performance. | Size, quality and configuration of components; hourly weather data. |
| SIMSHAC | Gearold R. Johnson Mechanical Engineering Colorado State Univ. Fort Collins, CO 80523 | To simulate solar systems for the dynamic performance. | Component description, parameters and geometry; building characteristics; and weather data. |

# 12

# Thermal Storage

This chapter discusses the following:

> Advantages of Thermal Storage
> Basic Definitions
> Status of Energy Storage Technology
>> Water Tanks
>> Ice Storage
>> Phase Change Materials
>> Rock Beds
>> Ground Coupled Storage

## INTRODUCTION

Thermal storage can be defined as the temporary storage of energy at high or low temperature for use when it is needed. Most of the energy waste occurs because, in many situations, there is a time gap between energy availability and energy use. For example, in large buildings with interior zones that require year-round cooling, the energy typically exhausted to the atmosphere through cooling towers in the winter could be stored in water for heating during the nighttime and unoccupied periods. Similarly, solar energy could be stored during the daytime for night heating.

Unfortunately, the words "heat storage" for storage of energy at high temperature and "cool storage" for storage at low temperature have become common usage. This practice is contradictory to the concepts stressed in thermodynamics, where "heat" is used for energy in transition and not in storage. Similarly, "hot/cold" terms refer to temperature and not to heat.

Finally, the "energy availability" term used above refers to the time of availability and not to the quality of energy (useful work) as used in thermodynamics. For our purpose in this chapter, the words "thermal storage" will be used for energy in the form of internal/kinetic/potential energy stored on a temporary basis at a time when it is not needed for later use when it is needed. Similarly, "cool" and "heat" storage will mean storage of energy in the form of internal energy of a medium at a temperature of 20°F-70°F and 70°F-160°F, respectively.

## ADVANTAGES OF THERMAL STORAGE

The concept of energy storage is not new and has been used for centuries in the design and operation of occupied spaces. In fact, a study of the design of Taj Mahal, one of the most well-known buildings in the world, shows that the concepts of thermal storage were used in its design to make the building energy-efficient and comfortable. The building, which is a 17th-century specimen of Mughal architecture that was built by Emperor Shah Jahan, uses high values of thermal mass to bridge the time gap between the extreme day and night temperature swings in the local climate throughout the year.

Currently, thermal storage techniques are of great interest to an energy engineer because of the following possible advantages:

a) Savings in the first and operating costs of the equipment are possible, especially in the design of new facilities;

b) Use of alternative energy sources like solar and wind energy becomes feasible;

c) Off-Peak rates from the utility company can be utilized.

d) Energy use efficiency can be improved through the use of technologies like heat reclamation.

Significant savings in the first cost of the equipment are possible because thermal storage permits the installation of smaller-size heating or cooling equipment than would otherwise be required. Conditions favoring thermal storage are:

a) High loads of relatively short duration

b) High electric power demand charges

c) Low-cost electrical energy during off-peak hours

d) Building expansion (installation of storage may eliminate the need for heating/cooling equipment addition);

e) Need to provide cooling for small-duration heavy loads such as cooling of restaurants and computer rooms;

f) Need to supplement a limited-capacity cogeneration plant,

and so on.

## Example 12-1

As an example of the possible benefits from thermal storage, let us consider a hypothetical case of a building having a cooling requirement of 500 tons for 8 hours a day (9 a.m.-5 p.m.) and a second cooling load of 1000 tons for 2 hours a day (5 p.m.-7 p.m.). The designer could have different choices for installing a chiller as shown in Table 12.1.

### Table 12-1. Choices for Chiller

|  | Choice No. 1 | Choice No. 2 | Choice No. 3 |
|---|---|---|---|
| Chiller capacity | 1000 Tons | 500 Tons | 250 Tons |
| Thermal storage capacity | — | 1000 Ton hrs | 3500 Ton hrs |
| First cost of chiller | $400,000 | $250,000 | $150,000 |
| First cost of storage | — | $60,000 | $100,000 |
| Total cost | $400,000 | $310,000 | $250,000 |
| Savings |  | $90,000 | $150,000 |

The above example clearly demonstrates how the installation of thermal storage can reduce the first cost of the cooling equipment. It may be pointed out that, if the designer chooses a chiller of 500 or 250 tons, it will cut down the demand charges also. The actual savings from the reduction in the demand charges will depend on the location, type of building and the rate structure of the utility.

It may also be observed in the above example that, if the durations of the cooling requirements for either load are likely to increase in the future, a 250-ton chiller will not serve the purpose even with thermal storage. In that situation, Choice No. 2 may be the only alternative available, as long as a 500-ton chiller can meet the storage cooling requirements during the non-occupied periods.

## BASIC DEFINITIONS

Thermal storage could be accomplished through *sensible heat storage* or as *latent heat storage.* In sensible heat storage, storage is accomplished by raising or lowering the temperature of the storage medium like water, sand, oil, rock beds, ceramic bricks, concrete slabs, etc. In latent heat storage, storage is accomplished by a change in the physical state of the storage medium, with or without a change in its temperature. Change of state may be from solid to liquid or vice versa. Examples of the phase change materials (PCM) used for energy storage are ice (water), salt hydrates and some organic materials. Phase change materials can store larger amounts of energy per unit mass compared to the sensible heat storage media and hence result in smaller and lighter storage devices with lower storage losses and higher storage efficiencies.

*Storage efficiency* is defined as the energy delivered by the storage system divided by the energy delivered to the storage system. Storage efficiencies on the order of 90 percent have been reported for well-stratified water tanks which are having daily duty cycling.

Thermal storage systems may be classified as *full storage* systems or *partial storage* systems. In full storage systems, the equivalent of the entire cooling or heating load energy for the design day is stored in the storage off-peak and used during the following peak period. In partial storage systems, only a portion of the daily load is generated during the preceding off-peak period and stored. In partial storage, the load is satisfied by simultaneous operation of the cooling or heating equipment and withdrawal from storage during the peak period.

## STATUS OF ENERGY STORAGE TECHNOLOGY

Currently, the following methods are used for accomplishing thermal storage:

    a)  Water tanks

b) Ice storage
c) Phase change materials
d) Rock beds
e) Ground-coupled storage

A brief description of each one of the methods is given below.

## a) Water tanks

Water tanks are the most commonly used method of thermal storage, and ASHRAE Standard 94.3-1986 has been developed which describes the procedures for measuring performance of water tanks. Water has the highest specific heat of all common materials, namely 1 Btu/lbm°F (4.18 KJ/Kgm°K), which makes it a very suitable medium for sensible heat thermal storage. Thermal storage in water has the following advantages:

(i) Interfacing the thermal storage with the HVAC equipment is easy.

(ii) With proper stratification, uniform discharge temperature is possible.

(iii) Water tanks may be located above ground or underground.

(iv) Simple control techniques can be used.

(v) Operating cost is very reasonable.

Water tanks may be made of steel for above-ground installations or concrete for burial in the ground. Geometrical shape for the tanks may be cylindrical or rectangular.

Stratification of the stored water has attracted considerable attention by researchers because it has a significant effect on the thermal performance of water tanks, especially when they are used for storing chilled water. Some methods used to reduce mixing of stored and returned water are as follows:

(i) *Temperature Stratification,* in which water stored in tanks is stratified thermally with the lighter, warmer water on top. However, in chilled water storage where water in most cases swings between 41°F (5°C) and 59°F (15°C), the density difference is too small to use this method.

(ii) *Empty Tank Method* uses two tanks, and the return water is pumped into a second tank. If more than two tanks are used, the equivalent of one tank should be empty all the time. In situations where hot and chilled water are to be stored simultaneously, the equivalent of two tanks should be empty all the time.

(iii) *Labyrinth Method,* developed by Japanese designers, moves water through interconnecting cubicles, as shown in Figure 12-1.

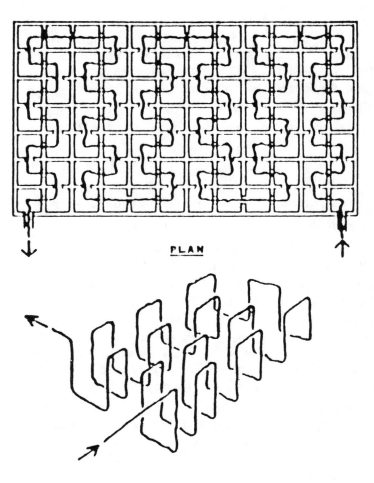

PLAN

**Figure 12-1. Labyrinth Concept**
(Source - HVAC Systems and Applications. ASHRAE Handbook 1987)

(iv) *Flexible Diaphragm* is still another method of temperature stratification, which is available under a license from Engineering Interface of Toronto. In this method, a sheet of coated fabric is attached at the middle of the tank and floats up and down as the volume of stored and return water in the tank varies.

## Storage Capacity of Water Tanks

Storage capacity C of water tanks in Btu's can be calculated using the relation

$$C = M \, Cp \, \Delta T \qquad\qquad \textit{Formula (12-1)}$$

Where  M  =  Mass of stored water (lbm)
    Cp  =  Specific heat of water (1 Btu/lbm° R)
    $\Delta$T  =  Temperature swing (° R )

## Example Problem 12-2

Calculate the volume of the water tank required for thermal storage for Choices No. 2 and No. 3 of Example 12.1.

*Choice No. 2*
Required thermal storage  =  1000 Ton hours
             =  12,000 X 1000 Btu's
             =  12 X $10^6$ Btu's

Assuming $\Delta$T = 20° R and taking Cp = 1 Btu/lbm ° R in Formula (12-1)

$$M = C/c_p\Delta T = \frac{12 \times 10^6 \ (Btu)}{1 \ (Btu/lbm°F) \ 20 \ (°R)} = 0.6 \times 10^6 \ lbm \ of \ water$$

Volume of water stored  =  $\dfrac{Mass}{Density}$

             =  $\dfrac{1.2 \times 10^6 \ (lbm)}{62.5 \ (lbm/ft^3)}$

             =  9,600 ft$^3$

*Choice No. 3*
Required thermal storage  =  3500 Ton hours
             =  42 X $10^6$ Btu's

Assuming $\Delta T = 20°R$ and taking $c_p = 1$ Btu/lbm°R in Formula (12.1)

$$M = C/C_p \Delta T \qquad = \frac{42 \times 10^6 \ (Btu)}{1 \ (Btu/lbm°F) \ 20 \ (°R)}$$

$$= 2.1 \times 10^6 \text{ lbm of water}$$

Volume of water stored $= \dfrac{\text{Mass}}{\text{Density}}$

$$= \frac{2.1 \times 10^6 \text{ lbm}}{62.5 \ (lbm/ft^3)}$$

$$= 33,600 \text{ ft}^3$$

## b) Ice storage

Ice storage is based on the use of latent heat storage compared to the use of sensible heat storage in water tanks. Water also has one of the highest latent heats of fusion, namely 144 Btu/lbm. Ice storeage has the following advantages:

(i) Ice storage has smaller standby losses.

(ii) Ice storage needs smaller volume.

(iii) Ice storage can be easily interfaced with the existing refrigeration units.

Ice storage has lower refrigeration efficiencies compared with water storage because lower evaporator temperatures are needed to make ice than to chill water. In a direct expansion coil making ice, evaporator temperature may go as low as 10°F. This severely impacts the coefficient of performance of the system.

Ice storage could be accomplished using the following techniques:

(i) *Direct Expansion Coils.* In this method, a direct expansion coil is fitted inside a storage tank filled with water. Ice layers about 3 inches thick are formed around the coil. Stirrers could be used to aid uniform buildup of ice. Cool storage is utilized by circulating return water in the tank, which melts the ice and uses its latent heat of fusion for cooling. Commercial units having a capacity as large as 100,000/lbm of ice are available.

(ii) *Brine Coils.* In this method, a direct expansion coil is used to cool a brine solution (for example, a solution of ethylene glycol and water), which in turn is circulated through a coil in the storage tank. Brine has the advantage of lower refrigerant quantity requirements. Thickness of ice formed is generally lower in brine coils compared with the direct expansion coils.

(iii) *Plate Ice Makers.* In this method, the evaporators are vertical plates which are fitted above a water/ice storage tank. Water from the tank is pumped over these plates. When the water has been cooled to 32°F, it starts making thin films of ice on the plates. These layers of ice are discharged to the storage tank by circulating hot gas in the plates.

**Capacity of Ice Storage**

For operation at 32°F, the capacity of ice storage is equal to 144 Btu/lbm. For operation at higher temperatures, say 45°F, the storage capacity is increased by the sensible heat between 45°F and 32°F; i.e., capacity is $(144 + 1.0 \times [45-32]) = 157$ Btu/lbm. If a fraction of the tank is ice, a correction must be applied accordingly.

**c) Phase Change Materials**

Water is not the only medium which is used for latent heat storage. Several other materials, known as phase change materials (PCM), have been used where heat of fusion between solid and liquid phases has been used for thermal storage with the same advantages which were mentioned for ice storage. PCM are more appropriate compared with ice when the storage temperature desired is other than 32°F. For example, solar energy storage systems operate in the range of 80°F -124°F, and storage systems for off-peak electricity use can operate at temperatures well above 1000°F. Salt hydrates, organic materials and clathrates are the most frequently used PCM. Table 12.2 lists some important properties of such materials.

Salt hydrates are compounds of salt and water. Some salt hydrates have heats of fusion which are comparable to ice. Generally, these materials are non-toxic, non-flammable and not very expensive.

Table 12-2. Properties of Commercially Available Phase Change Materials (Lane 1985)*

| Material | Type | Melting Point, F° | C° | Heat of Fusion Btu/lb | kJ/kg | Latent Heat Btu/ft | MN/m³ | Cost |
|---|---|---|---|---|---|---|---|---|
| $MgCl_2.6H_2O$ | Quasi Congruent | 243 | 117 | 72.5 | 168.6 | 6500 | 242 | medium |
| $Na_4P_2O_7.10H_2O$ | Incongruent | 158 | 70 | – | – | – | – | medium |
| Sodium Acetate $.3H_2O$ | Incongruent | 136 | 58 | 97.0 | 225.6 | 7700 | 287 | medium |
| $MgCl_2.6H_2O/Mg(NO_3)_2.6H_2O$ | Eutectic | 136 | 58 | 56.9 | 132.2 | 5400 | 201 | medium |
| Paraffin Wax | Congruent | 122 | 50 | 114.3 | 265.9 | 5500 | 205 | low |
| $Na_2S_2O_3.5H_2O$(Hypo) | Semi-congruent | 118 | 48 | 86.4 | 201.0 | 9000 | 335 | medium |
| Neopentyl Glycol | Congruent[a] | 110 | 43 | 56.3 | 131.0 | – | – | very high |
| $CaBr_2.6H_2O$ | Congruent | 93 | 34 | 49.7 | 115.6 | 6000 | 224 | high |
| $Na_2SO_4.10H_2O$ | Incongruent | 89 | 32 | 108.0 | 251.2 | 9000 | 335 | low |
| $Na_2SO_4.10H_2O$ (with admixtures) | Incongruent | 89 | 32 | 89.0 | 207.0 | 7300 | 272 | low |
| $CaCl_2.6H_2O$ | Semi-congruent | 82 | 28 | 74.0 | 172.1 | 7000 | 261 | low |
| $CaCl_2.6H_2O$ (modified) | Congruent | 81 | 27 | 82.0 | 190.7 | 7800 | 291 | low |
| PE Olycol | Congruent | 74 | 23 | 64.7 | 150.5 | 4500 | 177 | high |
| Glauber's salt with various percentages of $NaCl. NH_4Cl$ and KCl | Incongruent | 70 to 38 | 21 to 3 | 70 to 41 | 163 to 95 | 7000 to 4000 | 260 to 150 | low |
| $CaBr_2.6H_2O/CaCl_2.6H_2O$ | Congruent[b] | 57 | 14 | 61.0 | 141.0 | 6800 | 253 | high |
| Water/Ice | Congruent | 32 | 0 | 144.0 | 355.0 | 9000 | 335 | very high |

[a]Phase change is solid/solid    [b]Isomorphous

*Source — ASHRAE HVAC Systems and Applications Handbook, 1987

Salt hydrates have one major disadvantage, namely they melt incongruently or semi-congruantly. They melt to a saturated aqueous phase and a solid phase, which is the anhydrous salt or a lower hydrate of the same salt.

Mixtures of two or more materials, mixed in a controlled ratio so that a minimum melting point is obtained, are called Eutectics. These materials melt completely at the melting point, and they have the same composition in the liquid and solid phases.

Organic materials, such as paraffin, have characteristics like low thermal conductivity, low density and contraction of volume on freezing. Lower densities result in larger storage volumes. However, they have few problems of thermal stratification, incongruent melting or subcooling.

Clathrates materials are those compounds in which one chemical entity is bonded inside the structure of another. For example, some commonly used refrigerants (R-11, R-12, R-22) can use water molecules for bonding to form clathrates. Some clathrates melt in the range of 40°F to 56°F.

ASHRAE Standard 94.1-1985 may be referenced for the procedures for testing the performance of PCM storage systems.

### d) Rock beds

Thermal capacity of rocks or pebbles can be used to store energy. These materials are packed in a container which contains some wire-screen to support the bed of the material. The container is provided with air plenums for inlet and outlet of the fluid, which is generally air. Air flows through the bed in one direction during the transfer of heat to the bed and in the opposite direction during transfer of heat from the bed. Major use of rock beds is for the storage of solar energy.

### e) Ground-coupled storage

Earth was the first natural thermal storage. Earth can be coupled to an HVAC system to function as a heat source or sink. There are two types of earth-coupled systems, namely direct heating/cooling systems and heat pump systems. In direct heating/cooling system coupling to the earth, energy is stored in a localized volume of earth and transferred back when needed for heating or cooling. In earth-coupled heat pump systems, heat is transferred to the earth in summer and transferred from the earth in winter for heating.

Coupling is accomplished by burying a large, closed-loop heat exchanger, either in a vertical or in a horizontal format. The medium used to transfer energy to or from the earth is a solution of ethylene glycol or calcium chloride in water.

## R EFERENCES

1 _____ *HVAC Systems and Applications Handbook*. ASHRAE. Atlanta, GA, 1987.

2 Smith, C.B. *Energy Management Principles*. Pergamon Press, 1981.

3 Tran, N., et al. "Field Measurement of Chilled Water Storage Thermal Performance." *ASHRAE Trans.* 95(1),1989

4 Sohn, C.W.; and Tomlinson, J.J. "Diurnal Ice Storage Cooling Systems in Army Facilities." *ASHRAE Trans.* 95(1), 1989

5 Wildin, M.W.; and Truman, C.R. "Performance of Stratified Vertical Cylindrical Thermal Storage Tanks, Part 1: Scale Model Tank."*ASHRAE Trans.* 95(1), 1989

6 Lane, G.A. "Congruent-Melting Phase-Change Heat Storage Materials." *ASHRAE Trans.* 88(2), 1982.

7 Lane, G.A. "PCM Science and Technology: The Essential Connection." *ASHRAE Trans.* 91(2), 1985.

8 Lorsch, Harold G. "Thermal Energy Storage for Solar Heating." *ASHRAE Journal,* Nov. 1975.

# 13

# Passive Solar
# Energy Systems*

## INTRODUCTION

Passive solar energy systems use natural, nonmechanical forces—including sunshine, shading, and breezes—to help heat and cool homes with little or no use of electrical controls, pumps, or fans.

It should not be assumed that passive solar building heating measures are less effective or that their application is less widespread than the active measures. In fact there are few, if any, buildings that would not benefit in comfort and energy cost savings by the use of one or more passive measures.

Passive heating systems often depend on "thermal storage mass" to absorb and store the sun's energy for use within a building. Storage mass includes materials such as rock, stone, concrete or water, which store large quantities of heat as they absorb the sun's energy. This storage is accompanied by a moderate rise in temperature—the larger the temperature rise, the greater the energy stored. Later, as the thermal storage mass cools, the stored heat is released to heat the living space. "Phase change" materials may also be used for storage.

New passive buildings are generally integrated into their sites. Their design takes advantage of the natural environment to heat and cool the building by natural means. For example, evergreen trees are planted on the north side for winter protection, and deciduous trees on the south side to heighten winter sun access. Hillsides are used as earth berms that insulate the building. Indigenous materials, such as

---

*Source: Residential Conservation Service Auditor Training Manual.

adobe or rocks are used for thermal mass and insulation. Southerly orientation and glazing contribute direct solar gain for heating. Thermal mass on or near south glazing contributes to heating as well. Shading devices and overhangs are used to keep out the high summer sun when cooling is needed, and the thermal mass absorbs heat, effectively cooling the structure. Many of these passive features can be retrofitted to appropriately oriented existing structures.

Passive systems may require some attention from building occupants to perform at their peak. For example, someone may have to slide insulation into place to close off windows at night or during periods of extended cloud cover to reduce heat losses. Also, vents may have to be opened to control heat flow.

Well-designed passive buildings incorporate energy-saving features beyond standard improvements (insulation, weatherstripping and caulking). Depending on the specific location and design of the building, a variety of methods are suitable for additional energy savings in retrofit situations. For example, vestibules, now called airlock entries, are reappearing as a popular way to keep incoming blasts of cold or hot outside air to a minimum. Effective placement of vegetation such as trees, shrubs and vines is a natural, simple way to protect a building from temperature extremes.

This chapter discusses five passive solar systems.

- Thermosyphon domestic hot water systems
- Direct gain glazing systems
- Indirect gain systems
- Sunspaces
- Window heat gain retardant devices

## THERMOSYPHON DOMESTIC
## HOT WATER SYSTEMS

Thermosyphon systems are generally considered to be passive measures most suitable for warm climates.

## DIRECT GAIN GLAZING SYSTEMS

In direct gain systems, the building itself is a solar collector. Direct gain in its simplest form is probably the easiest way to apply solar energy to any building. It simply involves letting the sun shine in through added south-facing windows, thereby heating space directly. Nearly all of the sunlight entering a room is immediately converted to heat.

Thermal mass for storing excess heat (such as a concrete floor) may be located with direct exposure to the sunlight or in some other part of the building. All materials can store heat, but some store more than others. Table 13-1 lists the heat capacities (ability to store heat) of some common materials. Of the materials listed, air holds the least amount of heat and water the most.

**Table 13-1. Heat Capacities of Common Materials**

| Material | Heat Capacity $(Btu/ft^3 - °F)$ |
|---|---|
| Air (75°F) | 0.018 |
| Clay | 13.9 |
| Sand | 18.1 |
| Gypsum | 20.3 |
| Limestone | 22.4 |
| Wood, oak | 26.8 |
| Glass | 27.7 |
| Brick | 28 |
| Concrete | 28 |
| Asphalt | 29 |
| Aluminum | 36.6 |
| Marble | 38 |
| Copper | 51.2 |
| Iron | 55 |
| Water | 62.5 |

To reduce heat loss and thus increase overall thermal performance, insulation may be placed next to the glass at night, either inside or outside. During the heating season, south-facing glass takes advantage of the sun's low position in the sky; in the summer when the sun is high in the sky, the glass may be shaded by overhangs or foliage. Both east and west glass, even in cold climates, can admit somewhat more solar energy than they lose if nighttime insulation is used. North glazing should be kept at a minimum.

Vertical glass admits almost as much heat during the winter as tilted glass (skylights) and is much easier to insulate and keep clean;

also, vertical glass does not break as easily. Many codes require tilted glass to be tempered, which is more expensive than regular window glass.

South-facing windows that are designed to distribute heat to as much of the building as possible are preferred. Clerestory windows with adequate overhangs should be considered before skylights.

If possible, south-facing windows that are exposed to the sun should be located so that the sunlight falls directly on thermal mass (heavy masonry fireplaces, masonry floors, water walls, etc.). Structures built with concrete floors (e.g., slab on grade) are frequently carpeted, eliminating the exposure of thermal mass. Replacing the carpeting with heavy ceramic tiles could result in greatly increased solar storage, reduced building heat load and lower heating bills.

Thermal mass provides nighttime heating by reradiating stored heat to living space. Thermal mass also tempers temperature swings that can occur. Some people will not want nighttime heating if it means altering their space to provide storage; others will not mind the changes necessary to provide some nighttime heating, to prevent overheating and to moderate temperature variations.

### Applications

Addition of south-facing glazing should be considered as a potential passive solar retrofit wherever south-facing walls are available and an outside view is desired. Direct gain systems help to heat buildings.

### Advantages and Disadvantages

Glass is relatively inexpensive, widely available and thoroughly tested. The overall direct gain system can be one of the least expensive means of solar heating, the simplest solar energy system to conceptualize and the easiest to build. In many instances, it can be achieved by simply enlarging existing windows. Direct gain systems, besides heating the interior, provide natural lighting and a view. To meet a small fraction of the heating needs of a building, direct gain systems do not necessarily need thermal storage.

On the other hand, ultraviolet radiation in the sunlight can degrade fabrics and photographs. If the desire is to achieve large energy savings, then relatively large glazing areas and correspondingly large amounts of thermal mass are required to decrease temperature swings. Thermal mass can be expensive, unless it serves a structural purpose. Interior daily temperature swings of 15 to 20 degrees are common even with thermal mass.

## Construction Terms

*Clerestory*—Vertical window placed high in wall near eaves; used for light, heat gain, and ventilation.

*Double Glazed*—A frame with two panes of transparent glazing with space between the panes.

*Fenestration*—The arrangement, proportioning, and design of windows or doors in a building.

*Glazing*—Transparent or translucent material, generally glass or plastic, used to cover a window opening in a building.

*Header*—A horizontal structural member over an opening used to support the load above the opening.

*Lite*—A single pane of glazing.

*Movable Insulation*—Insulation (such as shutters, panels, curtains, or reflective foil draperies) that can be moved manually or by mechanical means.

*Shading Coefficient*—The ratio of the solar heat gain through a specific glazing system to the total solar heat gain through a single layer of clear double-strength glass.

*Tempered Glazing*—Glazing that has been specially treated to resist breakage.

*Thermal Mass*—Any material used to store the sun's heat or the night's coolness. Water, concrete, and rock are common choices for thermal mass. In winter, thermal mass stores solar energy collected during the day and releases it during sunless periods (nights or cloudy days). In summer, thermal mass absorbs excess daytime heat, and ventilation allows it to be discharged to the outdoors at night.

*Thermal Storage Floors, Ceilings, and Interior Walls*—Floors, ceilings, and interior walls that contain thermal mass and are used to collect and/or store heat in solar energy systems. They can be exposed to sun directly or receive only indirect solar heat to be effective.

## INDIRECT GAIN GLAZING SYSTEMS

Any passive heating system that uses some intermediary material for heat collection and/or storage before passing that heat on to its desired place of use can be called an indirect passive system. The term "indirect gain systems" means the use of panels of insulated glass, fiberglass or other transparent substances that direct the sun's rays onto specially constructed thermal walls, ceilings, rockbeds, or containers of water or other fluids where heat is stored and radiated.

### Trombe Wall

The Trombe wall uses a heat storage mass placed between glass and the space to be heated. Per unit of thermal storage mass used, the Trombe wall makes the best use of the material. While the temperature swing in the material is great, the temperature variation in the heated space is small. See Figure 13-1 for some features of a Trombe wall.

### Water Wall

The water wall uses the same principle as the Trombe wall and involves replacing the existing wall, or parts of it, with containers that hold water. The water mass then stores heat during the day and releases the heat as needed.

### Thermosyphon Air Panels (TAP)

TAP systems are often called "day heaters" because of their effective use during the day, since they have no storage. They are similar in appearance to active flat-plate collectors and are often mounted vertically. A TAP system has one or more glazings of glass or plastic, an air space, an absorber, another air space and (often) an insulated backing. Air flows naturally up in front of, behind or through the absorber and re-enters the building through a vent at the top. Figure 13-2 displays structural features of TAP systems.

### Applications

Trombe walls, water walls and thermosyphon air panel systems all provide heat. Each system requires south-facing opaque walls with solar access.

**Figure 13-1.** **Trombe Wall: Heating and Cooling**

**Figure 13-2.** **Airflow Through a TAP System**

**Advantages and Disadvantages**

*Trombe and water walls* have several advantages:

- Glare and ultraviolet degradation of fabrics are not problems.
- Temperature swings in the living space are lower than with direct gain or convective loop systems.
- The time delay between the absorption of solar radiant energy by the surface and the delivery of the resulting heat to the space provides warmth in the evening.

Trombe and water walls have the following disadvantages:

- Two south walls (a glazed wall and a mass wall) are needed.
- Massive walls are not often found in residential construction (although thermal storage walls may be the least expensive way to achieve the required thermal storage since they are compactly located behind the glass).
- In cold climates, considerable heat is lost to the outside from the warm wall through the glazing unless the glazing is insulated at night.

*Thermosyphon Air Panels (TAP)* systems have the following advantages:

- Glare and ultraviolet degradation of fabrics are not problems.
- TAPs provide one of the least expensive ways to collect solar heat.
- To provide a small fraction of the heating needs of a building, thermal storage is not needed.
- TAPs are easily incorporated onto south facades.
- TAPs are readily adaptable to existing buildings and require less skill for resident installation.
- Because the collector can be thermally isolated from the building interior, night heat losses are lower than for any other uninsulated passive system.

TAP systems have the following disadvantages:

- The collector is an obvious add-on device.
- Both careful engineering and construction are required to

ensure proper airflows, air seals and adequate thermal iso-
lation at night.

- The thermal energy is delivered as warmed air—it is difficult
  to store this heat for later retrieval because air transfers heat
  poorly to other mass.
- It's often impractical to add much collector area.
- Occupant use patterns, and natural airflows within the structure will strongly affect the usability of energy delivered;
  siting should thus be thoughtful and creative.

**Construction Terms**

*Absorber*—The surface in a collector that absorbs solar radiation
and converts it to heat energy. Generally, matte black surfaces are
good absorbers and emitters of thermal radiation, while white and
metallic or shiny surfaces are not.

*Backdraft Damper*—A damper designed to allow air flow in only
one direction.

*Damper*—A device used to vary the volume of air passing through
an air outlet, inlet or duct.

*Drum Wall*—A type of thermal storage wall in which the thermal mass is large metal drums filled with a storage medium, usually
water.

*Masonry*—Stone, brick rammed earth, adobe, ceramic, hollow
tile, concrete block, gypsum block or other similar building units or
materials, or a combination thereof, bonded together with mortar to
form a wall, pier, floor, roof or similar form.

*Solid Masonry*—Masonry in which there are no voids; for instance, concrete block with filled cores.

*Thermal Lag*—The ability of materials to delay the transmission
of heat; can be used interchangeably with time lag.

*Trombe Wall (or Solar Mass Wall)*—A massive wall that absorbs
collected solar heat and holds it until it is needed to heat the interior.

## SOLARIA/SUNSPACE SYSTEMS

The terms "sunspaces" and "greenhouses" are both commonly
used to refer to this measure; they are used interchangeably in this
discussion. Sunspaces could be considered a form of direct gain sys-

tem. Sunlight is absorbed in the sunspace by thermal storage mass (such as bricks or water in containers), which then radiates heat. Existing doors and windows are frequently used to allow heated greenhouse air to flow into interior living spaces. The sunspace can be sealed off from the rest when too much heat is gathered or when there is insufficient sunlight to contribute to heating. Closing doors and windows between interior space and the sunspace at night or on particularly cloudy days may also be necessary.

Although single glazing for sunspaces will result in a maximum light transmission for plant growth and solar gain, single glass will permit a large amount of heat loss at night. In northern climates, double glazing will retard this heat loss. Additionally, movable insulation can be applied to prevent nighttime heat loss.

Thermal mass can be expensive. Therefore, if it can be reduced or eliminated and wider temperature fluctuations allowed, the cost of a sunspace will be substantially less. However, system efficiency will be reduced.

A greenhouse can be thought of as a buffer zone between the outside environment and the inside of the building. In this buffer zone, direct solar gain causes wide temperature fluctuations and a higher temperature indoors than outdoors at virtually all times of the day. These higher indoor temperatures buffer the adjacent living area, reduce building heat loss and can be used immediately to help heat living area.

The sunspace is a versatile passive solar measure. It can be added to many different architectural designs with pleasing results. Sunspaces are equally compatible with expensive and inexpensive homes. Figure 13-3 illustrates attached retrofit sunspaces. Figure 13-4 illustrates a retrofit greenhouse for a mobile home.

The four basic methods for transferring thermal energy from the greenhouse into interior living space are

- Direct solar transmission
- Direct air exchange
- Conduction through common walls
- Storage in and transfer from gravel beds or other thermal mass

Attached
Pit Greenhouse

**Figure 13-3. Attached Retrofit Greenhouses**

Figure 13-5 illustrates an example of thermal mass in the form of water drums. These storage methods are often used in combination. For example, in addition to a common heat storage wall to conduct heat from the greenhouse to the building, forced or natural air flow (direct air exchange) can also be used.

### Applications

A sunspace can provide heat and additional living space, as well as the opportunity to grow vegetables during most of the year, in nearly all climates. Building a sunspace is a good do-it-yourself project, especially for small groups of neighbors and friends.

**Figure 13-4. Attached Greenhouse on a Mobile Home**

**Figure 13-5. Example of Storage in an Attached Greenhouse**

### Advantages and Disadvantages

Greenhouses have several advantages:

- Temperature swings in adjacent living spaces are moderated.
- They provide space for growing food and other plants.
- They reduce heat loss from buildings by acting as a buffer zone.
- They are readily adaptable to most existing residential buildings.
- Since the greenhouse serves more than one function, it can be a natural and integrated part of the building design.

Because sunspaces are a form of direct gain system, their greatest disadvantage is that a poor design can lead to wide indoor temperature fluctuations. To combat this problem, indirect gain elements are often combined with direct gain, e.g., Trombe walls or water walls, to provide the tempering effect of thermal mass. In general, greenhouses are most economical when they have purposes in addition to providing heat and are built to a standard of quality that will enhance the functional and appraised value of the home.

A second consideration is having maximum sunlight penetration in the winter (for heat) and less penetration in the summer (to prevent overheating). To achieve this, a rule of thumb of latitude ±15° is applicable for optimal greenhouse glazing tilt, with appropriate shading.

Aesthetics can also be a concern with sunspaces. The architectural lines of a house need to be aesthetically matched. Sometimes this leads to an extreme tilt and sometimes no tilt. In each case, interior design (addition or subtraction of thermal mass, outside venting, movable insulation, etc.) has to compensate for poor design from a passive solar energy point of view.

A fourth concern is movable insulation, which is necessary in most cases for retention of winter heat gain.

## WINDOW HEAT GAIN RETARDANTS

Shading is the most effective method of reducing heat gain through transparent materials, and, ideally, a shading device should allow sunlight in during the winter. The term "window heat gain retardants" means mechanisms that significantly reduce summer heat gain through south-facing windows (± 45° of true south) by use of devices such as awnings, insulated rollup shades, metal or plastic solar screens, or movable rigid insulation. Included as window heat gain retardants (WHGR) are any devices that provide shading. Other examples of these devices are sunscreens, exterior roll blinds, exterior shutters, venetian blinds, film shades, opaque rolls and insulating shutters and blinds.

Internal shading can reduce the amount of heat dispersed within a space. The most common internal shading devices are venetian blinds, vertical blinds, shades, draperies, reflective films and shutters. These devices can reject up to 65% of the solar radiation that strikes the glass directly.

External shading is more effective than internal shading; it can keep up to 95% of the solar radiation from entering the building (100% if opaque shutters are used). Many devices are available for exterior shading. Horizontal overhangs, with fixed and movable elements, are very effective on south windows, because the sun is highest as it approaches due south during midday. The sun also is higher in summer than in winter, and the overhang can be positioned to screen the sun in summer but admit it in winter. On east or west elevations, however, the sun's angle is too low to be blocked out by horizontal overhangs, and properly oriented vertical louvers have proved more beneficial. If the louvers are movable, the user can control them to provide a better view or greater admission of light when the sun is located on the opposite face of the building. Trees, vines, and adjacent buildings may also provide shade, depending on their proximity, height and orientation. Many summer shading devices can also provide window *insulation* in the winter. Care should be taken in their application so that beneficial winter solar heat gain is not lost. As an example, heat reflecting material may work best year-round if it is mounted on a retractable shade.

### Applications

Window heat gain retardants help keep buildings cool in summer. They should be considered wherever summer solar heat gain through windows causes overheating. More particularly, substantial dollar (and resource) savings are possible when retardants are used to diminish expenses for air conditioning.

## Advantages and Disadvantages

Any heat gain retardant device has advantages and disadvantages and to list all of them for the wide range of shading devices would be impractical. Instead, here are a few key ones for each device.

For sunscreens, visibility can remain as high as 86% while providing daytime privacy. Sunscreens block solar radiation reflected from nearby objects more effectively than other measures. However, drapes or blinds must be used for privacy at night.

With an exterior roll blind, energy-conserving effects can be achieved year-round. In summer, shading is provided by day and ventilation by night (with the blind raised). In winter, insulation is provided by night with the blind lowered, and solar gain by day with the blind raised. However, the blind cannot be tilted to provide a view.

For exterior shutters and blinds, protection of windows is provided from storm damage, vandalism, or intrusion. However, operation and maintenance of exterior devices can be time-consuming.

For awnings, ease of installation and attractiveness are advantages. However, they are subject to wind damage and they must be maintained. New fabric may be necessary every four to eight years if fabric awnings are used instead of more permanent wood or metal awnings.

## Construction Terms

*Awning*—A shading device, usually movable, used over the exterior of a window.

*Heat Reflective and Heat Absorbing Window or Door Materials* —Glazing, films or coating applied to existing windows or doors; they have exceptional heat absorbing or heat reflecting properties.

*Overhang*—A horizontal or vertical projection over or beside a window used to selectively shade the window or door on a yearly basis.

*Shading Device*—A covering that blocks the passage of solar radiation; common shading devices consist of awnings, overhangs, or trees.

*Shutter*—Movable cover or screen for a window or door.

# 14

# Energy Management

Energy management is the judicious and effective use of energy to maximize profits (minimize costs) and to enhance competitive positions. A successful energy management program is more than conservation. It is a total program that involves every area of a business. A comprehensive energy management program is not purely technical. It takes into account planning and communication as well as salesmanship and marketing. It affects the bottom line profits of every business; thus the individual who is assigned the role of "energy manager" has high visibility within the organization.

Energy management includes energy productivity, which is defined as reducing the amount of energy needed to produce one unit of output. Energy management includes energy awareness, which is essential in motivating employees to save energy. Probably the highest initial rate of return will occur through the establishment of a good maintenance management program.

This chapter reviews the basics of maintenance management and energy management organization.

## MAINTENANCE MANAGEMENT

There are obvious losses from poor maintenance such as steam and air leaks and uninsulated steam lines. There are also losses from less obvious areas. For example Tables 14-1 and 14-2 illustrate the hidden effect of dirty evaporators and condensers on equipment performance. These losses go undetected and result in decreased capacity of equipment and an increase in energy usage. In Table 14-1, for a reciprocating compressor under the dirty condenser and evaporator conditions, capacity is reduced 25.4% and the increase in brake horsepower per ton is 39%. In Table 14-2, for an absorption chiller under similar conditions, capacity is reduced 23.8% and power requirements are increased 7.5%.

**Table 14-1.** The Effects of Poor Maintenance on the Efficiency of a
Reciprocating Compressor, Nominal 15-Ton Capacity

| Conditions | (1) °F | (2) °F | (3) Tons | (4) % | (5) HP | (6) HP/T | (7) % |
|---|---|---|---|---|---|---|---|
| Normal | 45 | 105 | 17.0 | — | 15.9 | 0.93 | — |
| Dirty Condenser | 45 | 115 | 15.6 | 8.2 | 17.5 | 1.12 | 20 |
| Dirty Evaporator | 35 | 105 | 13.8 | 18.9 | 15.3 | 1.10 | 18 |
| Dirty Condenser and Evaporator | 35 | 115 | 12.7 | 25.4 | 16.4 | 1.29 | 39 |

(1) Suction Temp, °F
(2) Condensing Temp, °F
(3) Tons of refrigerant
(4) Reduction in capacity %
(5) Brake horsepower
(6) Brake horsepower per Ton
(7) Percent increase in compressor bh per/ton

**Table 14-2.** The Effects of Poor Maintenance on the Efficiency of
an Absorption Chiller, 520-Ton Capacity

| Conditions | Chilled Water °F | Tower Water °F | Tons | Reduction in Capacity % | Steam lb/ton/H | Percent |
|---|---|---|---|---|---|---|
| Normal | 44 | 85 | 520 | — | 18.7 | — |
| Dirty Condenser | 44 | 90 | 457 | 12 | 19.3 | 3 |
| Dirty Evaporator | 40 | 85 | 468 | 10 | 19.2 | 2.5 |
| Dirty Condenser and Evaporator | 40 | 90 | 396 | 23.8 | 20.1 | 7.5 |

A third major loss is in missing the opportunity to upgrade the facilities within the spare parts program. For example, when a motor burns out it is usually replaced with the same model or sent to a shop for rewinding. If the motor is sent to a rewind shop, it will usually have poorer efficiency and power factor characteristics than before. Thus the energy manager should consider upgrading the replacement with either a high-efficiency motor or requiring a higher specification from the shop rewinding the motor.

The following summarizes key elements of the maintenance management program.

## Preventive Maintenance Survey

This survey is made to establish a list of all equipment on the property that requires periodic maintenance and the maintenance that is required. The survey should list all items of equipment according to physical location. The survey sheet should list the following columns:

1. Item
2. Location of item
3. Frequency of maintenance
4. Estimated time required for maintenance
5. Time of day maintenance should be done
6. Brief description of maintenance to be done

## Preventive Maintenance Schedule

The preventive maintenance schedule is prepared from the information gathered during the survey. Items are to be arranged on schedule sheets according to physical location. The schedule sheet should list the following columns:

1. Item
2. Location of item
3. Time of day maintenance should be done
4. Weekly schedule with double columns for each day of the week (one column for "scheduled" and one for "completed")
5. Brief description of maintenance to be done
6. Maintenance mechanic assigned to do the work

*Use of Preventive Maintenance (PM) Schedule*—At some time before the beginning of the week, the supervisor of maintenance will take a copy of the schedule. The copy the supervisor prepares should be available in a three-ring notebook. He will go over the assignments in person with each mechanic.

After completing the work, the mechanic will note this on the schedule by placing a check under the "completed" column for that day and the index card system for cross-checking the PM program.

The supervisor of maintenance or the mechanic will check the schedule daily to determine that all work is being completed according to the plan. At the end of the week, the schedule will be removed from the book and checked to be sure that all work was completed. It will then be filed.

### Preventive Maintenance Training

The supervisor of maintenance or a mechanic is responsible for assisting department heads in the training of employees in handling, daily care and the use of equipment. When equipment is mishandled, he must take an active part in correcting this through training.

### Spare Parts

All too often equipment is replaced with the exact model as presently installed. Excellent energy conservation opportunities exist in upgrading a plant by installing more efficient replacement parts. Consideration should be given to the following:

- Efficient line motor to replace standard motors
- Efficient model burners to replace obsolete burners
- Upgrading lighting systems

### Leaks—Steam, Water and Air

The importance of leakage cannot be understated. If a plant has many leaks, this may be indicative of a low standard of operation involving the loss not only of steam but also of water, condensate, compressed air, etc.

If, for example, a valve spindle is worn, or badly packed, giving a clearance of 0.010 inch between the spindle, for a spindle of ¾-inch diameter the area of leakage will be equal to a 3/32-inch diameter hole. Table 14-3 illustrates fluid loss through small holes.

Although the plant may not be in full production for every hour of the entire year (i.e., 8760 hours), the boiler plant water systems and compressed air could be operable. Losses through leakage are usually, therefore, of a continuous nature.

### Thermal Insulation

Whatever the pipework system, there is one fundamental: It should be adequately insulated. Table 14-4 gives a guide to the de-

## Table 14-3. Fluid Loss Through Small Holes

| Diameter of Hole | Steam—lb/hour | | Water—gals/hour | | Air SCFM 80 psig |
| --- | --- | --- | --- | --- | --- |
| | 100 psig | 300 psig | 20 psig | 100 psig | |
| 1/16" | 14 | 33 | 20 | 45 | 4 |
| 1/8" | 56 | 132 | 80 | 180 | 26 |
| 3/16" | 126 | 297 | 180 | 405 | 36 |
| 1/4" | 224 | 528 | 320 | 720 | 64 |

## Table 14-4. Pipe Heat Losses

| Pipe Dia Inches | Surface Temp °F | Insulation Thickness Inches | Heat Loss (Btu/Ft/Hr) | | Insulation Efficiency |
| --- | --- | --- | --- | --- | --- |
| | | | Uninsulated | Insulated | |
| 4 | 200 | 1½ | 300 | 70 | 76.7 |
| | 300 | 2 | 800 | 120 | 85.0 |
| | 400 | 2½ | 1500 | 150 | 90.0 |
| 6 | 200 | 1½ | 425 | 95 | 78.7 |
| | 300 | 2 | 1300 | 180 | 85.8 |
| | 400 | 2½ | 2000 | 195 | 90.25 |
| 8 | 200 | 1½ | 550 | 115 | 79.1 |
| | 300 | 2 | 1500 | 200 | 86.7 |
| | 400 | 2½ | 2750 | 250 | 91.0 |

gree of insulation required. Obviously, there are a number of types of insulating materials with different properties and at different costs, each one of which will give a variance return on capital. Table 14-4 is based on magnesia insulation, but most manufacturers have cataloged data indicating various benefits and savings that can be achieved with their particular product.

## Steam Traps

The method of removing condensate is through steam trapping equipment. Most plants will have effective trapping systems. Others may have problems with both the type of traps and the effectiveness of the system.

The problems can vary from the wrong type of trap being installed to air locking or steam locking. A well-maintained trap system can be a great steam saver. A bad system can be a notorious steam waster, particularly where traps have to be bypassed or are leaking.

Therefore, the key to efficient trapping of most systems is good installation and maintenance. To facilitate the condensate removal, the pipes should slope in the direction of steam flow. This has two obvious advantages in relationship to the removal of condensate: One is the action of gravity, and the other the pushing action of the steam flow. Under these circumstances the strategic siting of the traps and drainage points is greatly simplified.

One common fault that often occurs at the outset is installing the wrong size traps. Traps are very often ordered by the size of the pipe connection. Unfortunately, the pipe connection size has nothing whatsoever to do with the capacity of the trap. The discharge capacity of the trap depends upon the area of the valve, the pressure drop across it and the temperature of the condensate.

It is therefore worth recapping exactly what a steam trap is. It is a device that distinguishes between steam and water and automatically opens a valve to allow the water to pass through but not the steam. There are numerous types of traps with various characteristics. Even within the same category of traps, e.g., ball floats or thermo-expansion traps, there are numerous designs, and the following guide is given for selection purposes:

1.  Where a small amount of condensate is to be removed, an expansion or thermostatic trap is preferred.
2.  Where intermittent discharge is acceptable and air is not a large problem, inverted bucket traps will adequately suffice.
3.  Where condensate must be continuously removed at steam temperatures, float traps must be used.
4.  When large amounts of condensate have to be removed, relay traps must be used. However, this type of steam trap is unlikely to be required for use in the food industry.

To insure that a steam trap is not stuck open, a weekly inspection should be made and corrective action taken. Steam trap testing can utilize several methods to insure proper operation:

- Install heat sensing tape on trap discharge. The color indicates proper operation.
- Place a screwdriver to the ear lobe with the other end on the trap. If the trap is a bucket-type, listen for the click of the trap operating.
- Use acoustical or infrared instruments to check operation.

## ENERGY MANAGEMENT ORGANIZATION

A common problem facing energy managers is that they have too much responsibility and very little authority to get the job done. A second problem is the lack of definition of the job. Probably the biggest problem is the lack of true commitment from top management. As pointed out in the text, an overall energy utilization program requires an "investment" in order to get the return desired.

The first phase of the program should start with top management establishing the organization, defining the goals and providing the resources for doing an effective job. Tables 14-5 and 14-6 illustrate a checklist for top management and typical energy conservation goals.

**Table 14-5. Checklist for Top Management**

| | |
|---|---|
| A. | Inform line supervisors of: |
| | 1. The economic reasons for the need to conserve energy |
| | 2. Their responsibility for implementing energy saving actions in the areas of their accountability |
| B. | Establish a team having the responsibility for formulating and conducting an energy conservation program and consisting of: |
| | 1. Representatives from each department in the plant |
| | 2. A coordinator appointed by and reporting to management |
| | NOTE: In smaller organizations, the manager and his staff may conduct energy conservation activities as part of their management duties. |
| C. | Provide the team with guidelines as to what is expected of them: |
| | 1. Plan and participate in energy saving surveys |
| | 2. Develop uniform record keeping, reporting, and energy accounting |
| | 3. Research and develop ideas on ways to save energy |
| | 4. Communicate these ideas and suggestions |
| | 5. Suggest tough, but achievable, goals for energy saving |
| | 6. Develop ideas and plans for enlisting employee support and participation |
| | 7. Plan and conduct a continuing program of activities to stimulate interest in energy conservation efforts |
| D. | Set goals in energy saving: |
| | 1. A preliminary goal at the start of the program |
| | 2. Later, a revised goal based on savings potential estimates from results of surveys |
| E. | Employ external assistance in surveying the plant and making recommendations, if necessary |
| F. | Communicate periodically to employees regarding management's emphasis on energy conservation action and report on progress |

Adapted from *NBS Handbook*, 115.

**Table 14-6. Typical Energy Conservation Goals**

---

1. Overall energy reduction goals
   - (a) Reduce yearly electrical bills by _____%.
   - (b) Reduce steam usage by _____ %.
   - (c) Reduce natural gas usage by _____%.
   - (d) Reduce fuel oil usage by _____ %.
   - (e) Reduce compressed air usage by _____%.
2. Return on investment goals for individual projects
   - (a) Minimum rate of return on investment before taxes is _____.
   - (b) Minimum payout period is _____.
   - (c) Minimum ratio of $\dfrac{\text{Btu/year savings}}{\text{capital cost}}$ is _____.
   - (d) Minimum rate of return on investment after taxes is _____.

---

A key element in any energy management program is "flexibility." For example, in one year natural gas prices may accelerate steeply. Unless an organization reacts quickly, it soon may be out of business. Every energy management program should have a contingency or backup plan. Periodically, the risk associated with each scenario should be reviewed. Risk assessment can help the energy manager determine the resources required to meet various emergency scenarios. A method of evaluating the cost to business for each scenario is illustrated in Formula 14-1.

$$C = P \times C_1 \qquad\qquad Formula\ (14\text{-}1)$$

Where

$C$ = cost as a result of emergency where no contingency plans are in effect.

$P$ = the probability or likelihood the emergency will occur.

$C_1$ = the loss in dollars as a result of an emergency where no plan is in effect.

## Implementing the Energy Management Program

The first phase of the energy management program involves the accumulation of data. Table 14-7 illustrates the minimum information required to evaluate the various energy utilization opportunities. The energy manager must also review government regulations which will affect the program.

**Table 14-7. Information Required to Set Energy Projects on the Same Base**

| Fuel | Cost At Present | Estimated Cost Escalation Per Year | Energy Equivalent |
|---|---|---|---|
| *1. Energy equivalents and costs for plant utilities.* | | | |
| Natural gas | $_____/1000 ft³ | $_____/1000 ft³ | _____Btu/ft³ |
| Fuel oil | $_____/gal | $_____/gal | _____Btu/gal |
| Coal | $_____/ton | $_____/ton | _____Btu/lb |
| Electric power | $_____/Kwh | $_____/Kwh | _____Btu/Kwh |
| Steam | | | |
| _____psig | $_____/1000 lb | $_____/1000 lb | _____Btu/1000 lb |
| _____psig | $_____/1000 lb | $_____/1000 lb | _____Btu/1000 lb |
| _____psig | $_____/1000 lb | $_____/1000 lb | _____Btu/1000 lb |
| Compressed air | $_____/1000 ft³ | $_____/1000 ft³ | _____Btu/1000 ft³ |
| Water | $_____/1000 lb | $_____/1000 lb | _____Btu/1000 lb |
| Boiler make-up water | $_____/1000 lb | $_____/1000 lb | _____Btu/1000 lb |

*2. Life Cycle Costing Equivalents*

After tax computations required _____

Depreciation method _____

Income tax bracket _____

Minimum rate of return _____

Economic life _____

Tax credit _____

Method of life cycle costing _____

(annual cost method, payout period, etc.)

All too often the energy manager is the last to find out information after the fact. For example, the procurement of motors may be the electrical department's responsibility. The energy manager must get his or her input into the evaluation process prior to purchase. One way is to establish a review process where important documents are initialled by the energy manager before they are issued. Key documents include

• Bid summaries for major equipment

- One-line diagrams
- Process flow diagrams
- Heat and material balances
- Plot plan
- Piping and instrument diagrams

It is much easier to add local instrumentation to the piping and instrument diagram prior to construction. Another way of insuring that the energy manager gets timely input is to incorporate these activities into the overall planning document, such as the Critical Path or PERT Schedules. Figure 14-1 illustrates a planning schedule incorporating input from the energy manager. Figure 14-2 illustrates how energy management activities interact with various departments.

Richard L. Aspenson, director of Facilities Engineering and Real Estate at 3M, summarized* the areas a good energy manager must master as follows:

### Technical Expertise

Energy management normally begins with a solid technical background —preferably in mechanical, electrical or plant engineering. Managers will need a good grasp of both design theory and the nuts-and-bolts details of conservation programs. This includes a thorough understanding of the company's processes, products, maintenance procedures and facilities.

### Communication

In the course of a single week, energy managers might find themselves dealing with lawyers, engineers, accountants, financial planners, public relations specialists, government officials, and even journalists and legislators. A good energy manager has to be able to communicate clearly and persuasively with all of these people—in *their* language. Above all, energy managers must be able to sell the benefits of their programs to top management.

### Financial Understanding

To enlist the support of top management, energy managers will have to develop and present their programs as investments, with predictable returns, instead of as unrecoverable costs. They will have to demonstrate what kind of returns—in both energy and cost savings—can be expected

---

*"The Skills of the Energy Manager," *Energy Economics, Policy and Management,* Vol. 2, No. 2, 1982.

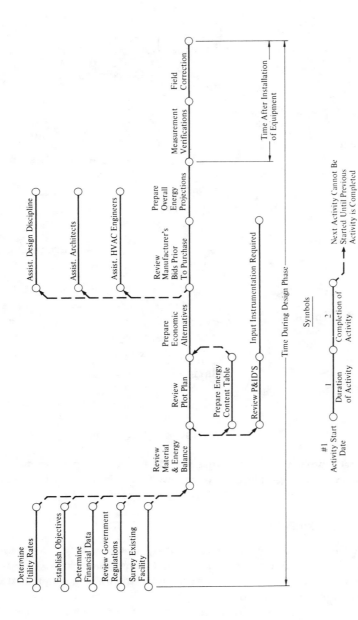

**Figure 14-1. Schedule of Energy Management Activities**

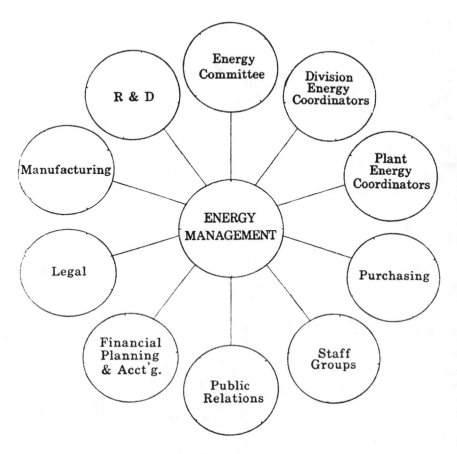

Figure 14-2. Energy Management Interactions

from each project, and over what period of time. This means first of all developing some credible way to measure returns—a method that will be understood and accepted by the financial officers of the company.

**Planning and Strategy**

A basic part of energy management is forecasting future energy supplies and costs with reasonable accuracy. This means coming to grips with the complexities of worldwide supply, market trends, demand projections, and the international political climate. There is no way, of course, to predict all of these things with certainty, but every business—especially energy-intensive ones—will need some kind of reliable forecasting from now on.

**Community Relations and Public Policy**

Energy managers have some responsibility to go outside their companies to share their ideas and experience with a variety of publics. Trade and professional associations can become clearing houses for new ideas. Legislators and government agencies need, and often welcome, expert help in setting standards and policies. And the public needs help in understanding what is at stake in learning to use limited supplies of energy wisely and efficiently.

Energy management can be an exciting, challenging and rewarding job if you put your skills to work.

We as Energy Managers are a special breed of people who have unlimited opportunities if we develop the skills and implement the programs that will improve the profitability of our respective businesses. We can and must influence public policy decision by using our technical knowledge in a constructive and objective way.

**Pointers for the Energy Manager**

1.  Be aggressive. Learn how to say *no* in a diplomatic way.
2.  Energy is not a soft sell. Stick your neck out.
3.  Be sure you anticipate questions before you request approval on energy expenditures. Prepare factual data—think ahead.
4.  Be creative. Identify needs. Prioritize your action plan tasks.
5.  Be a positive person—one with perpetual motion—a catalyst for action.
6.  Establish a free-thinking environment. Be a good listener.
7.  Develop a five-year plan. Update yearly. Include a strategy for implementation—action plans.
8.  Establish credibility through an accurate energy accountability system.
9.  Efficient use of energy is an evolutionary process. Be patient but demanding. Follow through on commitments.
10.  Give individual recognition for achievements.

Energy managers can turn the problems of a changing energy situation into opportunities. At the same time, we can grow to become better corporate citizens of our society.

# 15

# Competitive Power[1]

## COMPETITION VS REGULATION

For decades, electric utilities have been regulated monopolies. State regulators and their utilities operate under a "regulatory compact." The state grants the utility a monopoly franchise; within a certain geographic area, the utility has the exclusive right to sell electricity. In return for that franchise, the utility has an obligation to provide service to any qualified customer that wants service.

Two factors supported the monopoly status. First, the generation, transmission, and distribution of electricity was believed to be a natural monopoly, such that provision of service by two or more entities would increase costs significantly. Second, the supply of electricity was considered a necessity, "clothed with the public interest" (California Public Utilities Commission 1993).

During the past several years, more and more questions have been raised about both utility regulation and its monopoly status. Are utilities still natural monopolies? Is state regulation continuing to serve the broad public interest? Might customers be better served by companies that had to compete with each other? How can small electricity consumers, for whom competition may offer few benefits, be protected? How might a competitive electric-utility industry best meet environmental-protection goals?

Competition is increasingly discussed these days as a way to deal with a variety of problems. Several industries, including telecommunications, banking, airlines, and natural gas have been or are in the process of being deregulated and made more competitive. The pros and cons of meeting society's wants and needs by competition vs regulation vs managed competition are shaped largely by individual values. Those who believe that free markets work best tend to oppose government regulation because such regulations restrict individual freedom and impose inefficiencies on the economy. Those who believe that government programs can work favor government oversight because markets may yield outcomes that are inequitable and inconsistent with society's long-run objectives.

Although competition offers many benefits, it may also create some problems (Table 15-1). On the plus side, competition stimulates the introduction of new products and services and the demise of products and services that no longer meet customer needs. Although nothing inherent in competition favors the short-term over the long-term, competitive markets appear to price outputs to optimize short-term economic efficiency. On the other hand, societal values, such as environmental quality, long-term conservation of nonrenewable resources, research to develop new energy sources, and equity concerns (e.g., protection of low-income households), are largely ignored in a competitive environment.

Regulation, on the other hand, can protect these broader societal interests. However, it does so at the cost of timely and efficient decision making. Consider the complexity, expense, and amount of time required to reach decisions at state regulatory commissions. Also, whether regulatory decisions truly reflect customer interests is often unclear. If certain interests "capture" the regulators, their decisions may not benefit customers broadly.

Because of the limitations in competitive markets, all markets are bound by government or industry rules and procedures. As Reich (1991) noted:

> The idea of a "free market" apart from the laws and political decisions that create it is pure fantasy anyway. The market was not created by God on any of the first six days (at least, not directly), nor is it maintained by divine will. It is a human artifact, the shifting sum of a set of judgments about individual rights and responsibilities....

> ...In modern nations, government is the principal agency by which the society deliberates, defines, and enforces the norms that organize the market. Judges and legislators, as well as government executives and administrators, endlessly alter and adapt the rules of the game— usually tacitly, often unintentionally, always under the watchful eye and sometimes under the guiding hand of interests with clear stakes

**Table 15-1. Characteristics of competitive vs regulated markets**

| Issue | Competition | Regulation |
|---|---|---|
| Time focus | Short-term | Long-term |
| Efficiency | Economic (low cost, high value) | Societal (economic, environment, equity) |
| Products/services | Many | Few |
| Price differentia-tion | Substantial | Limited to a few customer classes |
| Recognition of differences among customers | Yes | No |
| Decision making | Fast | Slow, litigated |

in the outcomes of particular decisions. To the extent that rhetoric frames the issue as one grand choice between government and market, it befogs our view of the series of smaller choices about an endless set of alternative ways to structure the rules of ownership and exchange.

We will return to Reich's point about the role of government in establishing, maintaining, and conditioning markets later in the discussion of retail competition.

## FORMS OF COMPETITION
## IN THE ELECTRIC-UTILITY INDUSTRY

The second background point concerns the forms of competition taking place in the utility industry today (Fig. 15-1). The *Public Utility Regulatory Policies Act* (U.S. Congress 1978) helped to create new nonutility companies that build and operate power plants. The *Energy Policy Act of 1992* (U.S. Congress 1992) furthered that process by creating a class of generation companies exempt from the details of state and federal regulation. Thus, competition at the production end already exists.

If existing utilities are able to design, build, and operate power plants at costs below those of independent power producers (IPPs), then utilities may continue to be major players on the generation side. On the other hand, if the IPPs are more efficient and can build and operate power plants more cheaply and as reliably as today's utilities, then the utility role in generation will decline over time (assuming that utilities do not unfairly control access to transmission). Even here, the utility role is likely to be important in life-extension and repowering of existing power plants.

Transmission and distribution (T&D; the wires) are natural monopolies. However, tough questions remain about what parties should operate and control today's system and build new systems. The *Energy Policy Act* recognized the importance of transmission access to making generation truly competitive. It gave the Federal Energy Regulatory Commission (FERC) the authority to order a transmission-owning utility to provide access to another party wanting to move electricity from a power plant to a wholesale customer. But much is unresolved concerning access to, pricing of, and expansion of transmission services.

To illustrate, FERC can order a utility to construct new transmission facilities to meet the needs of another party. However, state siting and regulatory authorities determine whether and when those facilities are actually built. Resolving this potential state/federal jurisdictional issue may be difficult.

Utilities might continue to dominate transmission markets, perhaps through the creation of regional transmission groups. Alternatively, regional entities (analogous to natural-gas pipelines) might arise whose sole function is to build and operate the T&D systems. Hogan (1993) suggests splitting the transmission monopoly into two parts. The first, which he calls Gridco, would be responsible for the construction, operation, and maintenance of the transmission system. The second, which he calls Poolco, would manage the minute-by-minute dispatch of power plants and use of the transmission grid to transport power from generating units to load centers.

Small modular power plants (diesel generators, small combustion turbines, fuel cells, and photovoltaics) may become more cost-effective. If that happens, generating units can be located close to load centers, reducing the need for transmission of bulk power supplies. Although transmission would remain a monopoly, its importance would diminish in this scenario.

Although transmission is likely to be regulated primarily by FERC, state PUCs will retain regulatory control over distribution. Therefore, PUCs could have a substantial effect on the extent and nature of retail competition. Such competition, probably the most controversial element of the debates on the future of the utility industry, involves competition for retail customers among:

- Alternative fuels (e.g., switching from electricity to natural gas for certain end uses or industrial use of new electrotechnologies);

- Customer cogeneration or self-generation;

- Actions taken by customers, either on their own or with assistance from energy-service companies, to reduce peak demands and electricity use (thereby cutting electricity bills);

- Utilities seeking to encourage large customers to move from one local service area to another (utility to utility competition); and

- Alternative suppliers of electricity (i.e., retail wheeling).

The first four forms of retail competition already exist, although to varying degrees around the country. To the extent that customers believe they can meet their energy-service needs better and less expensively without the local utility, they can switch fuels, cogenerate or self-generate to meet their own electric needs, adopt energy-efficiency measures to reduce their electric bills, or move outside the local utility's service area. Whether customers take such actions depends, in part, on the level of retail electricity prices relative to the costs of these alternatives.

Retail wheeling (under which end users are free to choose their electricity suppliers) engenders the loudest and strongest debates, probably because of its effects on the traditional regulated monopoly franchise. If regulators allow retail customers to choose their electricity supplier, implicitly at least, they are taking away the local utility's franchise. In return, the local utility would (or at least should) no longer have any obligation to serve.

Proponents of retail wheeling believe it is crucial for customers, especially large industrials, to be free to choose their own electricity supplier (Hughes 1993). Absent such freedom, they will be unable to compete effectively in international markets, and local utilities will not be forced to improve efficiency.

Opponents believe that retail wheeling will lower costs only for large customers at the expense of utility shareholders and the remaining residential and commercial (core) customers (Cavanagh 1994). In addition to their worry about stranded investment, retail-wheeling opponents are concerned about the loss of the environmental benefits of DSM and renewables and the elimination of the utility's role as a portfolio manager on behalf of all retail customers.

Figure 15-1 shows several possible outcomes for competition at each of the three levels discussed here: generation, T&D, and retail. Different combinations from each column, in a mix-and-match mode, seem feasible.

To the extent that competition at the wholesale level lowers the costs of electricity production, the avoided costs against which DSM must compete will go down. Therefore, less DSM will be cost-effective. During much of the 1970s and 1980s, long-run marginal costs were above average costs. Thus, consumers faced price signals that, on average, encouraged them to use more electricity than was economically efficient. To the extent that long-run costs are now below average costs (and average rates), price signals encourage consumers to overinvest in energy efficiency. This change in the relationship between long-run marginal and average costs weakens the rationale for utility DSM programs. However, many important nonprice barriers to customer adoption of energy-efficiency measures remain, barriers that utilities can help overcome (Levine et al. 1994).

Under one scenario, at the retail level the franchise monopoly will hold, and today's utilities will continue to serve most of their existing customers, retaining the obligation to serve in return for which they retain their monopoly rights (top right box in Fig. 15-1). Only a few large industrials cogenerate, self-generate, or work out special deals that allow them to leave the system.

At the other end of the retail-competition spectrum, the franchise monopoly no longer exists (bottom right box in Fig. 15-1. All customers are free to choose their own supplier. Several utilities, as well as independent brokers, vie for retail customers. Brokers, who may own only telephones, computers, modems, and faxes, make "deals" in which they aggregate retail customers and purchase power from a combination of suppliers and arrange for T&D delivery of that power to customers (Bretz 1994).

If the first retail scenario prevails, the effects on utility DSM programs are likely to be minimal. On the other hand, if there is considerable competition for end-use customers, utility DSM programs may change dramatically. In particular, utilities will no longer be able to charge all customers for the costs of these programs because prices will be set in a competitive market and not in a regulatory proceeding. Utilities will (1) either charge participating customers for their DSM services or (2) be able to spread the costs of DSM programs across only the remaining core customers. Therefore, utilities will run fewer programs that offer rebates to customers that install energy-efficient equipment; such programs will be aimed only at core customers. Instead, utilities will sell information, contracting services, measures and their installation, and performance guarantees to those customers choosing to buy these DSM services. And the emerging brokers may also offer DSM services to retail customers through energy-service companies.

The future of utility DSM programs is largely independent of the extent and form of competition at either the generation or transmission levels, although its success will depend on the costs of generation and transmission. Utility DSM programs are likely to be affected primarily by competition at the retail level.

## STATE REGULATION OF RETAIL COMPETITION

It seems likely that FERC will play a stronger role than the state PUCs in shaping wholesale competition. But at the retail level, state PUCs are likely to be much more influential than FERC (Table 15-2), consistent with the role of government espoused by Reich above. Commissions may choose to shape the extent, pace, and beneficiaries of competition for retail electricity customers for three reasons.

First, it may be economically inefficient to allow customers to arbitrarily select their electricity suppliers. This situation could occur if the utility's retail prices do not fully reflect the costs of providing services to

**Fig. 15-1. The forms that competition might take in generation, transmission and distribution, and retail services in the electric-utility industry.**

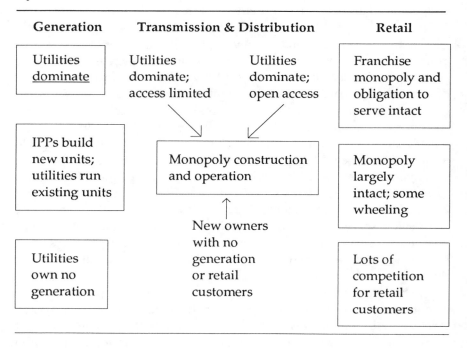

| Generation | Transmission & Distribution | | Retail |
|---|---|---|---|
| Utilities <u>dominate</u> | Utilities dominate; access limited | Utilities dominate; open access | Franchise monopoly and obligation to serve intact |
| IPPs build new units; utilities run existing units | Monopoly construction and operation | | Monopoly largely intact; some wheeling |
| | New owners with no generation or retail customers | | |
| Utilities own no generation | | | Lots of competition for retail customers |

**Table 15-2. Possible PUC rules to structure retail wheeling**

Define core vs noncore customers

Set terms and conditions for noncore customers
- Exit fee
- Reentry fee
- Advance notice to exit or reenter
- T&D wheeling charge
- Prices for other services (e.g., backup, power quality)

some customers and these customers then choose other options (e.g., cogeneration) that reduce their costs but not the overall costs to society.

Second, it may be inequitable to allow such freedom. Customers that leave the utility system may create a "stranded investment" problem for the remaining customers and the utility's shareholders. Although this is primarily a transition problem, the amount of money at stake (as much as 10 to 20% of utility assets) is large enough to warrant careful attention.*

Third, in a fully competitive retail market, utilities may want to drop the societal functions they currently furnish, including provision of demand-side services to customers. For various reasons, including environmental quality, economic productivity, and the provision of capacity and energy resources, commissions may want to be sure that such services are still available to customers.

Discussing the steps that PUCs could take to shape retail competition is important because these actions could have major effects on future utility DSM programs. In addition, a recent survey found that PUCs are not yet ready to deal with increased retail competition, lacking policies that require, encourage, permit, discourage, or prohibit retail wheeling (R.J. Rudden Associates and Fitch Investors Service 1993). According to that survey, only 19% of the responding PUCs had policies in place to deal with recovery of stranded investment, and only 31% had policies to protect core customers from higher prices.

PUCs could decide which customers are eligible to become noncore. In some jurisdictions, PUCs have set minimum size limits for noncore customers of natural-gas utilities. PUC decisions on size might reflect their judgments on which customers are most able to negotiate well with suppliers and which customers can most benefit from that choice. On both scores, large industrial customers may be those most likely to be able to select their own electricity suppliers.

Although the California PUC began by setting minimum size limits for noncore customers of the local gas utilities, these limits soon dropped and were then replaced by unbundled rates. These rates covered separately the commodity, capacity, transportation, and storage costs of delivering gas to customers. On the electric side, Commissions could establish

---

*One way to address this problem would he to allow utilities to write up their T&D assets by the same amount that they write down their generating assets. This procedure would rationalize prices for both generation and T&D and keep utility shareholders and core customers whole.

separate rate schedules for each utility service, including energy, capacity, transmission service, distribution service, reserve margin, power-factor correction, power quality, and so on.

For potential noncore customers, PUCs could establish requirements for exit and/or reentry. To illustrate, the PUC could require exiting customers to reimburse the utility (i.e., remaining customers) for the undepreciated investments made by the utility on behalf of the departing customer. Or the Commission could allow the utility to charge the departing customer the difference between the revenues that customer would likely pay the utility over a set number of future years and the utility's year-to-year avoided costs (Fig. 15-2). If the utility has considerable excess capacity and therefore low avoided costs for several years, the exit fee could be substantial. On the other hand, if the utility faced rapid load growth and needed capacity soon, the exit fee could be much smaller or even zero. In such a case, the departing customer would lower long-term costs for the remaining customers.

Such a requirement would have two powerful effects. First, it would ensure that neither utility shareholders nor remaining customers would have to pay for "stranded" investment. Second, the charge might discourage customers from pursuing uneconomic retail-wheeling options.

Commissions could require noncore customers to provide advance notice if they wish to return to core (full service) status; TransAlta, Canada's largest investor-owned utility proposed just such a five-year advance-notice requirement (Huntingford and Nelson 1993).

Considerable work remains to determine the economic rationale for either exit or reentry fees. In particular, commissions need to establish the relationship between such fees and the utility's current retail tariff and its short- and long-run marginal costs.

Commissions might impose on noncore customers a distribution-system fee, the proceeds of which could be used to pay for the functions that the utility undertakes on behalf of society in general and certain core customers in particular. (FERC could impose a similar fee on transmission services.) The fee, set on a ¢/kWh or $/kW basis, could be justified in different ways. One approach would set the fee equal to the difference between the utility's current retail rate and its marginal supply cost (Moskovitz et al. 1993). Setting the fee in this way would ensure that retail customers do not make uneconomic purchase decisions.

The fees collected in this way, beyond the utility's direct costs for

**Fig. 15-2.** Schematic showing exit fees for large customers that leave the local utility's system in (1) 1995 or (2) in 1998. In each case, the exiting customer pays the utility the net present worth of the difference between avoided cost and retail rate for five years.

building and operating its distribution system, could be used for various social purposes. These purposes include:

- lifeline rates for low-income customers,

- low-income weatherization,

- acquisition of renewable resources that are not yet cost-effective,

- acquisition of DSM resources that provide broad economic and environmental benefits but that raise prices, and

- research and development on advanced energy-supply and DSM technologies.

These functions might be especially important to implement the carbon-dioxide reductions called for by the national *Climate Change Action Plan.*
    Massachusetts Electric Company (MECO) filed a request along these

lines with FERC (Corcoran 1994b). The case involves the Massachusetts Bay Transportation Authority (MBTA), which until recently was a MECO customer but now is requesting only wheeling services from MECO. MECO's request includes, in addition to the usual charges for transmission and distribution services, a stranded-investment charge based on its power-supply costs. MECO claims that it arranged for power supplies on the expectation that the MBTA would remain a customer. The proposed stranded-investment charge will automatically be reduced to reflect actual load growth that, presumably, will use the capacity made excess by the MBTA's departure from the MECO system.

Commissions may want to expand utility flexibility in negotiating prices with the larger customers that have choices. Commissions can require utilities to offer energy-efficiency services to such customers in negotiating lower prices and costs. For example, Southern California Edison's (1993) self-generation deferral rate is intended to prevent uneconomic bypass. It permits the utility to offer large customers a rate lower than its standard industrial tariff if the customer can "demonstrate serious intent and feasibility to self-generate." The PUC guidelines for this rate require the utility to offer a conservation alternative to the special rate. This alternative involves utility financing of an investment in energy-efficient equipment at the customer's facility.

To illustrate, Southern California Edison recently agreed to install various measures at the Eisenhower Medical Center under the self-generation deferral rate. The measures include efficient chillers, a hydronic economizer, and high-efficiency lamps and ballasts. These measures will cost about $600,000 and should reduce utility bills by about $900,000/year.

Finally, PUCs will likely define the utility's "obligation to serve" differently for core and noncore customers. The utility's obligations to provide reserve margin, power quality, and other services to noncore customers may be minimal or zero. On the other hand, PUCs may require utilities to offer backup services to such customers, the pricing of which would affect the attractiveness of retail wheeling.

Whether any of these mechanisms are legal and politically feasible will vary from state to state. Large industrial customers (those most likely to benefit from easy access to other suppliers) will argue strongly against such "obstacles" to competition. In addition, substantial practical problems may arise in designing equitable mechanisms. For example, what criteria would be used to decide which customers would pay an exit fee or a reentry fee? Would it be based on size and, if so, on annual revenue, sales, or peak demand? Could the exit fee be applied to industrial customers that physically leave the utility's service area or only to those customers that switch suppliers? Would the reentry fee be applied to new customers moving into the service area or only to customers previously served by the local utility? If the latter, would the charge be a function of the number of years that the customer had been off the system?

A 1993 New York Public Service Commission decision provides precedent for the kinds of requirements discussed above (Kelleher and Wood 1994). The Commission approved an industrial Subscription Option program that allows large commercial and industrial customers to "opt out" of the standard DSM programs offered by Niagara Mohawk. Customers that chose to opt out are no longer eligible for participation in these programs and no longer have to pay for part of those DSM-program costs. However, as a condition for opting out, the industrial customer has to pay for a detailed energy audit at its facility and provide that audit to Niagara Mohawk. (Initial results from these audits show average potential energy savings of 7% with a less-than-four-year payback.) The utility offers a shared-savings program to help these customers finance installation of measures recommended in the energy audit. This decision shows the feasibility of imposing DSM-related requirements on customers as a condition for certain treatment (e.g., access to retail wheeling).

The California PUC (1993), in its review of regulatory reforms, suggested four strategies for state regulation of electric utilities. Although much broader than this focus on retail competition, the PUC's strategies contain some of the same elements:

- Limited reform, in which current cost-of-service regulation is retained, but rate cases are held every year, and most balancing accounts are eliminated.


- Price-cap model, in which the regulatory compact remains, but the utilities gain considerable pricing flexibility, and the link between utility rates and costs is broken.

- Limited customer choice, in which a limited segment of customers is given access to competitive markets. The remaining (core) customers are served under traditional cost-of-service regulation, but the utility's obligation to serve noncore customers is substantially reduced.

- Restructured utility industry, in which the regulated utilities divest themselves of all generating assets and become common-carrier transmission and distribution companies. Utilities no longer enjoy a monopoly retail franchise, but in return they no longer have an obligation to serve those customers that choose other suppliers.

The role of utility DSM programs would likely differ substantially across these four strategies. Under limited reform, such programs could continue with little change from today. Moving down the list, the programs would be progressively smaller and shift more of the costs to participating customers.

## RETAIL COMPETITION, PRICES, AND DSM

A key question concerning retail competition is the role of price. Will utilities and others compete for retail customers on the basis of unit price (¢/kWh) alone? Do consumers view electricity as a homogeneous product for which price is paramount? Or do they view electricity as a service with many attributes? Although we do not now know the answers to these questions, it seems likely that customers differ, with some emphasizing price and some emphasizing service.

Although much of the rhetoric suggests that price will dominate, it is helpful to consider the role that price plays in other competitive markets. Consider the extent to which people buy cars on the basis of unit price ($/pound), rather than on the basis of total cost, comfort, reliability, and other attributes. Similarly, do people go to movies on the basis of their unit cost in ¢/minute; do they buy books on the basis of ¢/page? These examples suggest that consumers will choose utilities on the basis of price, cost (the utility bill, which is the product of price and quantity), convenience, reliability, safety, and various consumer services, such as energy efficiency.

Regardless of whether price is the only factor on which suppliers will compete, price will surely be important. Industrial groups are aggressively comparing industrial electricity prices of their utility with those of surrounding utilities, and Wall Street financial advisers are doing the same. For example, the Tennessee Valley Industrial Committee, an association of large industrial customers of the Tennessee Valley Authority, commissioned a study to compare TVA industrial rates with those of other southeastern utilities (Drazen-Brubaker & Associates 1993). Rates varied by a factor of two, from 2.6 to 5.5¢/kWh (Utilities with high retail rates might use DSM as a way to compete with lower-cost providers.)

Regardless of whether customers focus on price, many utilities are acting as if they do. This means that utilities will look for more and more ways to control costs. Indeed, much of the recent downsizing at utilities is driven by management interest in lowering costs to become more competitive.

Just how important is DSM to a utility's prices? Answering this question requires information on the utility's DSM-program costs, the supply costs the utility would avoid because of the energy and demand reductions caused by the DSM programs, and the lost revenues caused by DSM-induced electricity savings. As explained by Hirst and Blank (1993), improvements in customer energy efficiency that lower sales also cut utility revenues. While utility costs go down also, the short-term cost reductions (primarily for fuel) are much less than the revenue reductions. In the long term, if DSM programs defer construction of new resources, costs may decline more than revenues do.

This difference between revenue and short-term cost reductions, called net lost revenues, is the fixed-cost component of electricity price, typically 50 to 75 % of the total cost of electricity. In many states, utilities recover these lost revenues from their customers. In other states, utility shareholders bear these losses until the next rate case.

In general, the effects of DSM on electricity price are small, on the order of a few percent (Faruqui and Chamberlin 1993). A forthcoming study, based on 66 data points, finds that the median price increase caused by utility DSM programs is 2.7% (Pye and Nadel 1994). Even in New York, which has aggressive utility DSM programs and low avoided costs, average electricity prices in the year 2000 will be only 6% higher than they would be without DSM programs (Gallagher 1994).

Cost overruns at nuclear plants and excess capacity are much more important factors in increasing electricity prices. On the other hand, the costs of these plants are sunk, whereas DSM program costs are at the margin and therefore discretionary. However, tree trimming along distribution lines to reduce the frequency of outages, maintaining inventories of transformers and cable to reduce the duration of outages, and providing telephone customer-assistance centers to improve customer service all cost money and do not increase electricity sales. Thus, these activities, like DSM, raise short-term prices. It is unclear whether these expenditures are receiving the same scrutiny that DSM-program costs are.

If electricity prices were set equal to short-term variable costs, then the lost-revenue problem would disappear. However, because short-term costs are generally below long-term marginal costs, basing prices on short-term costs would send misleading signals to customers, encouraging them to ignore the long-term consequences of their energy-related decisions. In particular, such pricing would lead them to underinvest in energy-efficiency measures. On the other hand, if nonprice market barriers are the primary obstacle to customer adoption of cost-effective energy-efficiency measures, setting prices equal to short-term marginal costs and having utilities offer DSM programs may be the preferred route to achieving the benefits of improved energy efficiency. Also, pricing at short-term marginal cost is what a competitive power market is likely to yield (Hogan 1993).

To explore the effects of utility DSM programs on retail electricity prices, consider a program that runs from 1995 through 1997. The utility pays the full cost of installing DSM measures. The program cuts overall demand by 1% as of 1998 and has a benefit/cost ratio (based on the total resource cost test) of 1.6. If program costs are ratebased, rather than expensed, net-lost revenues might account for about one-third of the average price increase (Fig. 15-3).

If short-term variable costs are lower than in this example, the lost-revenue component and total price impact would be greater. For example, Public Service Company of Colorado (1993) proposed a set of DSM programs that provide substantial system benefits. Over the 20-year analysis period, however, these programs are likely to raise electricity prices by 3%. Net-lost revenues contribute almost 60% to this price increase, with program costs accounting for the other 40%. Thus, the rate impacts of DSM programs depend both on program costs and on the differences over time between avoided costs and retail prices.

This discussion suggests that utilities in the future are likely to design and operate their DSM programs with greater attention to the rate-

**Fig. 15-3. Effects of a hypothetical utility DSM program on electricity prices, showing the relative effects of program costs and lost revenues.**

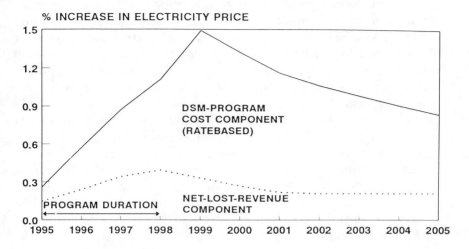

impact measure (RIM) and less attention to the total-resource-cost (TRC) test (California Commissions 1987). The two tests differ primarily in their treatment of lost revenues. The TRC test, which includes both participating and nonparticipating customers, ignores lost revenues. The RIM test, which focuses on nonparticipants, treats lost revenues as a cost.

## DESIGNING DSM PROGRAMS TO LOWER RATE IMPACTS

The more market research utilities conduct (to better understand what customers want and what utilities can profitably provide), the more successful their DSM programs will be. Success includes increased participation rates and lower utility costs. If utilities can better understand their customers, they may be able to offer DSM programs that both improve customer service and provide the resource benefits of "traditional" DSM programs. As Stone (1993) notes, "Utilities that practice DSM aggressively have a sophisticated understanding of critical end-use technologies, of customers' decision processes and the factors that influence them, and of how their own cost [and price] structures position them competitively."

It will be interesting to see whether utilities use the opportunities provided by current regulation (i.e., mandates to run DSM programs that are paid for by all customers) to develop the marketing skills they need to deliver DSM in a competitive environment. In particular, will utilities be able to overcome their tradition of long-leadtime projects that required decision making based on a hierarchical structure and careful analysis? Or will they be able to shift to the rapid, flexible, decentralized decision making needed to market DSM programs to customers? An equally interesting issue is whether PUCs will allow utilities such flexibility in designing, running, and modifying their DSM programs.

A recent study suggests that some utilities, at least, are learning how to market skillfully to their customers (Keneipp et al. 1993). Pacific Gas & Electric has corporate account managers who work closely with customers that spend more than $0.5 million/year on PG&E utility bills. Each manager has 25 to 30 such accounts and has frequent personal contact with them. Similarly, Wisconsin Electric has an account executive who works with "chains" (e.g., supermarkets, department stores, discount stores, and fast-food stores) for which decisions may be made at the regional or national level. These skills are important because the largest potential savings are in the commercial/industrial (C/I) sector, and these customers are most at risk.

The keys to running successful DSM programs seem to be:

• Keep the program simple.

• Update the program to recognize changing market dynamics (e.g., that would allow use of a lower rebate), but do so only once a year.

• Work closely with trade allies (but remember that their interests may not match the utilities').

• Use financial incentives to motivate utility marketing staff.

• Train in-house and trade-ally staff on the program and its measures/technologies.

Learning how to market effectively will be increasingly important if utilities can no longer use rebates (paid for by customers in general) to entice customers to participate in DSM programs. On the contrary, utilities may have to sell DSM to customers and convince them that such services are worth their full cost.

Increasingly, utilities will unbundle their services so that each customer is offered (and purchases) an appropriate mix of kWh, kW, reliability, power quality, and DSM at an appropriate price. For those customers that focus on price, fewer services will likely be selected. Other customers may select DSM options if they are more concerned with bills than prices and if the value of these DSM services exceed their costs. Finally, utilities will increasingly use DSM as an alternative to price breaks.

Consider how a utility might respond to the "threat" of an industrial cogeneration project in a competitive environment. Traditionally, the utility might have (1) ignored the possibility and let the customer leave the system with no effort to retain that customer, (2) pointed out the complications and negative aspects of cogeneration in an effort to dissuade the customer from installing such a system, or (3) offered a discount rate to retain the customer.

In a more competitive environment, the utility might help analyze the economics of a cogeneration system. If the economics are attractive, the utility might offer to build and operate the system on behalf of the customer. The utility might benefit from owning and operating this system if it costs less than the net-revenue losses associated with the customer's departure from the utility system. Also, building and operating cogeneration systems might provide new business opportunities for utilities.

Alternatively, the utility might offer DSM as a way to cut the customer's utility bill. For example, Rochester Gas & Electric (RG&E) convinced the University of Rochester not to install a 10-MW cogeneration system (Demand-Side Report 1994). The key to this change of heart was the utility's offer of an $8-million program of energy-efficiency improvements. The university will pay about $6 million, the utility (through an energy-service company) will pay about $2 million, and the reduction in the university's utility bill should be about $2 million/year. In return, the university agreed to stay on the RG&E system for at least seven years. Southern California Edison's investment in efficiency improvements at the Eisenhower Medical Center, discussed earlier, is another example of this approach.

The likely key to success for DSM programs in the future will be increasing the participant contribution to program costs, thereby reducing the support for DSM from nonparticipating customers. Blank (1993) and Chamberlin, Herman, and Wikler (1993) suggest several possible ways to accomplish that objective.

Blank proposed a novel way to pay for DSM programs that he called the bonus-payment (BP) approach. The twin purposes of this approach are to reduce the rate impacts of DSM programs and to retain high participation levels. The utility, under this system, would pay all the up-front costs of the energy-efficiency measures plus a cash bonus. Thus, participants have no out-of-pocket costs when the measures are installed. In return, the utility recovers program costs and perhaps net lost revenues through the participants' monthly utility bills.

Blank's method is similar to shared-savings schemes tried in the past. However, these approaches may reduce participation in the program relative to what would occur if the utility offered rebates to participating customers. Blank's innovation is to have the utility provide an up-front bonus that, implicitly, compensates participants for the inconvenience associated with participation. This bonus is intended to be so attractive that it more than offsets the subsequent customer repayment.

From an analytical perspective, Blank's method is appealing if customers use discount rates in assessing energy-efficiency alternatives that are higher than the utility's discount rate (typically, the weighted average cost of capital).* If this condition applies, then customers gain the benefit of the bonus, which they value more highly than the (heavily discounted) monthly repayments. The utility, on the other hand, values the repayments more because of its lower discount rate.

Figures 15-4 and 15-5 compare the cash flows to participants and the utility, respectively, with traditional rebate programs and the BP approach. With a rebate, customers typically pay part of the cost of the DSM measures and their installation. On the other hand, customers keep all of the subsequent bill savings for themselves, as shown by the solid line in Fig. 15-4. With the BP approach, customers have a positive cash flow from the start; their subsequent monthly savings are much smaller because they are shared with the utility, as shown by the dotted line.

Figure 15-5 shows the cash flows from the two program designs for the utility. With the BP method, the utility's initial cost is higher, but its

---

*Considerable evidence exists that consumers *do* use (implicitly) higher discount rates in making energy-efficiency decisions than utilities use in making investment decisions (Ruderman, Levine, and McMahon 1987).

Figure 15-4. Annual cash flows seen by DSM-program participants under rebate and bonus-payment program designs.

Figure 15-5. Annual cash flows seen by a utility rebate and bonus-payment program designs.

later revenues are also greater. Given the assumed differences in discount rates between participants and the utility (acting on behalf of customers as a whole), the bonus payment method should create a win-win situation.

Some utilities, including PacifiCorp and Southern California Edison, are moving to this model of DSM program. Although their programs do not include the up-front bonus that Blank recommends, both utilities provide turnkey services to participating customers in the commercial and industrial sectors. These services include analysis of energy-efficiency opportunities (including, in some cases, fuel-switching possibilities) and other productivity improvements; installation of measures; verification that the measures were installed and are operating properly (commissioning); and financing. In return, participating customers repay all the utility's costs, including interest payments, on their monthly utility bills (Hassan 1994).

Chamberlin, Herman, and Wikler (1993) suggest several ways to reduce the price increases normally associated with DSM programs. These possibilities include:

- Accept only those programs that pass the RIM test. This restriction will severely limit acceptance of programs that reduce electricity consumption, and focus DSM programs on reductions in peak demand.

- Accept programs primarily on the basis of the TRC test and
    - Ensure that all customers have access to at least some DSM programs,
    - Charge participants for energy services rather than for kWh and kW (similar to Blank's approach minus the up-front bonus payment), and
    - Allocate DSM-program costs and lost revenues within each customer class.

- Accept a package of DSM programs that passes the RIM test and has the highest TRC benefit. This approach involves selection of programs that pass the RIM test plus enough of the programs with high TRC benefits so that there is no net effect on electricity prices.

- Use the TRC test for residential programs and the RIM test for programs aimed at the larger commercial and industrial customers, with each class paying only for its own programs. This approach makes sense if small customers are responsive primarily to bills and large customers respond most to prices.

The timing of DSM programs can have substantial effects on the costs a utility avoids and therefore on the programs' cost-effectiveness (Richer 1993). DSM programs designed and timed to defer construction of new generating capacity can mitigate or eliminate rate increases.* Alternatively, running ambitious DSM programs when the utility's avoided costs are low relative to retail prices can exacerbate rate impacts. In addition, utilities could target DSM programs to those geographic areas where investments in T&D (as well as investments in generation) could be deferred, thus increasing the benefits of DSM programs.

On a more general level, utilities will need to understand, in much greater detail than they do today, what it costs to serve different groups of customers. The cost-of-service studies that utilities routinely conduct for rate cases will find new uses in utility marketing departments. Increasingly, these studies will include geographic disaggregation to reveal the large variations in avoided T&D costs across a utility's service area.

Whether the utility expenses DSM-program costs or treats these costs as capital investments that are part of the ratebase affects the timing of electricity-price increases (Hirst 1992). Expensing programs causes short-term price hikes, whereas ratebasing spreads the price increases over the amortization period. In the long run, if DSM programs are running at roughly a constant level from year to year, the differences between expensing and ratebasing in their effects on retail prices are minor. On the one hand, because customers likely have higher discount rates than utilities do, ratebasing is preferred. On the other hand, ratebasing is more expensive because of income-tax effects.

To reduce program costs, utilities may move upstream of their retail customers and undertake market-transformation programs. In such programs, utilities work with retailers, wholesalers, and manufacturers in an

---

*PUCs may still want utilities to acquire "lost-opportunity" resources even if the utility's short-term avoided costs do not call for ambitious DSM programs. Lost opportunities are actions that are time sensitive, including the chance to improve energy efficiency in a new factory or during extensive remodeling of an office building. Failure to act at the right time means that the potential efficiency gains are lost or that they will be much more expensive to acquire later on.

effort to completely change the market. Such programs are likely to be much more cost-effective than are efforts aimed at individual customers because the utility deals with far fewer entities. For example, B.C. Hydro ran an industrial efficient-motors program that largely converted the market in British Columbia from inefficient to efficient motors in three years (Flanigan and Fleming 1993), primarily by working with distributors and manufacturers.

Although such market-transformation programs are likely to be very cost-effective from a broad societal perspective, the very reason for their success—utilities no longer have to deal with individual customers to promote DSM actions—argues against their use in a competitive environment. Why should utilities work with dealers and manufacturers when the benefits will flow to all retail consumers whether or not they purchase their electricity from the utility? This dichotomy between private and public interests argues for a strong PUC role to ensure that the benefits of market transformation are not lost in a retail-competitive world.

Utilities during the past few years shifted their DSM emphasis from the residential sector to the C/I sector. They made this shift in recognition of the larger potentials for improving efficiency in the C/I sectors and the greater cost-effectiveness of such programs. In particular, the transaction costs per kWh or kW saved can be much less than in the residential sector, because each commercial or industrial customer has much greater DSM potential than individual residential customers. Ironically, if retail wheeling becomes widespread and the large C/I customers leave the local utility, the utility may be left with only the less cost-effective DSM opportunities, those in the small-C/I and residential sectors. Here, too, PUCs may want to establish regulations to ensure that the most cost-effective DSM opportunities are captured.

Utilities will have to find suitable niches (combinations of DSM products and services aimed at particular groups of customers) for their DSM activities. These niches might include new services not now widely available, such as building commissioning. They might include services that the utility currently provides for itself but not for customers, such as analysis of distribution transformers. And utilities might continue to offer many of the traditional DSM products they have been offering, although packaged in ways that are more appealing to customers, including computerized home energy audits and commercial-lighting improvements.

High-efficiency transformers provide an interesting example of a service that utilities, perhaps uniquely, could provide their large industrial customers. Such transformers, those bought both by utilities and by large customers, can provide cost-effective savings, with paybacks ranging from 1.5 to 4 years. But "very few utilities treat distribution transformers as a DSM resource or actively help their customers to purchase more efficient models. Absent encouragement from utilities, end users typically overlook efficiency in their transformer purchasing policies, most often buying based solely on lowest first cost" (Howe 1993).

Utilities have unique access to detailed information on the electricity-use patterns of their customers. Utilities might be able to sell information services to their customers, based on these data. For example, utilities could disaggregate residential-customer electric bills to show the costs of operating various appliances. Surveys show that customers do not know how much electricity various end uses consume and would find such information useful (Hopkins, Paquette, and Sioshansi 1994). For large commercial customers, utilities could link real-time price signals with building-automation systems, allowing the building's heating, ventilating, and air-conditioning control systems to respond to hour-by-hour changes in electricity prices. Such a service would benefit both the utility and the customer (Flood, Carmichael, and Vold 1994).

In the future, utility DSM programs are likely to focus more on costumer service, with less emphasis on providing capacity and energy resources to the utility. Utilities will explore, identify, and implement a much wider range of energy-efficiency and load-management programs than exist today as they work with different market segments among their customers.

## CONCLUSIONS

A critical issue, not addressed in this report, concerns how today's retail monopoly franchise, cost-of-service regulation will be transformed to a more competitive environment. Utilities are concerned that PUCs may not provide them with enough flexibility to respond appropriately to competitive threats. Others are worried that residential and small commercial customers will subsidize the larger customers that have market power and choices. And environmentalists worry that DSM programs may be a victim of the rush to competition.

Another key issue concerns environmental quality. Environmental concerns may play an important role in determining both the extent of retail wheeling and the nature of utility DSM programs. If, as seems likely, society places stricter controls on emissions of air and water pollutants and remains concerned about possible electromagnetic-field effects of transmission and distribution lines, then state regulators will likely require continuation or expansion of utility DSM programs. The Climate Challenge (part of the *Climate Change Action Plan*), includes a prominent role for U.S. electric utilities to reduce $CO_2$ emissions. If the industry is largely fragmented, meeting the $CO_2$-emission-reduction goals will be very difficult. Utilities can also play an important role in helping their customers meet environmental regulations, as Southern California Edison and others are already doing.

As of early 1994, it is hard to know the extent to which competition will affect utility DSM programs. However, it seems clear that competition at the retail level would have much more effect on DSM programs than would wholesale-level competition. The extent to which retail competition occurs will depend strongly on the role that PUCs play. If PUCs encourage retail wheeling, utility DSM programs that are financed by customers as a whole would be severely restricted. Utilities in that kind of an environment could generally run only those programs that paid for themselves. In other words, individual customers would have to pay for the DSM services that the utility provides. And that payment might have to include both the program costs and the utility's lost revenues.

If, however, PUCs limit retail wheeling or impose a fee on use of the local utility's distribution system, DSM programs could continue much the same as today. Customers covered by such programs would diminish because noncore customers would presumably not be eligible (or want) to participate in the utility's DSM programs. But utilities could continue to rely primarily on the TRC test rather than the RIM test in choosing DSM programs. Even here, though, utilities may likely face pressure to keep rates low. This pressure will require utilities to look for least-cost ways to implement DSM programs; i.e., the rate impact measure will be important here, although less so than in a full blown retail-wheeling scenario.

Utility DSM programs will
- Focus more on customer service and less on the system-resource benefits.

- Emphasize capacity reductions more and energy savings less. This shift will occur as utilities seek to minimize the lost revenues associated with DSM.

- Become more cost-effective as utilities identify better ways to deliver DSM services at lower cost.

- Decrease as utilities increasingly seek to have individual customers pay for their own DSM services.

Regardless of the extent of retail competition, the role of utility DSM will depend in large part on the ability of utilities to deliver DSM, to become more customer driven, and to lower the costs of DSM. Utilities will package DSM with other customer services, including pricing alternatives, reliability levels, power quality, and productivity improvements. All these DSM-program changes will require large changes in utility structure, operation, and culture. The challenge to regulators will be to retain the public benefits of traditional DSM programs in a more competitive environment.

[1]Based on the report "Electric-Utility DSM Programs in a Competitive Market" by Eric Hirst.

Sponsored by Office of Energy Efficiency and Renewable Energy, U.S. Department of Energy

Oak Ridge National Laboratory, Oak Ridge, Tennessee 37831, managed by Martin Marietta Energy Systems, Inc. under contract No. DE-AC05-840R21400 for U.S. Department of Energy


## ACKNOWLEDGMENTS

Douglas Bauer, Eric Blank, David Berry, Ralph Cavanagh, John Chamberlin, Shel Feldman, Jane Finleon, James Gallagher, John Hartnett, Patricia Herman, Lynn Hobbie, David Meyer, Diane Pirkey, Ralph Prahl, Dan Quigley, John Rowe, Richard Scheer, Martin Schweitzer, Bruce Tonn, Michael Yokell, Fred O'Hara, and Ethel Schorn.

## REFERENCES

E. Blank 1993, "Minimizing Non-Participant DSM Rate Impacts—Without Harming Participation," *The Electricity Journal* 6(4), 32-41.

E. A. Bretz 1994, "Power Marketing: Wave of the Future?" *Electrical World* 208(2), 16, February.

California Public Utilities Commission and California Energy Commission 1987, *Standard Practice Manual, Economic Analysis of Demand-Side Management Programs*, San Francisco and Sacramento, CA, December.

California Public Utilities Commission 1993, *California's Electric Services Industry: Perspectives on the Past, Strategies for the Future*, Division of Strategic Planning, San Francisco, CA, February.

R. Cavanagh 1994, *The Great "Retail Wheeling" Illusion—And More Productive Energy Futures*, Natural Resources Defense Council, New York, February.

J. Chamberlin, P. Herman, and G. Wikler 1993, "Mitigating Rate Impacts of DSM Programs," *The Electricity Journal* 6(9), 46-56, November.

W.J. Clinton and A. Gore, Jr. 1993, *The Climate Change Action Plan*, The White House, Washington, DC, October.

P. Corcoran 1994a, "The Conowingo Power Stranded Investment Case," *Edison Times*, Edison Electric Institute, Washington, DC, January.

P. Corcoran 1994b, "FERC Sets Transit Agency Case for Hearing," *Edison Times*, Edison Electric Institute, Washington, DC, February.

Demand-Side Report 1994, "Noresco Program for University of Rochester Heads Off Plans for Cogeneration Unit," 5, January 6.

Drazen-Brubaker & Associates, Inc. 1993, "Survey of Power Costs in the Southeastern United States," prepared for the Tennessee Valley Industrial Committee, October.

Electric Power Research Institute 1994, *Proceedings: 1994 Innovative Electricity Pricing*, EPRI TR-103629, Palo Alto, CA, February.

# 16

# Financing Energy Projects

## INTRODUCTION

Every day spent by an organization without having installed the appropriate energy efficiency measures means lost savings and lost opportunities. Performance contracting is a sophisticated solution to this problem. As with any sophisticated system, there are elements of complexity to be managed. Therefore, communication, knowledge and experience are essential for successful project completion.

## DEFINITIONS AND CLARIFICATIONS

For the purposes of this chapter, it is assumed that the energy services company (ESCO) is providing the energy conservation measures (ECMs) such as audit, design, installation, monitoring and maintenance for the customer and that a separate third party, such as a bank or investment company, is providing the capital for the project. Often, energy services companies market themselves to customers as providing financing. In many cases, there is an independent financing source involved in the background. Alternatively, an ESCO can be a utility subsidiary which uses the utility's shareholder money to finance projects. For simplicity's sake, we will treat the ultimate source of capital for projects as a separate lender with its own guidelines.

## FINANCING ALTERNATIVES

There is a variety of options for financing energy conservation projects. Some of the most common are:

## General Obligation Bond

Specifically applicable to municipalities, these bonds are based on the general credit of a state or local government. The process is long and complicated, but interest rates are low.

## Municipal Lease

Specifically applicable to municipalities, state entities and local entities. The lessor earns tax-exempt credit and the borrowing entity pays low interest rates.

## Commercial Loan

This is a loan to the customer from a conventional bank based on the customer's assets and credit quality. This form of financing is rarely offered by typical finance companies or financial institutions for energy conservation projects requiring less than $5 million of capital.

## Taxable Lease

There are a number of leasing vehicles with a variety of names such as: operating lease, capital lease, guideline lease, tax-oriented lease, and non-tax-oriented lease. The tax ramifications of leasing are often not well understood. Figure 16-1 will help clarify some of those distinctions. Figure 16-2 will show the dynamics of the parties involved.

It is important to note that in all of the categories listed above, the customer is directly obligated to make payments relating to the installed energy-saving measures REGARDLESS OF PERFORMANCE of either the equipment or the ESCO. The customer may well have recourse under a separate contract to the service provider or equipment manufacturer, but will still owe under the financing instrument. The interest rate and any additional loan terms are based almost exclusively on the credit-worthiness of the customer.


## RELATIVE BENEFITS OF PROJECT FINANCING

Many companies and government entities are undergoing severe budget cutbacks. When these entities are approached by an ESCO offering virtually no up-front investment and a guaranteed amount of savings, new possibilities are created for energy projects to be developed

## Figure 16-1. Leasing Options

**Guideline Lease/True Lease/Tax-Oriented Lease**

Compliance with all of the IRS guidelines including those listed below is required:

(a) The total lease term (including extensions and renewals at a predetermined, fixed rate) must not exceed 80% of the estimated useful life of the equipment at the start of the lease, i.e., at the end of the lease the equipment must have an estimated remaining useful life equal to at least 20% of its originally estimated useful life. Also, this remaining useful life must not be less than one year thereby limiting the maximum term of the lease.

(b) The equipment's estimated residual value at the expiration of the lease term must equal at least 20% of its value at the start of the lease. This requirement limits the maximum lease term and the type of equipment to be leased.

(c) No bargain purchase option.

(d) The lessee cannot make any investment in the equipment.

(e) The equipment must not be "limited use" property. Equipment is "limited use" property if no one other than the lessee or a related party has a use for it at the end of the lease.

(f) Tax-oriented leasing is 100% financing. Guideline leases (tax-oriented leases) may be either a capital lease or an operating lease for reporting purposes under Financial Accounting Standards Board Rule #13.

**Capital lease**

A capital lease is one that fulfills any ONE of the following criteria:

(a) The lease transfers ownership of the property to the lessee by the end of the lease term.

(b) The lease contains a bargain purchase option (less than fair market value).

(c) The lease term is equal to 75% or more of the estimated economic life of the leased property.

(d) The present value of the minimum lease payments equals or exceeds 90% of the fair value of the leased property.

The capital lease shows up on the lessee's balance sheet as an asset and a liability.

**Operating lease**

An operating lease does not meet ANY of the above criteria for a capital lease.

An operating lease is not booked on the lessee's balance sheet but is recorded as a periodic expense on the income statement.

**Sources**

*Accounting For Leases,* Financial Accounting Standards Board
*Leasing and Tax Reform: A Guide Through The Maze,* General Electric Credit Corporation
*Handbook of Leasing,* General Electric Credit Corporation

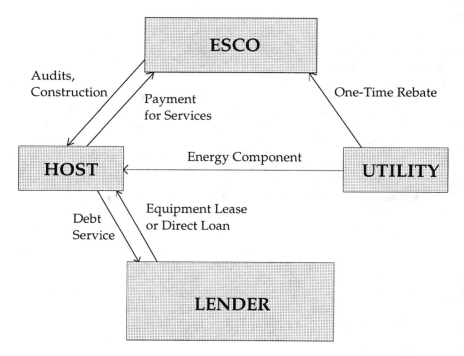

**Figure 16-2. Equipment Lease/Direct Financing**

and completed. If structured properly, performance contract financing has the benefits to the customer of an operating lease, i.e., off-balance-sheet treatment, with the added benefit of being non-recourse to the customer. By maintaining this financing out of its balance sheet, the customer can retain the use of available credit for expansion, research and development, additional inventory or business emergencies.

## BASIC STRUCTURE

Performance contract financing from a lender's perspective is a challenging combination of business credit analysis and project evaluation. The project financing is structured as follows:

(1)    The ESCO contracts with the customer to provide energy conservation measures (lighting, variable speed drives, etc.) as well as ongo-

ing services which may include warranties, handling and disposal of wastes, operation and maintenance of installed equipment, repair and replacement of measures, and measurement, monitoring and verification of savings. The Energy Services Agreement (ESA) is the foundational document upon which a lender will rely to confirm that the customer and ESCO have a clear understanding of all aspects of the ECMs being installed. The ESA incorporates a payment from the customer which is designed to cover the costs to finance the project as well as paying the ESCO for the services provided. (A sample ESA is included in Figure 16-3. Note that it is a guideline only. Be sure to have your legal counsel review any document before you authorize it.)

(2)  The lender lends directly to the ESCO. The ESCO uses the funds to recoup its project development expenses and purchase and install the equipment. Repayment to the lender is made by the ESCO out of the funds paid to it by the customer. One variation is to create a single-purpose entity exclusively to hold the assets of the project financing. Because the loan is technically made to this entity, it has the added benefit of keeping the transaction off the balance sheet of the ESCO as well as the customer (see Figure 16-4). In addition, the lender can lend up to 95% of the project amount.

## PROPOSAL REVIEW

The lender will analyze a proposed project with these questions in mind:

*   Will the revenues generated by the energy savings, utility rebates and customer payments be sufficient to cover debt service?

*   Can the ESCO perform as required under its contract with the customer and/or the utility for the term contract?

*   Are the risk allocations among all parties fair from both business and legal perspectives?

- Will the customer and/or the utility be able to meet its payment obligations for the term of the financing?

The lender will want to see three years of audited financial statements: balance sheet, income statement and cash flow statement with explanatory notes from the customer. These will be used to make trend and financial ratio analyses. The ESCO's financial statements are also important and will receive a similar review.

With utility payments and/or customer payments based on actual savings, the lender needs to make a thorough analysis of the projected savings of the energy conservation measures installed. The project figures (simple payback, types of equipment, maintenance savings, if any, construction period) will be carefully reviewed to determine how the savings will cover the loan payments and to determine the effects of factors such as energy price fluctuations and inflation.

Any information which the parties can provide early on in the proposal negotiation to make the lender feel more comfortable with these issues will save a tremendous amount of time, money and effort to all parties.

## CONCLUSION

The key point here is that if an event of default occurs in the ESA between the ESCO and the customer which gives the customer the right to reduce or cease its payment obligation, it is the lender which is most at risk of suffering a loss. The lender must either be comfortable that the ESCO will cure the default or be confident that it can hire another ESCO to cure the project default and force the customer to resume payments. The lender is also at risk of the energy savings not being estimated or measured properly. Despite all of this, good projects are being funded and customers are extremely satisfied with the resulting benefits.

## Figure 16-3. Sample Energy Services Agreement

*Note: Consult your legal counsel before authorizing any legal agreement.*

### I. TERM SHEET.

The purpose of the Agreement is the evaluation, engineering, design, procurement, installation, financing and monitoring by **ESCo** of Energy Conservation Measures ("ECMs") at **Customer's** facility(ies) identified in Appendix A attached hereto ("Premises").

**ESCo** and **Customer** (the "Parties") agree to the following terms pursuant to which this Agreement shall be performed:

1. EXECUTION DATE: _____

2. TERM OF AGREEMENT: _____

_____
      (years after ECM Commencement Date)

3. LOCATION: _____

4. OWNERSHIP OF PREMISES: _____

_____

### PROJECT FIGURES

| | Preliminary Estimates | Final Installed Figures |
|---|---|---|
| 5. **ESCo** PERCENTAGE OF ENERGY SAVINGS: | ———— | ———— |
| 6. PROJECT TOTAL CAPITAL COST ($ 000): | ———— | ———— |
| 7. ECM COMMENCEMENT DATE: | ———— | ———— |
| 8. VALUE OF FIRST YEAR ENERGY SAVINGS ($): | ———— | ———— |

9. **CUSTOMER** FIRST YEAR
   PAYMENT TO **ESCo**
   (line 5 X line 8) ($):                    _____          _____

10. **CUSTOMER** FIRST YEAR
    MONTHLY PAYMENT
    (line 9 12) ($):                         _____          _____

11. **CUSTOMER** FIXED MONTHLY
    PAYMENT (OPTIONAL):

                                             _____          _____

                              Please Initial:        ESCo  _____

                              CUSTOMER             _____

Final Installed Figures appearing in the right-hand column above shall
be completed in accordance with Section V and the Parties shall then re-
execute the Agreement below and enter their initials above.

As indicated in term #11 above, upon completion of any determi-
nations required by Section V, **Customer** shall have the option to fix its
monthly payment to **ESCo**, as required under Section VI, by multiplying
One Hundred and _____ Percent (1____%) of **ESCo**'s Percentage of
Energy Savings by the Final Installed Figure for the Value of First Year
Energy Savings and paying such product to **ESCo** each year in twelve
(12) monthly payments ("**Customer** Fixed Monthly Payment"). **Customer** shall indicate its exercise of such option by initialing the appropriate line below. Except as set forth in this Agreement, such monthly payments shall be due and payable each month of this Agreement from and
after the month in which the Commencement Date occurs and shall be
made by **Customer** without regard to the amount of Energy Savings in
any such month or year. Any excess of Energy Savings over the Final
Estimated Value of First Year Energy Savings shall be retained by **Customer. Customer**'s Fixed Monthly Payment shall not be revised except as
may be specifically required in accordance with the terms of this Agreement.

| ESCo: | Customer: |
|-------|-----------|
| _____ | _____ |
| (business organization) | (business organization) |

By:————————————————————————————
By:————————————————————————————
      (entity)                              (entity)

By:————————————————————————————
By:————————————————————————————
      (name)                              (name)

Its:————————————————————————————
Its:————————————————————————————
      (title)                              (title)

## ACCEPTANCE OF FINAL INSTALLED FIGURES

| ESCo: | Customer: |
|-------|-----------|
| _____ | _____ |
| (business organization) | (business organization) |

By:————————————————————————————
By:————————————————————————————
      (entity)                              (entity)

By:————————————————————————————
By:————————————————————————————
      (name)                              (name)

Its:————————————————————————————
Its:————————————————————————————
      (title)                              (title)

### CUSTOMER ACCEPTANCE OF FIXED MONTHLY PAYMENT

Accepted                          Not Accepted

——————                    ——————— (please initial)

### II. DEFINITIONS.
When used in this Agreement, the following terms shall have the meaning specified:

2.1 **Agreement:** This Agreement between **Customer** and **ESCo.**

2.2 **ESCo's Percentage of Energy Savings:** The percentage of Energy Savings **ESCo** shall receive as compensation for its services under this Agreement, subject to **Customer's** option to fix monthly payments set forth in Section I, paid to **ESCo** in accordance with Section VI and Section VII.

2.3 **Current Market Value of Energy Savings:** The total market value rate expressed in r/kWh of electrical energy use and/or $/kW of electrical demand imposed by the Utility company in the current monthly period then occurring, or in any future monthly period then being considered, including applicable taxes, surcharges and franchise fees, if applicable. The **Current Market Value of Energy Savings** shall be determined in accordance with Appendix C.

2.4 **ECM Commencement Date:** The ECM Commencement Date shall be the date on which the installation of the ECMs is substantially complete. Prior thereto, **ESCo** shall give **Customer** a Notice of Substantial Completion and shall therein identify the ECM Commencement Date, which shall occur no sooner than fifteen (15) days after such notice.

2.5 **Energy Audit Report ("EAR"):** The analysis performed by **ESCo** of the electric energy use by **Customer** at the Premises, and the potential for electric energy savings. Such analysis includes, without limitation, the ECMs recommended by **ESCo** and agreed to by the Parties for installation at the Premises and the Measurement Plan for measuring the savings estimated to result from such ECMs, all as attached hereto and made a part hereof as Appendix B.

**2.6 Energy Conservation Measure ("ECM"):** The various items of equipment, devices, materials and/or software as installed by **ESCo** at the Premises, or as repaired or replaced by **Customer** hereunder, for the purpose of improving the efficiency of electric consumption, or otherwise to reduce the electric utility costs of the Premises.

**2.7 Energy Savings:** Electric energy reduction (expressed in kilowatt-hours of electric energy and/or kilowatts of electric demand and measured in accordance with the Measurement Plan) achieved through the more efficient utilization of electricity resulting from the installation of the ECMs agreed to by the Parties under this Agreement.

**2.8 Measurement Plan:** The plan for measuring Energy Savings under this Agreement, which shall be in accordance with the requirements of the Utility Agreement and shall be a part of the Energy Audit Report attached as Appendix B hereto.

**2.9 Monthly Period:** A span of time covering approximately 30 days per month, corresponding to **Customer**'s billing period from its electric utility.

**2.10 Party: Customer** or **ESCo. Parties** means **Customer** and **ESCo.**

**2.11 Premises:** The buildings, facilities and equipment used by **Customer**, as identified in Section I and as more fully described in the attached Appendix A, where **ESCo** shall implement the Project under this Agreement.

**2.12 Project:** The complete range of services provided by **ESCo** pursuant to this Agreement, including evaluation, engineering, procurement, installation, financing and monitoring of ECMs at the Premises.

**2.13 Uncontrollable Circumstances:** Any event or condition having a material adverse effect on the rights, duties or obligations of **ESCo**, or materially adversely affecting the Project, if such event or condition is beyond the reasonable control, and not the result of willful or negligent action or omission or a lack of reasonable diligence, of **ESCo**; provided, however, that the contesting by **ESCo** in good faith of any event or condition constituting a Change in Law shall not constitute or be construed as a willful or negligent action, or a lack of reasonable diligence. Such events or conditions may include, but shall not be limited to, circumstances of the following kind:

      a. An act of God, epidemic, landslide, lightning, hurricane, earthquake, fire, explosion, storm, flood or similar occurrence, an equipment failure or outage, an interruption in supply, an act or omission by persons or entities other than a Party, an act of war, effects of nuclear radiation, blockade, insurrection, riot, civil disturbance or similar occurrences, or damage, interruption or interference to the Project caused by hazardous waste stored on or existing at the Project site;

      b. strikes, lockouts, work slowdowns or stoppages, or similar labor difficulties, affecting or impacting the performance of **ESCo** or its contractors and suppliers;

      c. a change in law or regulation or an act by a governmental agency or judicial authority.

**2.14 Utility Agreement:** The agreement entered into by **ESCo** with _____, a _____ public utility company ("Utility"), pursuant to which **ESCo** is required to install certain ECMs at facilities such as **Customer**'s Premises and in accordance with the terms of which **ESCo** has entered into this Agreement.

**III. ECM COMMENCEMENT DATE AND TERM OF AGREEMENT.** The term of this Agreement shall commence as of the date on which this Agreement is executed and shall continue, unless sooner terminated in accordance with the terms hereof, for the period of years after the ECM Commencement Date set forth in Section I.

      Upon receipt of the Notice of Substantial Completion identifying the ECM Commencement Date, **Customer** shall provide **ESCo**, within fifteen (15) days, any comments and requests for work or corrections. **ESCo** shall make all commercially reasonable efforts to respond to such comments and requests within thirty (30) days of the ECM Commencement Date, which shall occur on the identified date.

      Upon the expiration or termination of this Agreement the provisions of this Agreement that may reasonably be interpreted or construed as surviving the expiration or termination of this Agreement shall survive the expiration or termination for such period as may be necessary to effect the intent of this Agreement.

At the end of the term of this Agreement, **Customer** shall purchase the ECMs and **ESCo** shall transfer title to **Customer**, free and clear of all liens and encumbrances, all as set forth in Section 6 of the General Terms and Conditions.

## IV. SCOPE OF ESCO'S SERVICES.

Subject to and in accordance with the terms and conditions of this Agreement, **ESCo** shall provide the evaluation, engineering, design, procurement, installation, financing and monitoring of the ECMs set forth in the Energy Audit Report. The ECMs in the EAR may be modified pursuant to Section V. **ESCo** shall use reasonable commercial efforts to achieve the ECM Commencement Date estimated in Section I, line 7, column entitled, "Engineering Estimates."

ESCo agrees to extend its funds and install such ECMs in return for **Customer**'s agreement to perform hereunder and in particular, to pay **ESCo**'s Percentage of Energy Savings pursuant to Section VI. The Parties anticipate the measurement of Energy Savings, but notwithstanding this expectation or any provision of this Agreement which may suggest to the contrary, in light of the factors affecting savings which are beyond **ESCo**'s reasonable control, **ESCo** assumes no obligation that any particular level of savings shall materialize due to its services hereunder.

ESCo warrants that the ECMs which have a lifetime greater than one (1) year shall be, and shall remain, free of defects for one (1) year after the ECM Commencement Date. In addition, **ESCo** agrees to assign all manufacturers warranties for such ECMs to **Customer** for the period of manufacturer warranty, subject to all exclusions and limitations as may be set forth therein.

## V. DETERMINATION OF FINAL TERMS.

After the execution of this Agreement, if the Parties agree to revise the ECMs listed in the EAR, and to make all associated revisions to this Agreement, including, without limitation, to the numbers or dates, as the case may be, entered in the column entitled, "Engineering Estimates" appearing in lines 5 through 11 of Section I, the Parties shall, upon such agreement, make any associated revision and enter in the column entitled, "Final Installed Figures" the agreed-to numbers and/or dates. The Parties shall then re-execute Section I upon the entry of such numbers and/or dates and **ESCo** shall revise the Termination Values in Appendix D consistent with Appendix D and the Final Installed Project Total Capital Cost then appearing in line 6 of Section I above.

All such revisions shall be voluntary and the Parties shall not be required, absent mutual consent, to revise the ECMs listed in the EAR after the execution hereof; provided, however, **ESCo** shall not be required to install ECMs affected by Uncontrollable Circumstances. Such ECMs shall be deleted from the Project unless the Parties agree to all necessary changes to the Project numbers and/or dates required to adjust to such circumstances. Changes to Project numbers and/or dates in connection with Uncontrollable Circumstances shall be entered by **ESCo** in the column entitled, "Final Installed Figures," in lines 5 through 11 of Section I. The Parties shall then re-execute Section I and **ESCo** shall revise the Termination Values set forth in Appendix D consistent with Appendix D and the Final Installed Project Total Capital Cost then appearing in line 6 of Section I.

### VI. COMPENSATION.

From and after the ECM Commencement Date, except as provided in Sections I and VII with respect to **Customer**'s option to fix monthly payments, **Customer** shall pay **ESCo** an amount equal to **ESCo**'s Percentage of Energy Savings, as set forth in Section I above, multiplied by the applicable Current Market Value of Energy Savings, all as determined pursuant to Appendix C.

**ESCo** shall prepare and send **Customer**, and **Customer** agrees to pay, a monthly invoice calculated pursuant to Section VII below.

### VII. BILLING.

**ESCo** will submit monthly invoices to **Customer** in amounts determined in accordance with Section VI. In the event **Customer** exercises its option under Section I to fix its monthly payments to **ESCo**, **Customer** shall pay to **ESCo** the **Customer** Fixed Monthly Payment defined in Section I. If **Customer** does not exercise such option, monthly payments shall be estimated as set forth in this Section VII and paid on such estimated basis, subject to reconciliation as provided hereinafter. Monthly invoices shall be paid within thirty (30) calendar days following receipt. Reconciliation payments, or refunds, as the case may be, shall be due within thirty (30) calendar days of receipt of a reconciliation invoice.

Subject to reconciliation, invoice amounts shall be estimated for each year following the ECM Commencement Date. In the first such year, monthly payments shall be 1/12 of the product of **ESCo**'s Percentage of Energy Savings and the Current Market Value of Energy Savings expected in such first year as set forth in line 8 of the column entitled, "Final Installed Figures" in Section I above. Such fixed amounts shall be reconciled and adjusted as necessary every six (6) months based on the difference between the Current Market Value of Energy Savings measured pursuant to Appendix C and the estimated value then applicable. **ESCo** shall prepare and submit to **Customer** a reconciliation invoice within thirty (30) calendar days of the expiration of each such six-month period. The estimate of the value of Energy Savings then in effect at the end of any year shall continue until replaced with reconciled amounts in the succeeding year. In the second year and following years, the estimate of the Current Market Value of Energy Savings expected in such year shall equal the expected value (in nominal dollars for the billing year in question) of the Energy Savings actually delivered in the prior year.

## VIII. NOTICES.
All notices to be given by either Party to the other shall be in writing and must be delivered or mailed by registered or certified mail, return receipt requested, or sent by a courier service which renders a receipt upon delivery addressed as set forth above in Section I or such other addresses as either Party may hereinafter designate by a Notice to the other. Notices are deemed delivered or given and become effective upon mailing if mailed as aforesaid and upon actual receipt if otherwise delivered to the addresses set forth in Section I.

## IX. APPLICABLE LAW.
This Agreement and the construction and enforceability thereof shall be interpreted under the laws of the state of _____.

## X. FINAL AGREEMENT.
This Agreement, together with its appendices and attachments, shall constitute the full and final Agreement between the Parties, shall supersede all prior agreements, communications and understandings regarding the subject matter hereof and shall not be amended, modified or revised except in writing. **This Agreement shall bind the Parties as of the date on which it was executed.** Any inconsistency in this Agreement shall be resolved by giving priority in the following order: (a) Amendments to the Agreement, in reverse chronological order; (b) the Agree-

ment, Sections I through XI; (c) General Terms and Conditions of the Agreement; (d) Appendices to the Agreement; and (e) Attachments to, or other documents incorporated into, the Agreement.

## XI. INCORPORATION OF GENERAL TERMS AND CONDITIONS:

**THIS AGREEMENT IS SUBJECT TO THE GENERAL TERMS AND CONDITIONS ATTACHED HERETO AND MADE A PART HEREOF. IN ADDITION, THE PERFORMANCE OF THIS AGREEMENT BY ESCo IS SUBJECT TO THE PROVISIONS OF THE UTILITY AGREEMENT.**

**IN WITNESS WHEREOF**, and intending to be legally bound, the Parties hereto subscribe their names to this instrument hereinabove in Section I as of the date of execution first written above.

## SUMMARY OF GENERAL TERMS AND CONDITIONS

| | |
|---|---|
| Section 1. | Operating the Premises and Maintaining the Use of the ECMs. |
| Section 2. | Equipment Location and Access. |
| Section 3. | Construction. |
| Section 4. | Ownership of ECMs. |
| Section 5. | Condition of ECMs. |
| Section 6. | Transfer of the ECMs. |
| Section 6.1. | Material Shortfall in Energy Savings. |
| Section 6.2. | Customer Purchase Option. |
| Section 6.3. | Purchase Upon Customer Default. |
| Section 6.4. | Expiration of the Term. |
| Section 6.5. | Transfer Without Encumbrance. |
| Section 7. | Requirements of Utility Agreement. |
| Section 8. | Insurance. |
| Section 8.1. | Insurance on the Premises and the ECMs. |
| Section 8.2. | Risk of Loss of the ECMs. |
| Section 8.3. | ESCo's Insurance Requirements. |
| Section 9. | Hazardous Materials and Activities. |
| Section 10. | Events of Default. |
| Section 10.1. | Events of Default by Customer. |
| Section 10.2. | Events of Default by ESCo. |
| Section 11. | Remedies Upon Default. |
| Section 11.1. | Remedies Upon Default by Customer. |
| Section 11.2. | Remedies Upon Default by ESCo. |
| Section 11.3. | Limitation of Remedies. |

Section 12.     Representations and Warranties of Both Parties.
Section 13.     Compliance With Law and Standard Practices.
Section 14.     Assignment.
Section 15.     Taxes.
Section 16.     Severability.
Section 17.     Effect of Waiver.
Section 18.     Usage of Customer's Records.
Section 19.     Air Emission Rights, Credits or Allowances.

**APPENDICES:**
"A"     Description and Address of Customer's Facility(ies).
"B"     Energy Audit Report and List of ECMs.
"C"     Method of Savings Calculation.
"D"     Termination Values.
"E"     Requirements of Utility Agreement.

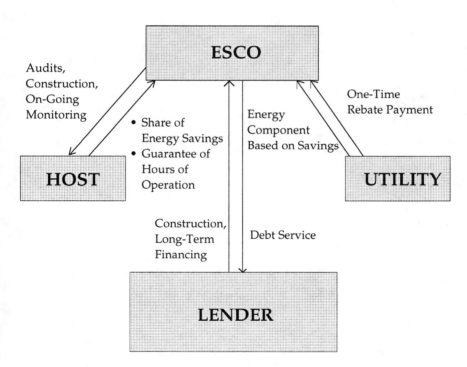

**Figure 16-4. Shared Savings/Performance Contracting**

# 17

# Appendix

## Table 15-1. 10% Interest Factor

| Period $n$ | Single-payment compound-amount (F/P) | Single-payment present-worth (P/F) | Uniform-series compound-amount (F/A) | Sinking-fund payment (A/F) | Capital recovery (A/P) | Uniform-series present-worth (P/A) |
|---|---|---|---|---|---|---|
| | Future value of \$1 $(1 + i)^n$ | Present value of \$1 $\dfrac{1}{(1 + i)^n}$ | Future value of uniform series of \$1 $\dfrac{(1 + i)^n - 1}{i}$ | Uniform series whose future value is \$1 $\dfrac{i}{(1 + i)^n - 1}$ | Uniform series with present value of \$1 $\dfrac{i(1 + i)^n}{(1 + i)^n - 1}$ | Present value of uniform series of \$1 $\dfrac{(1 + i)^n - 1}{i(1 + i)^n}$ |
| 1 | 1.100 | 0.9091 | 1.000 | 1.00000 | 1.10000 | 0.909 |
| 2 | 1.210 | 0.8264 | 2.100 | 0.47619 | 0.57619 | 1.736 |
| 3 | 1.331 | 0.7513 | 3.310 | 0.30211 | 0.40211 | 2.487 |
| 4 | 1.464 | 0.6830 | 4.641 | 0.21547 | 0.31547 | 3.170 |
| 5 | 1.611 | 0.6209 | 6.105 | 0.16380 | 0.26380 | 3.791 |
| 6 | 1.772 | 0.5645 | 7.716 | 0.12961 | 0.22961 | 4.355 |
| 7 | 1.949 | 0.5132 | 9.487 | 0.10541 | 0.20541 | 4.868 |
| 8 | 2.144 | 0.4665 | 11.436 | 0.08744 | 0.18744 | 5.335 |
| 9 | 2.358 | 0.4241 | 13.579 | 0.07364 | 0.17364 | 5.759 |
| 10 | 2.594 | 0.3855 | 15.937 | 0.06275 | 0.16275 | 6.144 |
| 11 | 2.853 | 0.3505 | 18.531 | 0.05396 | 0.15396 | 6.495 |
| 12 | 3.138 | 0.3186 | 21.384 | 0.04676 | 0.14676 | 6.814 |
| 13 | 3.452 | 0.2897 | 24.523 | 0.04078 | 0.14078 | 7.103 |
| 14 | 3.797 | 0.2633 | 27.975 | 0.03575 | 0.13575 | 7.367 |
| 15 | 4.177 | 0.2394 | 31.772 | 0.03147 | 0.13147 | 7.606 |
| 16 | 4.595 | 0.2176 | 35.950 | 0.02782 | 0.12782 | 7.824 |
| 17 | 5.054 | 0.1978 | 40.545 | 0.02466 | 0.12466 | 8.022 |
| 18 | 5.560 | 0.1799 | 45.599 | 0.02193 | 0.12193 | 8.201 |
| 19 | 6.116 | 0.1635 | 51.159 | 0.01955 | 0.11955 | 8.365 |
| 20 | 6.727 | 0.1486 | 57.275 | 0.01746 | 0.11746 | 8.514 |
| 21 | 7.400 | 0.1351 | 64.002 | 0.01562 | 0.11562 | 8.649 |
| 22 | 8.140 | 0.1228 | 71.403 | 0.01401 | 0.11401 | 8.772 |
| 23 | 8.954 | 0.1117 | 79.543 | 0.01257 | 0.11257 | 8.883 |
| 24 | 9.850 | 0.1015 | 88.497 | 0.01130 | 0.11130 | 8.985 |
| 25 | 10.835 | 0.0923 | 98.347 | 0.01017 | 0.11017 | 9.077 |
| 26 | 11.918 | 0.0839 | 109.182 | 0.00916 | 0.10916 | 9.161 |
| 27 | 13.110 | 0.0763 | 121.100 | 0.00826 | 0.10826 | 9.237 |
| 28 | 14.421 | 0.0693 | 134.210 | 0.00745 | 0.10745 | 9.307 |
| 29 | 15.863 | 0.0630 | 148.631 | 0.00673 | 0.10673 | 9.370 |
| 30 | 17.449 | 0.0573 | 164.494 | 0.00608 | 0.10608 | 9.427 |
| 35 | 28.102 | 0.0356 | 271.024 | 0.00369 | 0.10369 | 9.644 |
| 40 | 45.259 | 0.0221 | 442.593 | 0.00226 | 0.10226 | 9.779 |
| 45 | 72.890 | 0.0137 | 718.905 | 0.00139 | 0.10139 | 9.863 |
| 50 | 117.391 | 0.0085 | 1163.909 | 0.00086 | 0.10086 | 9.915 |
| 55 | 189.059 | 0.0053 | 1880.591 | 0.00053 | 0.10053 | 9.947 |
| 60 | 304.482 | 0.0033 | 3034.816 | 0.00033 | 0.10033 | 9.967 |
| 65 | 490.371 | 0.0020 | 4893.707 | 0.00020 | 0.10020 | 9.980 |
| 70 | 789.747 | 0.0013 | 7887.470 | 0.00013 | 0.10013 | 9.987 |
| 75 | 1271.895 | 0.0008 | 12708.954 | 0.00008 | 0.10008 | 9.992 |
| 80 | 2048.400 | 0.0005 | 20474.002 | 0.00005 | 0.10005 | 9.995 |
| 85 | 3298.969 | 0.0003 | 32979.690 | 0.00003 | 0.10003 | 9.997 |
| 90 | 5313.023 | 0.0002 | 53120.226 | 0.00002 | 0.10002 | 9.998 |
| 95 | 8556.676 | 0.0001 | 85556.760 | 0.00001 | 0.10001 | 9.999 |

## Table 15-2.  12% Interest Factor

| Period n | Single-payment compound-amount (F/P) | Single-payment present-worth (P/F) | Uniform-series compound-amount (F/A) | Sinking-fund payment (A/F) | Capital recovery (A/P) | Uniform-series present-worth (P/A) |
|---|---|---|---|---|---|---|
|  | Future value of $1 $(1 + i)^n$ | Present value of $1 $\dfrac{1}{(1 + i)^n}$ | Future value of uniform series of $1 $\dfrac{(1 + i)^n - 1}{i}$ | Uniform series whose future value is $1 $\dfrac{i}{(1 + i)^n - 1}$ | Uniform series with present value of $1 $\dfrac{i(1 + i)^n}{(1 + i)^n - 1}$ | Present value of uniform series of $1 $\dfrac{(1 + i)^n - 1}{i(1 + i)^n}$ |
| 1 | 1.120 | 0.8929 | 1.000 | 1.00000 | 1.12000 | 0.893 |
| 2 | 1.254 | 0.7972 | 2.120 | 0.47170 | 0.59170 | 1.690 |
| 3 | 1.405 | 0.7118 | 3.374 | 0.29635 | 0.41635 | 2.402 |
| 4 | 1.574 | 0.6355 | 4.779 | 0.20923 | 0.32923 | 3.037 |
| 5 | 1.762 | 0.5674 | 6.353 | 0.15741 | 0.27741 | 3.605 |
| 6 | 1.974 | 0.5066 | 8.115 | 0.12323 | 0.24323 | 4.111 |
| 7 | 2.211 | 0.4523 | 10.089 | 0.09912 | 0.21912 | 4.564 |
| 8 | 2.476 | 0.4039 | 12.300 | 0.08130 | 0.20130 | 4.968 |
| 9 | 2.773 | 0.3606 | 14.776 | 0.06768 | 0.18768 | 5.328 |
| 10 | 3.106 | 0.3220 | 17.549 | 0.05698 | 0.17698 | 5.650 |
| 11 | 3.479 | 0.2875 | 20.655 | 0.04842 | 0.16842 | 5.938 |
| 12 | 3.896 | 0.2567 | 24.133 | 0.04144 | 0.16144 | 6.194 |
| 13 | 4.363 | 0.2292 | 28.029 | 0.03568 | 0.15568 | 6.424 |
| 14 | 4.887 | 0.2046 | 32.393 | 0.03087 | 0.15087 | 6.628 |
| 15 | 5.474 | 0.1827 | 37.280 | 0.02682 | 0.14682 | 6.811 |
| 16 | 6.130 | 0.1631 | 42.753 | 0.02339 | 0.14339 | 6.974 |
| 17 | 6.866 | 0.1456 | 48.884 | 0.02046 | 0.14046 | 7.120 |
| 18 | 7.690 | 0.1300 | 55.750 | 0.01794 | 0.13794 | 7.250 |
| 19 | 8.613 | 0.1161 | 63.440 | 0.01576 | 0.13576 | 7.366 |
| 20 | 9.646 | 0.1037 | 72.052 | 0.01388 | 0.13388 | 7.469 |
| 21 | 10.804 | 0.0926 | 81.699 | 0.01224 | 0.13224 | 7.562 |
| 22 | 12.100 | 0.0826 | 92.503 | 0.01081 | 0.13081 | 7.645 |
| 23 | 13.552 | 0.0738 | 104.603 | 0.00956 | 0.12956 | 7.718 |
| 24 | 15.179 | 0.0659 | 118.155 | 0.00846 | 0.12846 | 7.784 |
| 25 | 17.000 | 0.0588 | 133.334 | 0.00750 | 0.12750 | 7.843 |
| 26 | 19.040 | 0.0525 | 150.334 | 0.00665 | 0.12665 | 7.896 |
| 27 | 21.325 | 0.0469 | 169.374 | 0.00590 | 0.12590 | 7.943 |
| 28 | 23.884 | 0.0419 | 190.699 | 0.00524 | 0.12524 | 7.984 |
| 29 | 26.750 | 0.0374 | 214.583 | 0.00466 | 0.12466 | 8.022 |
| 30 | 29.960 | 0.0334 | 241.333 | 0.00414 | 0.12414 | 8.055 |
| 35 | 52.800 | 0.0189 | 431.663 | 0.00232 | 0.12232 | 8.176 |
| 40 | 93.051 | 0.0107 | 767.091 | 0.00130 | 0.12130 | 8.244 |
| 45 | 163.988 | 0.0061 | 1358.230 | 0.00074 | 0.12074 | 8.283 |
| 50 | 289.002 | 0.0035 | 2400.018 | 0.00042 | 0.12042 | 8.304 |
| 55 | 509.321 | 0.0020 | 4236.005 | 0.00024 | 0.12024 | 8.317 |
| 60 | 897.597 | 0.0011 | 7471.641 | 0.00013 | 0.12013 | 8.324 |
| 65 | 1581.872 | 0.0006 | 13173.937 | 0.00008 | 0.12008 | 8.328 |
| 70 | 2787.800 | 0.0004 | 23223.332 | 0.00004 | 0.12004 | 8.330 |
| 75 | 4913.056 | 0.0002 | 40933.799 | 0.00002 | 0.12002 | 8.332 |
| 80 | 8658.483 | 0.0001 | 72145.692 | 0.00001 | 0.12001 | 8.332 |

## Table 15-3. 15% Interest Factors.

| Period n | Single-payment compound-amount (F/P) Future value of $1 $(1 + i)^n$ | Single-payment present-worth (P/F) Present value of $1 $\dfrac{1}{(1 + i)^n}$ | Uniform-series compound-amount (F/A) Future value of uniform series of $1 $\dfrac{(1 + i)^n - 1}{i}$ | Sinking-fund payment (A/F) Uniform series whose future value is $1 $\dfrac{i}{(1 + i)^n - 1}$ | Capital recovery (A/P Uniform series with present value of $1 $\dfrac{i(1 + i)^n}{(1 + i)^n - 1}$ | Uniform-series present-worth (P/A) Present value of uniform series of $1 $\dfrac{(1 + i)^n - 1}{i(1 + i)^n}$ |
|---|---|---|---|---|---|---|
| 1 | 1.150 | 0.8696 | 1.000 | 1.00000 | 1.15000 | 0.870 |
| 2 | 1.322 | 0.7561 | 2.150 | 0.46512 | 0.61512 | 1.626 |
| 3 | 1.521 | 0.6575 | 3.472 | 0.28798 | 0.43798 | 2.283 |
| 4 | 1.749 | 0.5718 | 4.993 | 0.20027 | 0.35027 | 2.855 |
| 5 | 2.011 | 0.4972 | 6.742 | 0.14832 | 0.29832 | 3.352 |
| 6 | 2.313 | 0.4323 | 8.754 | 0.11424 | 0.26424 | 3.784 |
| 7 | 2.660 | 0.3759 | 11.067 | 0.09036 | 0.24036 | 4.160 |
| 8 | 3.059 | 0.3269 | 13.727 | 0.07285 | 0.22285 | 4.487 |
| 9 | 3.518 | 0.2843 | 16.786 | 0.05957 | 0.20957 | 4.772 |
| 10 | 4.046 | 0.2472 | 20.304 | 0.04925 | 0.19925 | 5.019 |
| 11 | 4.652 | 0.2149 | 24.349 | 0.04107 | 0.19107 | 5.234 |
| 12 | 5.350 | 0.1869 | 29.002 | 0.03448 | 0.18448 | 5.421 |
| 13 | 6.153 | 0.1625 | 34.352 | 0.02911 | 0.17911 | 5.583 |
| 14 | 7.076 | 0.1413 | 40.505 | 0.02469 | 0.17469 | 5.724 |
| 15 | 8.137 | 0.1229 | 47.580 | 0.02102 | 0.17102 | 5.847 |
| 16 | 9.358 | 0.1069 | 55.717 | 0.01795 | 0.16795 | 5.954 |
| 17 | 10.761 | 0.0929 | 65.075 | 0.01537 | 0.16537 | 6.047 |
| 18 | 12.375 | 0.0808 | 75.836 | 0.01319 | 0.16319 | 6.128 |
| 19 | 14.232 | 0.0703 | 88.212 | 0.01134 | 0.16134 | 6.198 |
| 20 | 16.367 | 0.0611 | 102.444 | 0.00976 | 0.15976 | 6.259 |
| 21 | 18.822 | 0.0531 | 118.810 | 0.00842 | 0.15842 | 6.312 |
| 22 | 21.645 | 0.0462 | 137.632 | 0.00727 | 0.15727 | 6.359 |
| 23 | 24.891 | 0.0402 | 159.276 | 0.00628 | 0.15628 | 6.399 |
| 24 | 28.625 | 0.0349 | 184.168 | 0.00543 | 0.15543 | 6.434 |
| 25 | 32.919 | 0.0304 | 212.793 | 0.00470 | 0.15470 | 6.464 |
| 26 | 37.857 | 0.0264 | 245.712 | 0.00407 | 0.15407 | 6.491 |
| 27 | 43.535 | 0.0230 | 283.569 | 0.00353 | 0.15353 | 6.514 |
| 28 | 50.066 | 0.0200 | 327.104 | 0.00306 | 0.15306 | 6.534 |
| 29 | 57.575 | 0.0174 | 377.170 | 0.00265 | 0.15265 | 6.551 |
| 30 | 66.212 | 0.0151 | 434.745 | 0.00230 | 0.15230 | 6.566 |
| 35 | 133.176 | 0.0075 | 881.170 | 0.00113 | 0.15113 | 6.617 |
| 40 | 267.864 | 0.0037 | 1779.090 | 0.00056 | 0.15056 | 6.642 |
| 45 | 538.769 | 0.0019 | 3585.128 | 0.00028 | 0.15028 | 6.654 |
| 50 | 1083.657 | 0.0009 | 7217.716 | 0.00014 | 0.15014 | 6.661 |
| 55 | 2179.622 | 0.0005 | 14524.148 | 0.00007 | 0.15007 | 6.664 |
| 60 | 4383.999 | 0.0002 | 29219.992 | 0.00003 | 0.15003 | 6.665 |
| 65 | 8817.787 | 0.0001 | 58778.583 | 0.00002 | 0.15002 | 6.666 |

## Table 15-4.  20% Interest Factor

| Period $n$ | Single-payment compound-amount (F/P) | Single-payment present-worth (P/F) | Uniform-series compound-amount (F/A) | Sinking-fund payment (A/F) | Capital recovery (A/P) | Uniform-series present-worth (P/A) |
|---|---|---|---|---|---|---|
| | Future value of $1 $(1 + i)^n$ | Present value of $1 $\dfrac{1}{(1 + i)^n}$ | Future value of uniform series of $1 $\dfrac{(1 + i)^n - 1}{i}$ | Uniform series whose future value is $1 $\dfrac{i}{(1 + i)^n - 1}$ | Uniform series with present value of $1 $\dfrac{i(1 + i)^n}{(1 + i)^n - 1}$ | Present value of uniform series of $1 $\dfrac{(1 + i)^n - 1}{i(1 + i)^n}$ |
| 1 | 1.200 | 0.8333 | 1.000 | 1.00000 | 1.20000 | 0.833 |
| 2 | 1.440 | 0.6944 | 2.200 | 0.45455 | 0.65455 | 1.528 |
| 3 | 1.728 | 0.5787 | 3.640 | 0.27473 | 0.47473 | 2.106 |
| 4 | 2.074 | 0.4823 | 5.368 | 0.18629 | 0.38629 | 2.589 |
| 5 | 2.488 | 0.4019 | 7.442 | 0.13438 | 0.33438 | 2.991 |
| 6 | 2.986 | 0.3349 | 9.930 | 0.10071 | 0.30071 | 3.326 |
| 7 | 3.583 | 0.2791 | 12.916 | 0.07742 | 0.27742 | 3.605 |
| 8 | 4.300 | 0.2326 | 16.499 | 0.06061 | 0.26061 | 3.837 |
| 9 | 5.160 | 0.1938 | 20.799 | 0.04808 | 0.24808 | 4.031 |
| 10 | 6.192 | 0.1615 | 25.959 | 0.03852 | 0.23852 | 4.192 |
| 11 | 7.430 | 0.1346 | 32.150 | 0.03110 | 0.23110 | 4.327 |
| 12 | 8.916 | 0.1122 | 39.581 | 0.02526 | 0.22526 | 4.439 |
| 13 | 10.699 | 0.0935 | 48.497 | 0.02062 | 0.22062 | 4.533 |
| 14 | 12.839 | 0.0779 | 59.196 | 0.01689 | 0.21689 | 4.611 |
| 15 | 15.407 | 0.0649 | 72.035 | 0.01388 | 0.21388 | 4.675 |
| 16 | 18.488 | 0.0541 | 87.442 | 0.01144 | 0.21144 | 4.730 |
| 17 | 22.186 | 0.0451 | 105.931 | 0.00944 | 0.20944 | 4.775 |
| 18 | 26.623 | 0.0376 | 128.117 | 0.00781 | 0.20781 | 4.812 |
| 19 | 31.948 | 0.0313 | 154.740 | 0.00646 | 0.20646 | 4.843 |
| 20 | 38.338 | 0.0261 | 186.688 | 0.00536 | 0.20536 | 4.870 |
| 21 | 46.005 | 0.0217 | 225.026 | 0.00444 | 0.20444 | 4.891 |
| 22 | 55.206 | 0.0181 | 271.031 | 0.00369 | 0.20369 | 4.909 |
| 23 | 66.247 | 0.0151 | 326.237 | 0.00307 | 0.20307 | 4.925 |
| 24 | 79.497 | 0.0126 | 392.484 | 0.00255 | 0.20255 | 4.937 |
| 25 | 95.396 | 0.0105 | 471.981 | 0.00212 | 0.20212 | 4.948 |
| 26 | 114.475 | 0.0087 | 567.377 | 0.00176 | 0.20176 | 4.956 |
| 27 | 137.371 | 0.0073 | 681.853 | 0.00147 | 0.20147 | 4.964 |
| 28 | 164.845 | 0.0061 | 819.223 | 0.00122 | 0.20122 | 4.970 |
| 29 | 197.814 | 0.0051 | 984.068 | 0.00102 | 0.20102 | 4.975 |
| 30 | 237.376 | 0.0042 | 1181.882 | 0.00085 | 0.20085 | 4.979 |
| 35 | 590.668 | 0.0017 | 2948.341 | 0.00034 | 0.20034 | 4.992 |
| 40 | 1469.772 | 0.0007 | 7343.858 | 0.00014 | 0.20014 | 4.997 |
| 45 | 3657.262 | 0.0003 | 18281.310 | 0.00005 | 0.20005 | 4.999 |
| 50 | 9100.438 | 0.0001 | 45497.191 | 0.00002 | 0.20002 | 4.999 |

## Table 15-5. 25% Interest Factor

| Period $n$ | Single-payment compound-amount (F/P) | Single-payment present-worth (P/F) | Uniform-series compound-amount (F/A) | Sinking-fund payment (A/F) | Capital recovery (A/P) | Uniform-series present-worth (P/A) |
|---|---|---|---|---|---|---|
| | Future value of \$1 $(1 + i)^n$ | Present value of \$1 $\dfrac{1}{(1 + i)^n}$ | Future value of uniform series of \$1 $\dfrac{(1 + i)^n - 1}{i}$ | Uniform series whose future value is \$1 $\dfrac{i}{(1 + i)^n - 1}$ | Uniform series with present value of \$1 $\dfrac{i(1 + i)^n}{(1 + i)^n - 1}$ | Present value of uniform series of \$1 $\dfrac{(1 + i)^n - 1}{i(1 + i)^n}$ |
| 1 | 1.250 | 0.8000 | 1.000 | 1.00000 | 1.25000 | 0.800 |
| 2 | 1.562 | 0.6400 | 2.250 | 0.44444 | 0.69444 | 1.440 |
| 3 | 1.953 | 0.5120 | 3.812 | 0.26230 | 0.51230 | 1.952 |
| 4 | 2.441 | 0.4096 | 5.766 | 0.17344 | 0.42344 | 2.362 |
| 5 | 3.052 | 0.3277 | 8.207 | 0.12185 | 0.37185 | 2.689 |
| 6 | 3.815 | 0.2621 | 11.259 | 0.08882 | 0.33882 | 2.951 |
| 7 | 4.768 | 0.2097 | 15.073 | 0.06634 | 0.31634 | 3.161 |
| 8 | 5.960 | 0.1678 | 19.842 | 0.05040 | 0.30040 | 3.329 |
| 9 | 7.451 | 0.1342 | 25.802 | 0.03876 | 0.28876 | 3.463 |
| 10 | 9.313 | 0.1074 | 33.253 | 0.03007 | 0.28007 | 3.571 |
| 11 | 11.642 | 0.0859 | 42.566 | 0.02349 | 0.27349 | 3.656 |
| 12 | 14.552 | 0.0687 | 54.208 | 0.01845 | 0.26845 | 3.725 |
| 13 | 18.190 | 0.0550 | 68.760 | 0.01454 | 0.26454 | 3.780 |
| 14 | 22.737 | 0.0440 | 86.949 | 0.01150 | 0.26150 | 3.824 |
| 15 | 28.422 | 0.0352 | 109.687 | 0.00912 | 0.25912 | 3.859 |
| 16 | 35.527 | 0.0281 | 138.109 | 0.00724 | 0.25724 | 3.887 |
| 17 | 44.409 | 0.0225 | 173.636 | 0.00576 | 0.25576 | 3.910 |
| 18 | 55.511 | 0.0180 | 218.045 | 0.00459 | 0.25459 | 3.928 |
| 19 | 69.389 | 0.0144 | 273.556 | 0.00366 | 0.25366 | 3.942 |
| 20 | 86.736 | 0.0115 | 342.945 | 0.00292 | 0.25292 | 3.954 |
| 21 | 108.420 | 0.0092 | 429.681 | 0.00233 | 0.25233 | 3.963 |
| 22 | 135.525 | 0.0074 | 538.101 | 0.00186 | 0.25186 | 3.970 |
| 23 | 169.407 | 0.0059 | 673.626 | 0.00148 | 0.25148 | 3.976 |
| 24 | 211.758 | 0.0047 | 843.033 | 0.00119 | 0.25119 | 3.981 |
| 25 | 264.698 | 0.0038 | 1054.791 | 0.00095 | 0.25095 | 3.985 |
| 26 | 330.872 | 0.0030 | 1319.489 | 0.00076 | 0.25076 | 3.988 |
| 27 | 413.590 | 0.0024 | 1650.361 | 0.00061 | 0.25061 | 3.990 |
| 28 | 516.988 | 0.0019 | 2063.952 | 0.00048 | 0.25048 | 3.992 |
| 29 | 646.235 | 0.0015 | 2580.939 | 0.00039 | 0.25039 | 3.994 |
| 30 | 807.794 | 0.0012 | 3227.174 | 0.00031 | 0.25031 | 3.995 |
| 35 | 2465.190 | 0.0004 | 9856.761 | 0.00010 | 0.25010 | 3.998 |
| 40 | 7523.164 | 0.0001 | 30088.655 | 0.00003 | 0.25003 | 3.999 |

## Table 15-6.  30% Interest Factor

| Period $n$ | Single-payment compound-amount (F/P)<br><br>Future value of $1<br>$(1 + i)^n$ | Single-payment present-worth (P/F)<br><br>Present value of $1<br>$\dfrac{1}{(1 + i)^n}$ | Uniform-series compound-amount (F/A)<br><br>Future value of uniform series of $1<br>$\dfrac{(1 + i)^n - 1}{i}$ | Sinking-fund payment (A/F)<br><br>Uniform series whose future value is $1<br>$\dfrac{i}{(1 + i)^n - 1}$ | Capital recovery (A/P<br><br>Uniform series with present value of $1<br>$\dfrac{i(1 + i)^n}{(1 + i)^n - 1}$ | Uniform-series present-worth (P/A)<br><br>Present value of uniform series of $1<br>$\dfrac{(1 + i)^n - 1}{i(1 + i)^n}$ |
|---|---|---|---|---|---|---|
| 1 | 1.300 | 0.7692 | 1.000 | 1.00000 | 1.30000 | 0.769 |
| 2 | 1.690 | 0.5917 | 2.300 | 0.43478 | 0.73478 | 1.361 |
| 3 | 2.197 | 0.4552 | 3.990 | 0.25063 | 0.55063 | 1.816 |
| 4 | 2.856 | 0.3501 | 6.187 | 0.16163 | 0.46163 | 2.166 |
| 5 | 3.713 | 0.2693 | 9.043 | 0.11058 | 0.41058 | 2.436 |
| 6 | 4.827 | 0.2072 | 12.756 | 0.07839 | 0.37839 | 2.643 |
| 7 | 6.275 | 0.1594 | 17.583 | 0.05687 | 0.35687 | 2.802 |
| 8 | 8.157 | 0.1226 | 23.858 | 0.04192 | 0.34192 | 2.925 |
| 9 | 10.604 | 0.0943 | 32.015 | 0.03124 | 0.33124 | 3.019 |
| 10 | 13.786 | 0.0725 | 42.619 | 0.02346 | 0.32346 | 3.092 |
| 11 | 17.922 | 0.0558 | 56.405 | 0.01773 | 0.31773 | 3.147 |
| 12 | 23.298 | 0.0429 | 74.327 | 0.01345 | 0.31345 | 3.190 |
| 13 | 30.288 | 0.0330 | 97.625 | 0.01024 | 0.31024 | 3.223 |
| 14 | 39.374 | 0.0254 | 127.913 | 0.00782 | 0.30782 | 3.249 |
| 15 | 51.186 | 0.0195 | 167.286 | 0.00598 | 0.30598 | 3.268 |
| 16 | 66.542 | 0.0150 | 218.472 | 0.00458 | 0.30458 | 3.283 |
| 17 | 86.504 | 0.0116 | 285.014 | 0.00351 | 0.30351 | 3.295 |
| 18 | 112.455 | 0.0089 | 371.518 | 0.00269 | 0.30269 | 3.304 |
| 19 | 146.192 | 0.0068 | 483.973 | 0.00207 | 0.30207 | 3.311 |
| 20 | 190.050 | 0.0053 | 630.165 | 0.00159 | 0.30159 | 3.316 |
| 21 | 247.065 | 0.0040 | 820.215 | 0.00122 | 0.30122 | 3.320 |
| 22 | 321.184 | 0.0031 | 1067.280 | 0.00094 | 0.30094 | 3.323 |
| 23 | 417.539 | 0.0024 | 1388.464 | 0.00072 | 0.30072 | 3.325 |
| 24 | 542.801 | 0.0018 | 1806.003 | 0.00055 | 0.30055 | 3.327 |
| 25 | 705.641 | 0.0014 | 2348.803 | 0.00043 | 0.30043 | 3.329 |
| 26 | 917.333 | 0.0011 | 3054.444 | 0.00033 | 0.30033 | 3.330 |
| 27 | 1192.533 | 0.0008 | 3971.778 | 0.00025 | 0.30025 | 3.331 |
| 28 | 1550.293 | 0.0006 | 5164.311 | 0.00019 | 0.30019 | 3.331 |
| 29 | 2015.381 | 0.0005 | 6714.604 | 0.00015 | 0.30015 | 3.332 |
| 30 | 2619.996 | 0.0004 | 8729.985 | 0.00011 | 0.30011 | 3.332 |
| 35 | 9727.860 | 0.0001 | 32422.868 | 0.00003 | 0.30003 | 3.333 |

## Table 15-7. 40% Interest Factor

| Period $n$ | Single-payment compound-amount (F/P) Future value of $1 $(1 + i)^n$ | Single-payment present-worth (P/F) Present value of $1 $\dfrac{1}{(1 + i)^n}$ | Uniform-series compound-amount (F/A) Future value of uniform series of $1 $\dfrac{(1 + i)^n - 1}{i}$ | Sinking-fund payment (A/F) Uniform series whose future value is $1 $\dfrac{i}{(1 + i)^n - 1}$ | Capital recovery (A/P) Uniform series with present value of $1 $\dfrac{i(1 + i)^n}{(1 + i)^n - 1}$ | Uniform-series present-worth (P/A) Present value of uniform series of $1 $\dfrac{(1 + i)^n - 1}{i(1 + i)^n}$ |
|---|---|---|---|---|---|---|
| 1 | 1.400 | 0.7143 | 1.000 | 1.00000 | 1.40000 | 0.714 |
| 2 | 1.960 | 0.5102 | 2.400 | 0.41667 | 0.81667 | 1.224 |
| 3 | 2.744 | 0.3644 | 4.360 | 0.22936 | 0.62936 | 1.589 |
| 4 | 3.842 | 0.2603 | 7.104 | 0.14077 | 0.54077 | 1.849 |
| 5 | 5.378 | 0.1859 | 10.946 | 0.09136 | 0.49136 | 2.035 |
| 6 | 7.530 | 0.1328 | 16.324 | 0.06126 | 0.46126 | 2.168 |
| 7 | 10.541 | 0.0949 | 23.853 | 0.04192 | 0.44192 | 2.263 |
| 8 | 14.758 | 0.0678 | 34.395 | 0.02907 | 0.42907 | 2.331 |
| 9 | 20.661 | 0.0484 | 49.153 | 0.02034 | 0.42034 | 2.379 |
| 10 | 28.925 | 0.0346 | 69.814 | 0.01432 | 0.41432 | 2.414 |
| 11 | 40.496 | 0.0247 | 98.739 | 0.01013 | 0.41013 | 2.438 |
| 12 | 56.694 | 0.0176 | 139.235 | 0.00718 | 0.40718 | 2.456 |
| 13 | 79.371 | 0.0126 | 195.929 | 0.00510 | 0.40510 | 2.469 |
| 14 | 111.120 | 0.0090 | 275.300 | 0.00363 | 0.40363 | 2.478 |
| 15 | 155.568 | 0.0064 | 386.420 | 0.00259 | 0.40259 | 2.484 |
| 16 | 217.795 | 0.0046 | 541.988 | 0.00185 | 0.40185 | 2.489 |
| 17 | 304.913 | 0.0033 | 759.784 | 0.00132 | 0.40132 | 2.492 |
| 18 | 426.879 | 0.0023 | 1064.697 | 0.00094 | 0.40094 | 2.494 |
| 19 | 597.630 | 0.0017 | 1491.576 | 0.00067 | 0.40067 | 2.496 |
| 20 | 836.683 | 0.0012 | 2089.206 | 0.00048 | 0.40048 | 2.497 |
| 21 | 1171.356 | 0.0009 | 2925.889 | 0.00034 | 0.40034 | 2.498 |
| 22 | 1639.898 | 0.0006 | 4097.245 | 0.00024 | 0.40024 | 2.498 |
| 23 | 2295.857 | 0.0004 | 5737.142 | 0.00017 | 0.40017 | 2.499 |
| 24 | 3214.200 | 0.0003 | 8032.999 | 0.00012 | 0.40012 | 2.499 |
| 25 | 4499.880 | 0.0002 | 11247.199 | 0.00009 | 0.40009 | 2.499 |
| 26 | 6299.831 | 0.0002 | 15747.079 | 0.00006 | 0.40006 | 2.500 |
| 27 | 8819.764 | 0.0001 | 22046.910 | 0.00005 | 0.40005 | 2.500 |

## Table 15-8.  50% Interest Factor

| Period $n$ | Single-payment compound-amount (F/P) | Single-payment present-worth (P/F) | Uniform-series compound-amount (F/A) | Sinking-fund payment (A/F) | Capital recovery (A/P) | Uniform-series present-worth (P/A) |
|---|---|---|---|---|---|---|
| | Future value of \$1 $(1 + i)^n$ | Present value of \$1 $\dfrac{1}{(1 + i)^n}$ | Future value of uniform series of \$1 $\dfrac{(1 + i)^n - 1}{i}$ | Uniform series whose future value is \$1 $\dfrac{i}{(1 + i)^n - 1}$ | Uniform series with present value of \$1 $\dfrac{i(1 + i)^n}{(1 + i)^n - 1}$ | Present value of uniform series of \$1 $\dfrac{(1 + i)^n - 1}{i(1 + i)^n}$ |
| 1  | 1.500    | 0.6667 | 1.000     | 1.00000 | 1.50000 | 0.667 |
| 2  | 2.250    | 0.4444 | 2.500     | 0.40000 | 0.90000 | 1.111 |
| 3  | 3.375    | 0.2963 | 4.750     | 0.21053 | 0.71053 | 1.407 |
| 4  | 5.062    | 0.1975 | 8.125     | 0.12308 | 0.62308 | 1.605 |
| 5  | 7.594    | 0.1317 | 13.188    | 0.07583 | 0.57583 | 1.737 |
| 6  | 11.391   | 0.0878 | 20.781    | 0.04812 | 0.54812 | 1.824 |
| 7  | 17.086   | 0.0585 | 32.172    | 0.03108 | 0.53108 | 1.883 |
| 8  | 25.629   | 0.0390 | 49.258    | 0.02030 | 0.52030 | 1.922 |
| 9  | 38.443   | 0.0260 | 74.887    | 0.01335 | 0.51335 | 1.948 |
| 10 | 57.665   | 0.0173 | 113.330   | 0.00882 | 0.50882 | 1.965 |
| 11 | 86.498   | 0.0116 | 170.995   | 0.00585 | 0.50585 | 1.977 |
| 12 | 129.746  | 0.0077 | 257.493   | 0.00388 | 0.50388 | 1.985 |
| 13 | 194.620  | 0.0051 | 387.239   | 0.00258 | 0.50258 | 1.990 |
| 14 | 291.929  | 0.0034 | 581.859   | 0.00172 | 0.50172 | 1.993 |
| 15 | 437.894  | 0.0023 | 873.788   | 0.00114 | 0.50114 | 1.995 |
| 16 | 656.841  | 0.0015 | 1311.682  | 0.00076 | 0.50076 | 1.997 |
| 17 | 985.261  | 0.0010 | 1968.523  | 0.00051 | 0.50051 | 1.998 |
| 18 | 1477.892 | 0.0007 | 2953.784  | 0.00034 | 0.50034 | 1.999 |
| 19 | 2216.838 | 0.0005 | 4431.676  | 0.00023 | 0.50023 | 1.999 |
| 20 | 3325.257 | 0.0003 | 6648.513  | 0.00015 | 0.50015 | 1.999 |
| 21 | 4987.885 | 0.0002 | 9973.770  | 0.00010 | 0.50010 | 2.000 |
| 22 | 7481.828 | 0.0001 | 14961.655 | 0.00007 | 0.50007 | 2.000 |

## Table 15-9. Five-Year Escalation Table

| | Present Worth of a Series of Escalating Payments Compounded Annually Discount-Escalation Factors for $n$ = 5 Years | | | | | |
|---|---|---|---|---|---|---|
| | Annual Escalation Rate | | | | | |
| Discount Rate | 0.10 | 0.12 | 0.14 | 0.16 | 0.18 | 0.20 |
| 0.10 | 5.000000 | 5.279234 | 5.572605 | 5.880105 | 6.202627 | 6.540569 |
| 0.11 | 4.866862 | 5.136200 | 5.420152 | 5.717603 | 6.029313 | 6.355882 |
| 0.12 | 4.738562 | 5.000000 | 5.274242 | 5.561868 | 5.863289 | 6.179066 |
| 0.13 | 4.615647 | 4.869164 | 5.133876 | 5.412404 | 5.704137 | 6.009541 |
| 0.14 | 4.497670 | 4.742953 | 5.000000 | 5.269208 | 5.551563 | 5.847029 |
| 0.15 | 4.384494 | 4.622149 | 4.871228 | 5.131703 | 5.404955 | 5.691165 |
| 0.16 | 4.275647 | 4.505953 | 4.747390 | 5.000000 | 5.264441 | 5.541511 |
| 0.17 | 4.171042 | 4.394428 | 4.628438 | 4.873699 | 5.129353 | 5.397964 |
| 0.18 | 4.070432 | 4.287089 | 4.513947 | 4.751566 | 5.000000 | 5.259749 |
| 0.19 | 3.973684 | 4.183921 | 4.403996 | 4.634350 | 4.875619 | 5.126925 |
| 0.20 | 3.880510 | 4.084577 | 4.298207 | 4.521178 | 4.755725 | 5.000000 |
| 0.21 | 3.790801 | 3.989001 | 4.196400 | 4.413341 | 4.640260 | 4.877689 |
| 0.22 | 4.704368 | 3.896891 | 4.098287 | 4.308947 | 4.529298 | 4.759649 |
| 0.23 | 3.621094 | 3.808179 | 4.003835 | 4.208479 | 4.422339 | 4.645864 |
| 0.24 | 3.540773 | 3.722628 | 3.912807 | 4.111612 | 4.319417 | 4.536517 |
| 0.25 | 3.463301 | 3.640161 | 3.825008 | 4.018249 | 4.220158 | 4.431144 |
| 0.26 | 3.388553 | 3.560586 | 3.740376 | 3.928286 | 4.124553 | 4.329514 |
| 0.27 | 3.316408 | 3.483803 | 3.658706 | 3.841442 | 4.032275 | 4.231583 |
| 0.28 | 3.246718 | 3.409649 | 3.579870 | 3.757639 | 3.943295 | 4.137057 |
| 0.29 | 3.179393 | 3.338051 | 3.503722 | 3.676771 | 3.857370 | 4.045902 |
| 0.30 | 3.114338 | 3.268861 | 3.430201 | 3.598653 | 3.774459 | 3.957921 |
| 0.31 | 3.051452 | 3.201978 | 3.359143 | 3.523171 | 3.694328 | 3.872901 |
| 0.32 | 2.990618 | 3.137327 | 3.290436 | 3.450224 | 3.616936 | 3.790808 |
| 0.33 | 2.931764 | 3.074780 | 3.224015 | 3.379722 | 3.542100 | 3.711472 |
| 0.34 | 2.874812 | 3.014281 | 3.159770 | 3.311524 | 3.469775 | 3.634758 |

### Table 15-10. Ten-Year Escalation Table

Present Worth of a Series of Escalating Payments Compounded Annually
Discount-Escalation Factors for $n$ = 10 Years

| Discount Rate | Annual Escalation Rate | | | | | |
|---|---|---|---|---|---|---|
| | 0.10 | 0.12 | 0.14 | 0.16 | 0.18 | 0.20 |
| 0.10 | 10.000000 | 11.056250 | 12.234870 | 13.548650 | 15.013550 | 16.646080 |
| 0.11 | 9.518405 | 10.508020 | 11.613440 | 12.844310 | 14.215140 | 15.741560 |
| 0.12 | 9.068870 | 10.000000 | 11.036530 | 12.190470 | 13.474590 | 14.903510 |
| 0.13 | 8.650280 | 9.526666 | 10.498990 | 11.582430 | 12.786980 | 14.125780 |
| 0.14 | 8.259741 | 9.084209 | 10.000000 | 11.017130 | 12.147890 | 13.403480 |
| 0.15 | 7.895187 | 8.672058 | 9.534301 | 10.490510 | 11.552670 | 12.731900 |
| 0.16 | 7.554141 | 8.286779 | 9.099380 | 10.000000 | 10.998720 | 12.106600 |
| 0.17 | 7.234974 | 7.926784 | 8.693151 | 9.542653 | 10.481740 | 11.524400 |
| 0.18 | 6.935890 | 7.589595 | 8.312960 | 9.113885 | 10.000000 | 10.980620 |
| 0.19 | 6.655455 | 7.273785 | 7.957330 | 8.713262 | 9.549790 | 10.472990 |
| 0.20 | 6.392080 | 6.977461 | 7.624072 | 8.338518 | 9.128122 | 10.000000 |
| 0.21 | 6.144593 | 6.699373 | 7.311519 | 7.987156 | 8.733109 | 9.557141 |
| 0.22 | 5.911755 | 6.437922 | 7.017915 | 7.657542 | 8.363208 | 9.141752 |
| 0.23 | 5.692557 | 6.192047 | 6.742093 | 7.348193 | 8.015993 | 8.752133 |
| 0.24 | 5.485921 | 5.960481 | 6.482632 | 7.057347 | 7.690163 | 8.387045 |
| 0.25 | 5.290990 | 5.742294 | 6.238276 | 6.783767 | 7.383800 | 8.044173 |
| 0.26 | 5.106956 | 5.536463 | 6.008083 | 6.526298 | 7.095769 | 7.721807 |
| 0.27 | 4.933045 | 5.342146 | 5.790929 | 6.283557 | 6.824442 | 7.418647 |
| 0.28 | 4.768518 | 5.158489 | 5.585917 | 6.054608 | 6.568835 | 7.133100 |
| 0.29 | 4.612762 | 4.984826 | 5.392166 | 5.838531 | 6.327682 | 6.864109 |
| 0.30 | 4.465205 | 4.820429 | 5.209000 | 5.634354 | 6.100129 | 6.610435 |
| 0.31 | 4.325286 | 4.664669 | 5.035615 | 5.441257 | 5.885058 | 6.370867 |
| 0.32 | 4.192478 | 4.517015 | 4.871346 | 5.258512 | 5.681746 | 6.144601 |
| 0.33 | 4.066339 | 4.376884 | 4.715648 | 5.085461 | 5.489304 | 5.930659 |
| 0.34 | 3.946452 | 4.243845 | 4.567942 | 4.921409 | 5.307107 | 5.728189 |

## Table 15-11. Fifteen-Year Escalation Table

| | | | | | | |
|---|---|---|---|---|---|---|
| | Present Worth of a Series of Escalating Payments Compounded Annually Discount-Escalation Factors for $n$ = 15 years | | | | | |
| | Annual Escalation Rate | | | | | |
| Discount Rate | 0.10 | 0.12 | 0.14 | 0.16 | 0.18 | 0.20 |
| 0.10 | 15.000000 | 17.377880 | 20.199780 | 23.549540 | 27.529640 | 32.259620 |
| 0.11 | 13.964150 | 16.126230 | 18.690120 | 21.727370 | 25.328490 | 29.601330 |
| 0.12 | 13.026090 | 15.000000 | 17.332040 | 20.090360 | 23.355070 | 27.221890 |
| 0.13 | 12.177030 | 13.981710 | 16.105770 | 18.616160 | 21.581750 | 25.087260 |
| 0.14 | 11.406510 | 13.057790 | 15.000000 | 17.287320 | 19.985530 | 23.169060 |
| 0.15 | 10.706220 | 12.220570 | 13.998120 | 16.086500 | 18.545150 | 21.442230 |
| 0.16 | 10.068030 | 11.459170 | 13.088900 | 15.000000 | 17.244580 | 19.884420 |
| 0.17 | 9.485654 | 10.766180 | 12.262790 | 14.015480 | 16.066830 | 18.477610 |
| 0.18 | 8.953083 | 10.133630 | 11.510270 | 13.118840 | 15.000000 | 17.203010 |
| 0.19 | 8.465335 | 9.555676 | 10.824310 | 12.303300 | 14.030830 | 16.047480 |
| 0.20 | 8.017635 | 9.026333 | 10.197550 | 11.560150 | 13.148090 | 15.000000 |
| 0.21 | 7.606115 | 8.540965 | 9.623969 | 10.881130 | 12.343120 | 14.046400 |
| 0.22 | 7.227109 | 8.094845 | 9.097863 | 10.259820 | 11.608480 | 13.176250 |
| 0.23 | 6.877548 | 7.684317 | 8.614813 | 9.690559 | 10.936240 | 12.381480 |
| 0.24 | 6.554501 | 7.305762 | 8.170423 | 9.167798 | 10.320590 | 11.655310 |
| 0.25 | 6.255518 | 6.956243 | 7.760848 | 8.687104 | 9.755424 | 10.990130 |
| 0.26 | 5.978393 | 6.632936 | 7.382943 | 8.244519 | 9.236152 | 10.379760 |
| 0.27 | 5.721101 | 6.333429 | 7.033547 | 7.836080 | 8.757889 | 9.819020 |
| 0.28 | 5.481814 | 6.055485 | 6.710042 | 7.458700 | 8.316982 | 9.302823 |
| 0.29 | 5.258970 | 5.797236 | 6.410005 | 7.109541 | 7.909701 | 8.827153 |
| 0.30 | 5.051153 | 5.556882 | 6.131433 | 6.785917 | 7.533113 | 8.388091 |
| 0.31 | 4.857052 | 5.332839 | 5.872303 | 6.485500 | 7.184156 | 7.982019 |
| 0.32 | 4.675478 | 5.123753 | 5.630905 | 6.206250 | 6.860492 | 7.606122 |
| 0.33 | 4.505413 | 4.928297 | 5.405771 | 5.946343 | 6.559743 | 7.257569 |
| 0.34 | 4.345926 | 4.745399 | 5.195502 | 5.704048 | 6.280019 | 6.933897 |

### Table 15-12. Twenty-Year Escalation Table

Present Worth of a Series of Escalating Payments Compounded Annually
Discount-Escalation Factors for $n$ = 20 Years

| Discount Rate | Annual Escalation Rate | | | | | |
|---|---|---|---|---|---|---|
| | 0.10 | 0.12 | 0.14 | 0.16 | 0.18 | 0.20 |
| 0.10 | 20.000000 | 24.295450 | 29.722090 | 36.592170 | 45.308970 | 56.383330 |
| 0.11 | 18.213210 | 22.002090 | 26.776150 | 32.799710 | 40.417480 | 50.067940 |
| 0.12 | 16.642370 | 20.000000 | 24.210030 | 29.505400 | 36.181240 | 44.614710 |
| 0.13 | 15.259850 | 18.243100 | 21.964990 | 26.634490 | 32.502270 | 39.891400 |
| 0.14 | 14.038630 | 16.694830 | 20.000000 | 24.127100 | 29.298170 | 35.789680 |
| 0.15 | 12.957040 | 15.329770 | 18.271200 | 21.929940 | 26.498510 | 32.218060 |
| 0.16 | 11.995640 | 14.121040 | 16.746150 | 20.000000 | 24.047720 | 29.098950 |
| 0.17 | 11.138940 | 13.048560 | 15.397670 | 18.300390 | 21.894660 | 26.369210 |
| 0.18 | 10.373120 | 12.093400 | 14.201180 | 16.795710 | 20.000000 | 23.970940 |
| 0.19 | 9.686791 | 11.240870 | 13.137510 | 15.463070 | 18.326720 | 21.860120 |
| 0.20 | 9.069737 | 10.477430 | 12.188860 | 14.279470 | 16.844020 | 20.000000 |
| 0.21 | 8.513605 | 9.792256 | 11.340570 | 13.224610 | 15.527270 | 18.353210 |
| 0.22 | 8.010912 | 9.175267 | 10.579620 | 12.282120 | 14.355520 | 16.890730 |
| 0.23 | 7.555427 | 8.618459 | 9.895583 | 11.438060 | 13.309280 | 15.589300 |
| 0.24 | 7.141531 | 8.114476 | 9.278916 | 10.679810 | 12.373300 | 14.429370 |
| 0.25 | 6.764528 | 7.657278 | 8.721467 | 9.997057 | 11.533310 | 13.392180 |
| 0.26 | 6.420316 | 7.241402 | 8.216490 | 9.380883 | 10.778020 | 12.462340 |
| 0.27 | 6.105252 | 6.862203 | 7.757722 | 8.823063 | 10.096710 | 11.626890 |
| 0.28 | 5.816151 | 6.515563 | 7.339966 | 8.316995 | 9.480940 | 10.874120 |
| 0.29 | 5.550301 | 6.198027 | 6.958601 | 7.856833 | 8.922847 | 10.194520 |
| 0.30 | 5.305312 | 5.906440 | 6.609778 | 7.437339 | 8.416060 | 9.579437 |
| 0.31 | 5.079039 | 5.638064 | 6.289875 | 7.054007 | 7.954518 | 9.021190 |
| 0.32 | 4.869585 | 5.390575 | 5.995840 | 6.702967 | 7.533406 | 8.513612 |
| 0.33 | 4.675331 | 5.161809 | 5.725066 | 6.380829 | 7.148198 | 8.050965 |
| 0.34 | 4.494838 | 4.949990 | 5.475180 | 6.084525 | 6.795200 | 7.628322 |

# STEAM TABLES

## Table 15-13. Saturated Steam: Temperature

| Temp Fahr t | Abs Press. Lb per Sq In. p | Specific Volume | | | Enthalpy | | | Entropy | | | Temp Fahr t |
|---|---|---|---|---|---|---|---|---|---|---|---|
| | | Sat. Liquid $v_f$ | Evap $v_{fg}$ | Sat. Vapor $v_g$ | Sat. Liquid $h_f$ | Evap $h_{fg}$ | Sat. Vapor $h_g$ | Sat. Liquid $s_f$ | Evap $s_{fg}$ | Sat. Vapor $s_g$ | |
| 32.0* | 0.08859 | 0.016022 | 3304.7 | 3304.7 | 0.0179 | 1075.5 | 1075.5 | 0.0000 | 2.1873 | 2.1873 | 32.0* |
| 34.0 | 0.09600 | 0.016021 | 3061.9 | 3061.9 | 1.996 | 1074.4 | 1076.4 | 0.0041 | 2.1762 | 2.1802 | 34.0 |
| 36.0 | 0.10395 | 0.016020 | 2839.0 | 2839.0 | 4.008 | 1073.2 | 1077.2 | 0.0081 | 2.1651 | 2.1732 | 36.0 |
| 38.0 | 0.11249 | 0.016019 | 2634.1 | 2634.2 | 6.018 | 1072.1 | 1078.1 | 0.0122 | 2.1541 | 2.1663 | 38.0 |
| 40.0 | 0.12163 | 0.016019 | 2445.8 | 2445.8 | 8.027 | 1071.0 | 1079.0 | 0.0162 | 2.1432 | 2.1594 | 40.0 |
| 42.0 | 0.13143 | 0.016019 | 2272.4 | 2272.4 | 10.035 | 1069.8 | 1079.9 | 0.0202 | 2.1325 | 2.1527 | 42.0 |
| 44.0 | 0.14192 | 0.016019 | 2112.8 | 2112.8 | 12.041 | 1068.7 | 1080.7 | 0.0242 | 2.1217 | 2.1459 | 44.0 |
| 46.0 | 0.15314 | 0.016020 | 1965.7 | 1965.7 | 14.047 | 1067.6 | 1081.6 | 0.0282 | 2.1111 | 2.1393 | 46.0 |
| 48.0 | 0.16514 | 0.016021 | 1830.0 | 1830.0 | 16.051 | 1066.4 | 1082.5 | 0.0321 | 2.1006 | 2.1327 | 48.0 |
| 50.0 | 0.17796 | 0.016023 | 1704.8 | 1704.8 | 18.054 | 1065.3 | 1083.4 | 0.0361 | 2.0901 | 2.1262 | 50.0 |
| 52.0 | 0.19165 | 0.016024 | 1589.2 | 1589.2 | 20.057 | 1064.2 | 1084.2 | 0.0400 | 2.0798 | 2.1197 | 52.0 |
| 54.0 | 0.20625 | 0.016026 | 1482.4 | 1482.4 | 22.058 | 1063.1 | 1085.1 | 0.0439 | 2.0695 | 2.1134 | 54.0 |
| 56.0 | 0.22183 | 0.016028 | 1383.6 | 1383.6 | 24.059 | 1061.9 | 1086.0 | 0.0478 | 2.0593 | 2.1070 | 56.0 |
| 58.0 | 0.23843 | 0.016031 | 1292.2 | 1292.2 | 26.060 | 1060.8 | 1086.9 | 0.0516 | 2.0491 | 2.1008 | 58.0 |
| 60.0 | 0.25611 | 0.016033 | 1207.6 | 1207.6 | 28.060 | 1059.7 | 1087.7 | 0.0555 | 2.0391 | 2.0946 | 60.0 |
| 62.0 | 0.27494 | 0.016036 | 1129.2 | 1129.2 | 30.059 | 1058.5 | 1088.6 | 0.0593 | 2.0291 | 2.0885 | 62.0 |
| 64.0 | 0.29497 | 0.016039 | 1056.5 | 1056.5 | 32.058 | 1057.4 | 1089.5 | 0.0632 | 2.0192 | 2.0824 | 64.0 |
| 66.0 | 0.31626 | 0.016043 | 989.0 | 989.1 | 34.056 | 1056.3 | 1090.4 | 0.0670 | 2.0094 | 2.0764 | 66.0 |
| 68.0 | 0.33889 | 0.016046 | 926.5 | 926.5 | 36.054 | 1055.2 | 1091.2 | 0.0708 | 1.9996 | 2.0704 | 68.0 |
| 70.0 | 0.36292 | 0.016050 | 868.3 | 868.4 | 38.052 | 1054.0 | 1092.1 | 0.0745 | 1.9900 | 2.0645 | 70.0 |
| 72.0 | 0.38844 | 0.016054 | 814.3 | 814.3 | 40.049 | 1052.9 | 1093.0 | 0.0783 | 1.9804 | 2.0587 | 72.0 |
| 74.0 | 0.41550 | 0.016058 | 764.1 | 764.1 | 42.046 | 1051.8 | 1093.8 | 0.0821 | 1.9708 | 2.0529 | 74.0 |
| 76.0 | 0.44420 | 0.016063 | 717.4 | 717.4 | 44.043 | 1050.7 | 1094.7 | 0.0858 | 1.9614 | 2.0472 | 76.0 |
| 78.0 | 0.47461 | 0.016067 | 673.8 | 673.9 | 46.040 | 1049.5 | 1095.6 | 0.0895 | 1.9520 | 2.0415 | 78.0 |
| 80.0 | 0.50683 | 0.016072 | 633.3 | 633.3 | 48.037 | 1048.4 | 1096.4 | 0.0932 | 1.9426 | 2.0359 | 80.0 |
| 82.0 | 0.54093 | 0.016077 | 595.5 | 595.5 | 50.033 | 1047.3 | 1097.3 | 0.0969 | 1.9334 | 2.0303 | 82.0 |
| 84.0 | 0.57702 | 0.016082 | 560.3 | 560.3 | 52.029 | 1046.1 | 1098.2 | 0.1006 | 1.9242 | 2.0248 | 84.0 |
| 86.0 | 0.61518 | 0.016087 | 527.5 | 527.5 | 54.026 | 1045.0 | 1099.0 | 0.1043 | 1.9151 | 2.0193 | 86.0 |
| 88.0 | 0.65551 | 0.016093 | 496.8 | 496.8 | 56.022 | 1043.9 | 1099.9 | 0.1079 | 1.9060 | 2.0139 | 88.0 |
| 90.0 | 0.69813 | 0.016099 | 468.1 | 468.1 | 58.018 | 1042.7 | 1100.8 | 0.1115 | 1.8970 | 2.0086 | 90.0 |
| 92.0 | 0.74313 | 0.016105 | 441.3 | 441.3 | 60.014 | 1041.6 | 1101.6 | 0.1152 | 1.8881 | 2.0033 | 92.0 |
| 94.0 | 0.79062 | 0.016111 | 416.3 | 416.3 | 62.010 | 1040.5 | 1102.5 | 0.1188 | 1.8792 | 1.9980 | 94.0 |
| 96.0 | 0.84072 | 0.016117 | 392.8 | 392.9 | 64.006 | 1039.3 | 1103.3 | 0.1224 | 1.8704 | 1.9928 | 96.0 |
| 98.0 | 0.89356 | 0.016123 | 370.9 | 370.9 | 66.003 | 1038.2 | 1104.2 | 0.1260 | 1.8617 | 1.9876 | 98.0 |

*The states shown are metastable

| Temp Fahr t | Abs Press. Lb per Sq In. p | Specific Volume Sat. Liquid $v_f$ | Evap $v_{fg}$ | Sat. Vapor $v_g$ | Enthalpy Sat. Liquid $h_f$ | Evap $h_{fg}$ | Sat. Vapor $h_g$ | Entropy Sat. Liquid $s_f$ | Evap $s_{fg}$ | Sat. Vapor $s_g$ | Temp Fahr t |
|---|---|---|---|---|---|---|---|---|---|---|---|
| 100.0 | 0.94924 | 0.016130 | 350.4 | 350.4 | 67.999 | 1037.1 | 1105.1 | 0.1295 | 1.8530 | 1.9825 | 100.0 |
| 102.0 | 1.00789 | 0.016137 | 331.1 | 331.1 | 69.995 | 1035.9 | 1105.9 | 0.1331 | 1.8444 | 1.9775 | 102.0 |
| 104.0 | 1.06965 | 0.016144 | 313.1 | 313.1 | 71.992 | 1034.8 | 1106.8 | 0.1366 | 1.8358 | 1.9725 | 104.0 |
| 106.0 | 1.1347 | 0.016151 | 296.16 | 296.18 | 73.99 | 1033.6 | 1107.6 | 0.1402 | 1.8273 | 1.9675 | 106.0 |
| 108.0 | 1.2030 | 0.016158 | 280.28 | 280.30 | 75.98 | 1032.5 | 1108.5 | 0.1437 | 1.8188 | 1.9626 | 108.0 |
| 110.0 | 1.2750 | 0.016165 | 265.37 | 265.39 | 77.98 | 1031.4 | 1109.3 | 0.1472 | 1.8105 | 1.9577 | 110.0 |
| 112.0 | 1.3505 | 0.016173 | 251.37 | 251.38 | 79.98 | 1030.2 | 1110.2 | 0.1507 | 1.8021 | 1.9528 | 112.0 |
| 114.0 | 1.4299 | 0.016180 | 238.21 | 238.22 | 81.97 | 1029.1 | 1111.0 | 0.1542 | 1.7938 | 1.9480 | 114.0 |
| 116.0 | 1.5133 | 0.016188 | 225.84 | 225.85 | 83.97 | 1027.9 | 1111.9 | 0.1577 | 1.7856 | 1.9433 | 116.0 |
| 118.0 | 1.6009 | 0.016196 | 214.20 | 214.21 | 85.97 | 1026.8 | 1112.7 | 0.1611 | 1.7774 | 1.9386 | 118.0 |
| 120.0 | 1.6927 | 0.016204 | 203.25 | 203.26 | 87.97 | 1025.6 | 1113.6 | 0.1646 | 1.7693 | 1.9339 | 120.0 |
| 122.0 | 1.7891 | 0.016213 | 192.94 | 192.95 | 89.96 | 1024.5 | 1114.4 | 0.1680 | 1.7613 | 1.9293 | 122.0 |
| 124.0 | 1.8901 | 0.016221 | 183.23 | 183.24 | 91.96 | 1023.3 | 1115.3 | 0.1715 | 1.7533 | 1.9247 | 124.0 |
| 126.0 | 1.9959 | 0.016229 | 174.08 | 174.09 | 93.96 | 1022.2 | 1116.1 | 0.1749 | 1.7453 | 1.9202 | 126.0 |
| 128.0 | 2.1068 | 0.016238 | 165.45 | 165.47 | 95.96 | 1021.0 | 1117.0 | 0.1783 | 1.7374 | 1.9157 | 128.0 |
| 130.0 | 2.2230 | 0.016247 | 157.32 | 157.33 | 97.96 | 1019.8 | 1117.8 | 0.1817 | 1.7295 | 1.9112 | 130.0 |
| 132.0 | 2.3445 | 0.016256 | 149.64 | 149.66 | 99.95 | 1018.7 | 1118.6 | 0.1851 | 1.7217 | 1.9068 | 132.0 |
| 134.0 | 2.4717 | 0.016265 | 142.40 | 142.41 | 101.95 | 1017.5 | 1119.5 | 0.1884 | 1.7140 | 1.9024 | 134.0 |
| 136.0 | 2.6047 | 0.016274 | 135.55 | 135.57 | 103.95 | 1016.4 | 1120.3 | 0.1918 | 1.7063 | 1.8980 | 136.0 |
| 138.0 | 2.7438 | 0.016284 | 129.09 | 129.11 | 105.95 | 1015.2 | 1121.1 | 0.1951 | 1.6986 | 1.8937 | 138.0 |
| 140.0 | 2.8892 | 0.016293 | 122.98 | 123.00 | 107.95 | 1014.0 | 1122.0 | 0.1985 | 1.6910 | 1.8895 | 140.0 |
| 142.0 | 3.0411 | 0.016303 | 117.21 | 117.22 | 109.95 | 1012.9 | 1122.8 | 0.2018 | 1.6834 | 1.8852 | 142.0 |
| 144.0 | 3.1997 | 0.016312 | 111.74 | 111.76 | 111.95 | 1011.7 | 1123.6 | 0.2051 | 1.6759 | 1.8810 | 144.0 |
| 146.0 | 3.3653 | 0.016322 | 106.58 | 106.59 | 113.95 | 1010.5 | 1124.5 | 0.2084 | 1.6684 | 1.8769 | 146.0 |
| 148.0 | 3.5381 | 0.016332 | 101.68 | 101.70 | 115.95 | 1009.3 | 1125.3 | 0.2117 | 1.6610 | 1.8727 | 148.0 |
| 150.0 | 3.7184 | 0.016343 | 97.05 | 97.07 | 117.95 | 1008.2 | 1126.1 | 0.2150 | 1.6536 | 1.8686 | 150.0 |
| 152.0 | 3.9065 | 0.016353 | 92.66 | 92.68 | 119.95 | 1007.0 | 1126.9 | 0.2183 | 1.6463 | 1.8646 | 152.0 |
| 154.0 | 4.1025 | 0.016363 | 88.50 | 88.52 | 121.95 | 1005.8 | 1127.7 | 0.2216 | 1.6390 | 1.8606 | 154.0 |
| 156.0 | 4.3068 | 0.016374 | 84.56 | 84.57 | 123.95 | 1004.6 | 1128.6 | 0.2248 | 1.6318 | 1.8566 | 156.0 |
| 158.0 | 4.5197 | 0.016384 | 80.82 | 80.83 | 125.96 | 1003.4 | 1129.4 | 0.2281 | 1.6245 | 1.8526 | 158.0 |
| 160.0 | 4.7414 | 0.016395 | 77.27 | 77.29 | 127.96 | 1002.2 | 1130.2 | 0.2313 | 1.6174 | 1.8487 | 160.0 |
| 162.0 | 4.9722 | 0.016406 | 73.90 | 73.92 | 129.96 | 1001.0 | 1131.0 | 0.2345 | 1.6103 | 1.8448 | 162.0 |
| 164.0 | 5.2124 | 0.016417 | 70.70 | 70.72 | 131.96 | 999.8 | 1131.8 | 0.2377 | 1.6032 | 1.8409 | 164.0 |
| 166.0 | 5.4623 | 0.016428 | 67.67 | 67.68 | 133.97 | 998.6 | 1132.6 | 0.2409 | 1.5961 | 1.8371 | 166.0 |
| 168.0 | 5.7223 | 0.016440 | 64.78 | 64.80 | 135.97 | 997.4 | 1133.4 | 0.2441 | 1.5892 | 1.8333 | 168.0 |
| 170.0 | 5.9926 | 0.016451 | 62.04 | 62.06 | 137.97 | 996.2 | 1134.2 | 0.2473 | 1.5822 | 1.8295 | 170.0 |
| 172.0 | 6.2736 | 0.016463 | 59.43 | 59.45 | 139.98 | 995.0 | 1135.0 | 0.2505 | 1.5753 | 1.8258 | 172.0 |
| 174.0 | 6.5656 | 0.016474 | 56.95 | 56.97 | 141.98 | 993.8 | 1135.8 | 0.2537 | 1.5684 | 1.8221 | 174.0 |
| 176.0 | 6.8690 | 0.016486 | 54.59 | 54.61 | 143.99 | 992.6 | 1136.6 | 0.2568 | 1.5616 | 1.8184 | 176.0 |
| 178.0 | 7.1840 | 0.016498 | 52.35 | 52.36 | 145.99 | 991.4 | 1137.4 | 0.2600 | 1.5548 | 1.8147 | 178.0 |

## Table 15-13. (Continued)

| Temp °F | P | vf | vfg | vg | hf | hfg | hg | sf | sfg | sg |
|---|---|---|---|---|---|---|---|---|---|---|
| 180.0 | 7.5110 | 0.016510 | 50.21 | 50.22 | 148.00 | 990.2 | 1138.2 | 0.2631 | 1.5480 | 1.8111 |
| 182.0 | 7.850 | 0.016522 | 48.172 | 48.189 | 150.01 | 989.0 | 1139.0 | 0.2662 | 1.5413 | 1.8075 |
| 184.0 | 8.203 | 0.016534 | 46.232 | 46.249 | 152.01 | 987.8 | 1139.8 | 0.2694 | 1.5346 | 1.8040 |
| 186.0 | 8.568 | 0.016547 | 44.383 | 44.400 | 154.02 | 986.5 | 1140.5 | 0.2725 | 1.5279 | 1.8004 |
| 188.0 | 8.947 | 0.016559 | 42.621 | 42.638 | 156.03 | 985.3 | 1141.3 | 0.2756 | 1.5213 | 1.7969 |
| 190.0 | 9.340 | 0.016572 | 40.941 | 40.957 | 158.04 | 984.1 | 1142.1 | 0.2787 | 1.5148 | 1.7934 |
| 192.0 | 9.747 | 0.016585 | 39.337 | 39.354 | 160.05 | 982.8 | 1142.9 | 0.2818 | 1.5082 | 1.7900 |
| 194.0 | 10.168 | 0.016598 | 37.808 | 37.824 | 162.05 | 981.6 | 1143.7 | 0.2848 | 1.5017 | 1.7865 |
| 196.0 | 10.605 | 0.016611 | 36.348 | 36.364 | 164.06 | 980.4 | 1144.4 | 0.2879 | 1.4952 | 1.7831 |
| 198.0 | 11.058 | 0.016624 | 34.954 | 34.970 | 166.08 | 979.1 | 1145.2 | 0.2910 | 1.4888 | 1.7798 |
| 200.0 | 11.526 | 0.016637 | 33.622 | 33.639 | 168.09 | 977.9 | 1146.0 | 0.2940 | 1.4824 | 1.7764 |
| 204.0 | 12.512 | 0.016664 | 31.135 | 31.151 | 172.11 | 975.4 | 1147.5 | 0.3001 | 1.4697 | 1.7698 |
| 208.0 | 13.568 | 0.016691 | 28.862 | 28.878 | 176.14 | 972.8 | 1149.0 | 0.3061 | 1.4571 | 1.7632 |
| 212.0 | 14.696 | 0.016719 | 26.782 | 26.799 | 180.17 | 970.3 | 1150.5 | 0.3121 | 1.4447 | 1.7568 |
| 216.0 | 15.901 | 0.016747 | 24.878 | 24.894 | 184.20 | 967.8 | 1152.0 | 0.3181 | 1.4323 | 1.7505 |
| 220.0 | 17.186 | 0.016775 | 23.131 | 23.148 | 188.23 | 965.2 | 1153.4 | 0.3241 | 1.4201 | 1.7442 |
| 224.0 | 18.556 | 0.016805 | 21.529 | 21.545 | 192.27 | 962.6 | 1154.9 | 0.3300 | 1.4081 | 1.7380 |
| 228.0 | 20.015 | 0.016834 | 20.056 | 20.073 | 196.31 | 960.0 | 1156.3 | 0.3359 | 1.3961 | 1.7320 |
| 232.0 | 21.567 | 0.016864 | 18.701 | 18.718 | 200.35 | 957.4 | 1157.8 | 0.3417 | 1.3842 | 1.7260 |
| 236.0 | 23.216 | 0.016895 | 17.454 | 17.471 | 204.40 | 954.8 | 1159.2 | 0.3476 | 1.3725 | 1.7201 |
| 240.0 | 24.968 | 0.016926 | 16.304 | 16.321 | 208.45 | 952.1 | 1160.6 | 0.3533 | 1.3609 | 1.7142 |
| 244.0 | 26.826 | 0.016958 | 15.243 | 15.260 | 212.50 | 949.5 | 1162.0 | 0.3591 | 1.3494 | 1.7085 |
| 248.0 | 28.796 | 0.016990 | 14.264 | 14.281 | 216.56 | 946.8 | 1163.4 | 0.3649 | 1.3379 | 1.7028 |
| 252.0 | 30.883 | 0.017022 | 13.358 | 13.375 | 220.62 | 944.1 | 1164.7 | 0.3706 | 1.3266 | 1.6972 |
| 256.0 | 33.091 | 0.017055 | 12.520 | 12.538 | 224.69 | 941.4 | 1166.1 | 0.3763 | 1.3154 | 1.6917 |
| 260.0 | 35.427 | 0.017089 | 11.745 | 11.762 | 228.76 | 938.6 | 1167.4 | 0.3819 | 1.3043 | 1.6862 |
| 264.0 | 37.894 | 0.017123 | 11.025 | 11.042 | 232.83 | 935.9 | 1168.7 | 0.3876 | 1.2933 | 1.6808 |
| 268.0 | 40.500 | 0.017157 | 10.358 | 10.375 | 236.91 | 933.1 | 1170.0 | 0.3932 | 1.2823 | 1.6755 |
| 272.0 | 43.249 | 0.017193 | 9.738 | 9.755 | 240.99 | 930.3 | 1171.3 | 0.3987 | 1.2715 | 1.6702 |
| 276.0 | 46.147 | 0.017228 | 9.162 | 9.180 | 245.08 | 927.5 | 1172.5 | 0.4043 | 1.2607 | 1.6650 |
| 280.0 | 49.200 | 0.017264 | 8.627 | 8.644 | 249.17 | 924.6 | 1173.8 | 0.4098 | 1.2501 | 1.6599 |
| 284.0 | 52.414 | 0.01730 | 8.1280 | 8.1453 | 253.3 | 921.7 | 1175.0 | 0.4154 | 1.2395 | 1.6548 |
| 288.0 | 55.795 | 0.01734 | 7.6334 | 7.6807 | 257.4 | 918.8 | 1176.2 | 0.4208 | 1.2290 | 1.6498 |
| 292.0 | 59.350 | 0.01738 | 7.2301 | 7.2475 | 261.5 | 915.9 | 1177.4 | 0.4263 | 1.2186 | 1.6449 |
| 296.0 | 63.084 | 0.01741 | 6.8259 | 6.8433 | 265.6 | 913.0 | 1178.6 | 0.4317 | 1.2082 | 1.6400 |
| **300.0** | **67.005** | **0.01745** | **6.4483** | **6.4658** | **269.7** | **910.0** | **1179.7** | **0.4372** | **1.1979** | **1.6351** |
| **304.0** | **71.119** | **0.01749** | **6.0955** | **6.1130** | **273.8** | **907.0** | **1180.9** | **0.4426** | **1.1877** | **1.6303** |
| **308.0** | **75.433** | **0.01753** | **5.7655** | **5.7830** | **278.0** | **904.0** | **1182.0** | **0.4479** | **1.1776** | **1.6256** |
| **312.0** | **79.953** | **0.01757** | **5.4566** | **5.4742** | **282.1** | **901.0** | **1183.1** | **0.4533** | **1.1676** | **1.6209** |
| **316.0** | **84.688** | **0.01761** | **5.1673** | **5.1849** | **286.3** | **897.9** | **1184.1** | **0.4586** | **1.1576** | **1.6162** |
| **320.0** | **89.643** | **0.01766** | **4.8961** | **4.9138** | **290.4** | **894.8** | **1185.2** | **0.4640** | **1.1477** | **1.6116** |
| **324.0** | **94.826** | **0.01770** | **4.6418** | **4.6595** | **294.6** | **891.6** | **1186.2** | **0.4692** | **1.1378** | **1.6071** |
| **328.0** | **100.245** | **0.01774** | **4.4030** | **4.4208** | **298.7** | **888.5** | **1187.2** | **0.4745** | **1.1280** | **1.6025** |
| **332.0** | **105.907** | **0.01779** | **4.1788** | **4.1966** | **302.9** | **885.3** | **1188.2** | **0.4798** | **1.1183** | **1.5981** |
| **336.0** | **111.820** | **0.01783** | **3.9681** | **3.9859** | **307.1** | **882.1** | **1189.1** | **0.4850** | **1.1086** | **1.5936** |

| Temp Fahr t | Abs Press. Lb per Sq In. p | Specific Volume Sat. Liquid $v_f$ | Specific Volume Evap $v_{fg}$ | Specific Volume Sat. Vapor $v_g$ | Enthalpy Sat. Liquid $h_f$ | Enthalpy Evap $h_{fg}$ | Enthalpy Sat. Vapor $h_g$ | Entropy Sat. Liquid $s_f$ | Entropy Evap $s_{fg}$ | Entropy Sat. Vapor $s_g$ | Temp Fahr t |
|---|---|---|---|---|---|---|---|---|---|---|---|
| 340.0 | 117.992 | 0.01787 | 3.7699 | 3.7878 | 311.3 | 878.8 | 1190.1 | 0.4902 | 1.0990 | 1.5892 | 340.0 |
| 344.0 | 124.430 | 0.01792 | 3.5834 | 3.6013 | 315.5 | 875.5 | 1191.0 | 0.4954 | 1.0894 | 1.5849 | 344.0 |
| 348.0 | 131.142 | 0.01797 | 3.4078 | 3.4258 | 319.7 | 872.2 | 1191.1 | 0.5006 | 1.0799 | 1.5806 | 348.0 |
| 352.0 | 138.138 | 0.01801 | 3.2423 | 3.2603 | 323.9 | 868.9 | 1192.7 | 0.5058 | 1.0705 | 1.5763 | 352.0 |
| 356.0 | 145.424 | 0.01806 | 3.0863 | 3.1044 | 328.1 | 865.5 | 1193.6 | 0.5110 | 1.0611 | 1.5721 | 356.0 |
| 360.0 | 153.010 | 0.01811 | 2.9392 | 2.9573 | 332.3 | 862.1 | 1194.4 | 0.5161 | 1.0517 | 1.5678 | 360.0 |
| 364.0 | 160.903 | 0.01816 | 2.8002 | 2.8184 | 336.5 | 858.6 | 1195.2 | 0.5212 | 1.0424 | 1.5637 | 364.0 |
| 368.0 | 169.113 | 0.01821 | 2.6691 | 2.6873 | 340.8 | 855.1 | 1195.9 | 0.5263 | 1.0332 | 1.5595 | 368.0 |
| 372.0 | 177.648 | 0.01826 | 2.5451 | 2.5633 | 345.0 | 851.6 | 1196.7 | 0.5314 | 1.0240 | 1.5554 | 372.0 |
| 376.0 | 186.517 | 0.01831 | 2.4279 | 2.4462 | 349.3 | 848.1 | 1197.4 | 0.5365 | 1.0148 | 1.5513 | 376.0 |
| 380.0 | 195.729 | 0.01836 | 2.3170 | 2.3353 | 353.6 | 844.5 | 1198.0 | 0.5416 | 1.0057 | 1.5473 | 380.0 |
| 384.0 | 205.294 | 0.01842 | 2.2120 | 2.2304 | 357.9 | 840.8 | 1198.7 | 0.5466 | 0.9966 | 1.5432 | 384.0 |
| 388.0 | 215.220 | 0.01847 | 2.1126 | 2.1311 | 362.2 | 837.2 | 1199.3 | 0.5516 | 0.9876 | 1.5392 | 388.0 |
| 392.0 | 225.516 | 0.01853 | 2.0184 | 2.0369 | 366.5 | 833.4 | 1199.9 | 0.5567 | 0.9786 | 1.5352 | 392.0 |
| 396.0 | 236.193 | 0.01858 | 1.9291 | 1.9477 | 370.8 | 829.7 | 1200.4 | 0.5617 | 0.9696 | 1.5313 | 396.0 |
| 400.0 | 247.259 | 0.01864 | 1.8444 | 1.8630 | 375.1 | 825.9 | 1201.0 | 0.5667 | 0.9607 | 1.5274 | 400.0 |
| 404.0 | 258.725 | 0.01870 | 1.7640 | 1.7827 | 379.4 | 822.0 | 1201.5 | 0.5717 | 0.9518 | 1.5234 | 404.0 |
| 408.0 | 270.600 | 0.01875 | 1.6877 | 1.7064 | 383.8 | 818.2 | 1201.9 | 0.5766 | 0.9429 | 1.5195 | 408.0 |
| 412.0 | 282.894 | 0.01881 | 1.6152 | 1.6340 | 388.1 | 814.2 | 1202.4 | 0.5816 | 0.9341 | 1.5157 | 412.0 |
| 416.0 | 295.617 | 0.01887 | 1.5463 | 1.5651 | 392.5 | 810.2 | 1202.8 | 0.5866 | 0.9253 | 1.5118 | 416.0 |
| 420.0 | 308.780 | 0.01894 | 1.4808 | 1.4997 | 396.9 | 806.2 | 1203.1 | 0.5915 | 0.9165 | 1.5080 | 420.0 |
| 424.0 | 322.391 | 0.01900 | 1.4184 | 1.4374 | 401.3 | 802.2 | 1203.5 | 0.5964 | 0.9077 | 1.5042 | 424.0 |
| 428.0 | 336.463 | 0.01906 | 1.3591 | 1.3782 | 405.7 | 798.0 | 1203.7 | 0.6014 | 0.8990 | 1.5004 | 428.0 |
| 432.0 | 351.00 | 0.01913 | 1.30266 | 1.32179 | 410.1 | 793.9 | 1204.0 | 0.6063 | 0.8903 | 1.4966 | 432.0 |
| 436.0 | 366.03 | 0.01919 | 1.24887 | 1.26806 | 414.6 | 789.7 | 1204.2 | 0.6112 | 0.8816 | 1.4928 | 436.0 |
| 440.0 | 381.54 | 0.01926 | 1.19761 | 1.21687 | 419.0 | 785.4 | 1204.4 | 0.6161 | 0.8729 | 1.4890 | 440.0 |
| 444.0 | 397.56 | 0.01933 | 1.14874 | 1.16806 | 423.5 | 781.1 | 1204.6 | 0.6210 | 0.8643 | 1.4853 | 444.0 |
| 448.0 | 414.09 | 0.01940 | 1.10212 | 1.12152 | 428.0 | 776.7 | 1204.7 | 0.6259 | 0.8557 | 1.4815 | 448.0 |
| 452.0 | 431.14 | 0.01947 | 1.05764 | 1.07711 | 432.5 | 772.3 | 1204.8 | 0.6308 | 0.8471 | 1.4778 | 452.0 |
| 456.0 | 448.73 | 0.01954 | 1.01518 | 1.03472 | 437.0 | 767.8 | 1204.8 | 0.6356 | 0.8385 | 1.4741 | 456.0 |
| 460.0 | 466.87 | 0.01961 | 0.97463 | 0.99424 | 441.5 | 763.2 | 1204.8 | 0.6405 | 0.8299 | 1.4704 | 460.0 |
| 464.0 | 485.56 | 0.01969 | 0.93588 | 0.95557 | 446.1 | 758.6 | 1204.7 | 0.6454 | 0.8213 | 1.4667 | 464.0 |
| 468.0 | 504.83 | 0.01976 | 0.89885 | 0.91862 | 450.7 | 754.0 | 1204.6 | 0.6502 | 0.8127 | 1.4629 | 468.0 |
| 472.0 | 524.67 | 0.01984 | 0.86345 | 0.88329 | 455.2 | 749.3 | 1204.5 | 0.6551 | 0.8042 | 1.4592 | 472.0 |
| 476.0 | 545.11 | 0.01992 | 0.82958 | 0.84950 | 459.9 | 744.5 | 1204.3 | 0.6599 | 0.7956 | 1.4555 | 476.0 |

## Table 15-13. (Continued)

| | | | | | | | | | | | |
|---|---|---|---|---|---|---|---|---|---|---|---|
| **480.0** | 566.15 | 0.02000 | 0.79716 | 0.81717 | 464.5 | 739.6 | 1204.1 | 0.6648 | 0.7871 | 1.4518 | **480.0** |
| **484.0** | 587.81 | 0.02009 | 0.76613 | 0.78622 | 469.1 | 734.7 | 1203.8 | 0.6696 | 0.7785 | 1.4481 | **484.0** |
| **488.0** | 610.10 | 0.02017 | 0.73641 | 0.75658 | 473.8 | 729.7 | 1203.5 | 0.6745 | 0.7700 | 1.4444 | **488.0** |
| **492.0** | 633.03 | 0.02026 | 0.70794 | 0.72820 | 478.5 | 724.6 | 1203.1 | 0.6793 | 0.7614 | 1.4407 | **492.0** |
| **496.0** | 656.61 | 0.02034 | 0.68065 | 0.70100 | 483.2 | 719.5 | 1202.7 | 0.6842 | 0.7528 | 1.4370 | **496.0** |
| **500.0** | 680.86 | 0.02043 | 0.65448 | 0.67492 | 487.9 | 714.3 | 1202.2 | 0.6890 | 0.7443 | 1.4333 | **500.0** |
| **504.0** | 705.78 | 0.02053 | 0.62938 | 0.64991 | 492.7 | 709.0 | 1201.7 | 0.6939 | 0.7357 | 1.4296 | **504.0** |
| **508.0** | 731.40 | 0.02062 | 0.60530 | 0.62592 | 497.5 | 703.7 | 1201.1 | 0.6987 | 0.7271 | 1.4258 | **508.0** |
| **512.0** | 757.72 | 0.02072 | 0.58218 | 0.60289 | 502.3 | 698.2 | 1200.5 | 0.7036 | 0.7185 | 1.4221 | **512.0** |
| **516.0** | 784.76 | 0.02081 | 0.55997 | 0.58079 | 507.1 | 692.7 | 1199.8 | 0.7085 | 0.7099 | 1.4183 | **516.0** |
| **520.0** | 812.53 | 0.02091 | 0.53864 | 0.55956 | 512.0 | 687.0 | 1199.0 | 0.7133 | 0.7013 | 1.4146 | **520.0** |
| **524.0** | 841.04 | 0.02102 | 0.51814 | 0.53916 | 516.9 | 681.3 | 1198.2 | 0.7182 | 0.6926 | 1.4108 | **524.0** |
| **528.0** | 870.31 | 0.02112 | 0.49843 | 0.51955 | 521.8 | 675.5 | 1197.3 | 0.7231 | 0.6839 | 1.4070 | **528.0** |
| **532.0** | 900.34 | 0.02123 | 0.47947 | 0.50070 | 526.8 | 669.6 | 1196.4 | 0.7280 | 0.6752 | 1.4032 | **532.0** |
| **536.0** | 931.17 | 0.02134 | 0.46123 | 0.48257 | 531.7 | 663.6 | 1195.4 | 0.7329 | 0.6665 | 1.3993 | **536.0** |
| **540.0** | 962.79 | 0.02146 | 0.44367 | 0.46513 | 536.8 | 657.5 | 1194.3 | 0.7378 | 0.6577 | 1.3954 | **540.0** |
| **544.0** | 995.22 | 0.02157 | 0.42677 | 0.44834 | 541.8 | 651.3 | 1193.1 | 0.7427 | 0.6489 | 1.3915 | **544.0** |
| **548.0** | 1028.49 | 0.02169 | 0.41048 | 0.43217 | 546.9 | 645.0 | 1191.9 | 0.7476 | 0.6400 | 1.3876 | **548.0** |
| **552.0** | 1062.59 | 0.02182 | 0.39479 | 0.41660 | 552.0 | 638.5 | 1190.6 | 0.7525 | 0.6311 | 1.3837 | **552.0** |
| **556.0** | 1097.55 | 0.02194 | 0.37966 | 0.40160 | 557.2 | 632.0 | 1189.2 | 0.7575 | 0.6222 | 1.3797 | **556.0** |
| **560.0** | 1133.38 | 0.02207 | 0.36507 | 0.38714 | 562.4 | 625.3 | 1187.7 | 0.7625 | 0.6132 | 1.3757 | **560.0** |
| **564.0** | 1170.10 | 0.02221 | 0.35099 | 0.37320 | 567.6 | 618.5 | 1186.1 | 0.7674 | 0.6041 | 1.3716 | **564.0** |
| **568.0** | 1207.72 | 0.02235 | 0.33741 | 0.35975 | 572.9 | 611.5 | 1184.5 | 0.7725 | 0.5950 | 1.3675 | **568.0** |
| **572.0** | 1246.26 | 0.02249 | 0.32429 | 0.34678 | 578.3 | 604.5 | 1182.7 | 0.7775 | 0.5859 | 1.3634 | **572.0** |
| **576.0** | 1285.74 | 0.02264 | 0.31162 | 0.33426 | 583.7 | 597.2 | 1180.9 | 0.7825 | 0.5766 | 1.3592 | **576.0** |
| **580.0** | 1326.17 | 0.02279 | 0.29937 | 0.32216 | 589.1 | 589.9 | 1179.0 | 0.7876 | 0.5673 | 1.3550 | **580.0** |
| **584.0** | 1367.7 | 0.02295 | 0.28753 | 0.31048 | 594.6 | 582.4 | 1176.9 | 0.7927 | 0.5580 | 1.3507 | **584.0** |
| **588.0** | 1410.0 | 0.02311 | 0.27608 | 0.29919 | 600.1 | 574.7 | 1174.8 | 0.7978 | 0.5485 | 1.3464 | **588.0** |
| **592.0** | 1453.3 | 0.02328 | 0.26499 | 0.28827 | 605.7 | 566.8 | 1172.6 | 0.8030 | 0.5390 | 1.3420 | **592.0** |
| **596.0** | 1497.8 | 0.02345 | 0.25425 | 0.27770 | 611.4 | 558.8 | 1170.2 | 0.8082 | 0.5293 | 1.3375 | **596.0** |
| **600.0** | 1543.2 | 0.02364 | 0.24384 | 0.26747 | 617.1 | 550.6 | 1167.7 | 0.8134 | 0.5196 | 1.3330 | **600.0** |
| **604.0** | 1589.7 | 0.02382 | 0.23374 | 0.25757 | 622.9 | 542.2 | 1165.1 | 0.8187 | 0.5097 | 1.3284 | **604.0** |
| **608.0** | 1637.3 | 0.02402 | 0.22394 | 0.24796 | 628.8 | 533.6 | 1162.4 | 0.8240 | 0.4997 | 1.3238 | **608.0** |
| **612.0** | 1686.1 | 0.02422 | 0.21442 | 0.23865 | 634.8 | 524.7 | 1159.5 | 0.8294 | 0.4896 | 1.3190 | **612.0** |
| **616.0** | 1735.9 | 0.02444 | 0.20516 | 0.22960 | 640.8 | 515.6 | 1156.4 | 0.8348 | 0.4794 | 1.3141 | **616.0** |
| **620.0** | 1786.9 | 0.02466 | 0.19615 | 0.22081 | 646.9 | 506.3 | 1153.2 | 0.8403 | 0.4689 | 1.3092 | **620.0** |
| **624.0** | 1839.0 | 0.02489 | 0.18737 | 0.21226 | 653.1 | 496.6 | 1149.8 | 0.8458 | 0.4583 | 1.3041 | **624.0** |
| **628.0** | 1892.4 | 0.02514 | 0.17880 | 0.20394 | 659.5 | 486.7 | 1146.1 | 0.8514 | 0.4474 | 1.2988 | **628.0** |
| **632.0** | 1947.0 | 0.02539 | 0.17044 | 0.19583 | 665.9 | 476.4 | 1142.2 | 0.8571 | 0.4364 | 1.2934 | **632.0** |
| **636.0** | 2002.8 | 0.02566 | 0.16226 | 0.18792 | 672.4 | 465.7 | 1138.1 | 0.8628 | 0.4251 | 1.2879 | **636.0** |
| **640.0** | 2059.9 | 0.02595 | 0.15427 | 0.18021 | 679.1 | 454.6 | 1133.7 | 0.8686 | 0.4134 | 1.2821 | **640.0** |
| **644.0** | 2118.3 | 0.02625 | 0.14644 | 0.17269 | 685.9 | 443.1 | 1129.0 | 0.8746 | 0.4015 | 1.2761 | **644.0** |
| **648.0** | 2178.1 | 0.02657 | 0.13876 | 0.16534 | 692.9 | 431.1 | 1124.0 | 0.8806 | 0.3893 | 1.2699 | **648.0** |
| **652.0** | 2239.2 | 0.02691 | 0.13124 | 0.15816 | 700.0 | 418.7 | 1118.7 | 0.8868 | 0.3767 | 1.2634 | **652.0** |
| **656.0** | 2301.7 | 0.02728 | 0.12387 | 0.15115 | 707.4 | 405.7 | 1113.1 | 0.8931 | 0.3637 | 1.2567 | **656.0** |

| Temp Fahr t | Abs Press. lb per Sq In. p | Specific Volume | | | Enthalpy | | | Entropy | | | Temp Fahr t |
|---|---|---|---|---|---|---|---|---|---|---|---|
| | | Sat. Liquid $v_f$ | Evap $v_{fg}$ | Sat. Vapor $v_g$ | Sat. Liquid $h_f$ | Evap $h_{fg}$ | Sat. Vapor $h_g$ | Sat. Liquid $s_f$ | Evap $s_{fg}$ | Sat. Vapor $s_g$ | |
| **660.0** | 2365.7 | 0.02768 | 0.11663 | 0.14431 | 714.9 | 392.1 | 1107.0 | 0.8995 | 0.3502 | 1.2498 | **660.0** |
| **664.0** | 2431.1 | 0.02811 | 0.10947 | 0.13757 | 722.9 | 377.7 | 1100.6 | 0.9064 | 0.3361 | 1.2425 | **664.0** |
| **668.0** | 2498.1 | 0.02858 | 0.10229 | 0.13087 | 731.5 | 362.1 | 1093.5 | 0.9137 | 0.3210 | 1.2347 | **668.0** |
| **672.0** | 2566.6 | 0.02911 | 0.09514 | 0.12424 | 740.2 | 345.7 | 1085.9 | 0.9212 | 0.3054 | 1.2266 | **672.0** |
| **676.0** | 2636.8 | 0.02970 | 0.08799 | 0.11769 | 749.2 | 328.5 | 1077.6 | 0.9287 | 0.2892 | 1.2179 | **676.0** |
| **680.0** | 2708.6 | 0.03037 | 0.08080 | 0.11117 | 758.5 | 310.1 | 1068.5 | 0.9365 | 0.2720 | 1.2086 | **680.0** |
| **684.0** | 2782.1 | 0.03114 | 0.07349 | 0.10463 | 768.2 | 290.2 | 1058.4 | 0.9447 | 0.2537 | 1.1984 | **684.0** |
| **688.0** | 2857.4 | 0.03204 | 0.06595 | 0.09799 | 778.8 | 268.2 | 1047.0 | 0.9535 | 0.2337 | 1.1872 | **688.0** |
| **692.0** | 2934.5 | 0.03313 | 0.05797 | 0.09110 | 790.5 | 243.1 | 1033.6 | 0.9634 | 0.2110 | 1.1744 | **692.0** |
| **696.0** | 3013.4 | 0.03455 | 0.04916 | 0.08371 | 804.4 | 212.8 | 1017.2 | 0.9749 | 0.1841 | 1.1591 | **696.0** |
| **700.0** | 3094.3 | 0.03662 | 0.03857 | 0.07519 | 822.4 | 172.7 | 995.2 | 0.9901 | 0.1490 | 1.1390 | **700.0** |
| **702.0** | 3135.5 | 0.03824 | 0.03173 | 0.06997 | 835.0 | 144.7 | 979.7 | 1.0006 | 0.1246 | 1.1252 | **702.0** |
| **704.0** | 3177.2 | 0.04108 | 0.02192 | 0.06300 | 854.2 | 102.0 | 956.2 | 1.0169 | 0.0876 | 1.1046 | **704.0** |
| **705.0** | 3198.3 | 0.04427 | 0.01304 | 0.05730 | 873.0 | 61.4 | 934.4 | 1.0329 | 0.0527 | 1.0856 | **705.0** |
| **705.47*** | 3208.2 | 0.05078 | 0.00000 | 0.05078 | 906.0 | 0.0 | 906.0 | 1.0612 | 0.0000 | 1.0612 | **705.47*** |

*Critical temperature

## Table 15-14. Saturated Steam: Pressure Table

| Abs Press. Lb/Sq In. p | Temp Fahr t | Specific Volume Sat. Liquid v_f | Evap v_fg | Sat. Vapor v_g | Enthalpy Sat. Liquid h_f | Evap h_fg | Sat. Vapor h_g | Entropy Sat. Liquid s_f | Evap s_fg | Sat. Vapor s_g | Abs Press. Lb/Sq In. p |
|---|---|---|---|---|---|---|---|---|---|---|---|
| 0.08865 | 32.018 | 0.016022 | 3302.4 | 3302.4 | 0.0003 | 1075.5 | 1075.5 | 0.0000 | 2.1872 | 2.1872 | 0.08865 |
| 0.25 | 59.323 | 0.016032 | 1235.5 | 1235.5 | 27.382 | 1060.1 | 1087.4 | 0.0542 | 2.0425 | 2.0967 | 0.25 |
| 0.50 | 79.586 | 0.016071 | 641.5 | 641.5 | 47.623 | 1048.6 | 1096.6 | 0.0925 | 1.9446 | 2.0370 | 0.50 |
| 1.0 | 101.74 | 0.016136 | 333.59 | 333.60 | 69.73 | 1036.1 | 1105.8 | 0.1326 | 1.8455 | 1.9781 | 1.0 |
| 5.0 | 162.24 | 0.016407 | 73.515 | 73.532 | 130.20 | 1000.9 | 1131.1 | 0.2349 | 1.6094 | 1.8443 | 5.0 |
| 10.0 | 193.21 | 0.016592 | 38.404 | 38.420 | 161.26 | 982.1 | 1143.3 | 0.2836 | 1.5043 | 1.7879 | 10.0 |
| 14.696 | 212.00 | 0.016719 | 26.782 | 26.799 | 180.17 | 970.3 | 1150.5 | 0.3121 | 1.4447 | 1.7568 | 14.696 |
| 15.0 | 213.03 | 0.016726 | 26.274 | 26.290 | 181.21 | 969.7 | 1150.9 | 0.3137 | 1.4415 | 1.7552 | 15.0 |
| 20.0 | 227.96 | 0.016834 | 20.070 | 20.087 | 196.27 | 960.1 | 1156.3 | 0.3358 | 1.3962 | 1.7320 | 20.0 |
| 30.0 | 250.34 | 0.017009 | 13.7266 | 13.7436 | 218.9 | 945.2 | 1164.1 | 0.3682 | 1.3313 | 1.6995 | 30.0 |
| 40.0 | 267.25 | 0.017151 | 10.4794 | 10.4965 | 236.1 | 933.6 | 1169.8 | 0.3921 | 1.2844 | 1.6765 | 40.0 |
| 50.0 | 281.02 | 0.017274 | 8.4967 | 8.5140 | 250.2 | 923.9 | 1174.1 | 0.4112 | 1.2474 | 1.6586 | 50.0 |
| 60.0 | 292.71 | 0.017383 | 7.1562 | 7.1736 | 262.2 | 915.4 | 1177.6 | 0.4273 | 1.2167 | 1.6440 | 60.0 |
| 70.0 | 302.93 | 0.017482 | 6.1875 | 6.2050 | 272.7 | 907.8 | 1180.6 | 0.4411 | 1.1905 | 1.6316 | 70.0 |
| 80.0 | 312.04 | 0.017573 | 5.4536 | 5.4711 | 282.1 | 900.9 | 1183.1 | 0.4534 | 1.1675 | 1.6208 | 80.0 |
| 90.0 | 320.28 | 0.017659 | 4.8779 | 4.8953 | 290.7 | 894.6 | 1185.3 | 0.4643 | 1.1470 | 1.6113 | 90.0 |
| 100.0 | 327.82 | 0.017740 | 4.4133 | 4.4310 | 298.5 | 888.6 | 1187.2 | 0.4743 | 1.1284 | 1.6027 | 100.0 |
| 110.0 | 334.79 | 0.01782 | 4.0306 | 4.0484 | 305.8 | 883.1 | 1188.9 | 0.4834 | 1.1115 | 1.5950 | 110.0 |
| 120.0 | 341.27 | 0.01789 | 3.7097 | 3.7275 | 312.6 | 877.8 | 1190.4 | 0.4919 | 1.0960 | 1.5879 | 120.0 |
| 130.0 | 347.33 | 0.01796 | 3.4364 | 3.4544 | 319.0 | 872.8 | 1191.7 | 0.4998 | 1.0815 | 1.5813 | 130.0 |
| 140.0 | 353.04 | 0.01803 | 3.2010 | 3.2190 | 325.0 | 868.0 | 1193.0 | 0.5071 | 1.0681 | 1.5752 | 140.0 |
| 150.0 | 358.43 | 0.01809 | 2.9958 | 3.0139 | 330.6 | 863.4 | 1194.1 | 0.5141 | 1.0554 | 1.5695 | 150.0 |
| 160.0 | 363.55 | 0.01815 | 2.8155 | 2.8336 | 336.1 | 859.0 | 1195.1 | 0.5206 | 1.0435 | 1.5641 | 160.0 |
| 170.0 | 368.42 | 0.01821 | 2.6556 | 2.6738 | 341.2 | 854.8 | 1196.0 | 0.5269 | 1.0322 | 1.5591 | 170.0 |
| 180.0 | 373.08 | 0.01827 | 2.5129 | 2.5312 | 346.2 | 850.7 | 1196.9 | 0.5328 | 1.0215 | 1.5543 | 180.0 |
| 190.0 | 377.53 | 0.01833 | 2.3847 | 2.4030 | 350.9 | 846.7 | 1197.6 | 0.5384 | 1.0113 | 1.5498 | 190.0 |
| 200.0 | 381.80 | 0.01839 | 2.2689 | 2.2873 | 355.5 | 842.8 | 1198.3 | 0.5438 | 1.0016 | 1.5454 | 200.0 |
| 210.0 | 385.91 | 0.01844 | 2.16373 | 2.18217 | 359.9 | 839.1 | 1199.0 | 0.5490 | 0.9923 | 1.5413 | 210.0 |
| 220.0 | 389.88 | 0.01850 | 2.06779 | 2.08629 | 364.2 | 835.4 | 1199.6 | 0.5540 | 0.9834 | 1.5374 | 220.0 |
| 230.0 | 393.70 | 0.01855 | 1.97991 | 1.99846 | 368.3 | 831.8 | 1200.1 | 0.5588 | 0.9748 | 1.5336 | 230.0 |
| 240.0 | 397.39 | 0.01860 | 1.89909 | 1.91769 | 372.3 | 828.4 | 1200.6 | 0.5634 | 0.9665 | 1.5299 | 240.0 |
| 250.0 | 400.97 | 0.01865 | 1.82452 | 1.84317 | 376.1 | 825.0 | 1201.1 | 0.5679 | 0.9585 | 1.5264 | 250.0 |
| 260.0 | 404.44 | 0.01870 | 1.75548 | 1.77418 | 379.9 | 821.6 | 1201.5 | 0.5722 | 0.9508 | 1.5230 | 260.0 |
| 270.0 | 407.80 | 0.01875 | 1.69137 | 1.71013 | 383.6 | 818.3 | 1201.9 | 0.5764 | 0.9433 | 1.5197 | 270.0 |
| 280.0 | 411.07 | 0.01880 | 1.63169 | 1.65049 | 387.1 | 815.1 | 1202.3 | 0.5805 | 0.9361 | 1.5166 | 280.0 |
| 290.0 | 414.25 | 0.01885 | 1.57597 | 1.59482 | 390.6 | 812.0 | 1202.6 | 0.5844 | 0.9291 | 1.5135 | 290.0 |
| 300.0 | 417.35 | 0.01890 | 1.52384 | 1.54274 | 394.0 | 808.9 | 1202.9 | 0.5882 | 0.9223 | 1.5105 | 300.0 |
| 350.0 | 431.73 | 0.01912 | 1.30642 | 1.32554 | 409.8 | 794.2 | 1204.0 | 0.6059 | 0.8909 | 1.4968 | 350.0 |
| 400.0 | 444.60 | 0.01934 | 1.14162 | 1.16095 | 424.2 | 780.4 | 1204.6 | 0.6217 | 0.8630 | 1.4847 | 400.0 |

| Abs Press. Lb/Sq In. p | Temp Fahr t | Specific Volume | | | Enthalpy | | | Entropy | | | Abs Press. Lb/Sq In. p |
|---|---|---|---|---|---|---|---|---|---|---|---|
| | | Sat. Liquid $v_f$ | Evap $v_{fg}$ | Sat. Vapor $v_g$ | Sat. Liquid $h_f$ | Evap $h_{fg}$ | Sat. Vapor $h_g$ | Sat. Liquid $s_f$ | Evap $s_{fg}$ | Sat. Vapor $s_g$ | |
| 450.0 | 456.28 | 0.01954 | 1.01224 | 1.03179 | 437.3 | 767.5 | 1204.8 | 0.6360 | 0.8378 | 1.4738 | 450.0 |
| 500.0 | 467.01 | 0.01975 | 0.90787 | 0.92762 | 449.5 | 755.1 | 1204.7 | 0.6490 | 0.8148 | 1.4639 | 500.0 |
| 550.0 | 476.94 | 0.01994 | 0.82183 | 0.84177 | 460.9 | 743.3 | 1204.3 | 0.6611 | 0.7936 | 1.4547 | 550.0 |
| 600.0 | 486.20 | 0.02013 | 0.74962 | 0.76975 | 471.7 | 732.0 | 1203.7 | 0.6723 | 0.7738 | 1.4461 | 600.0 |
| 650.0 | 494.89 | 0.02032 | 0.68811 | 0.70843 | 481.9 | 720.9 | 1202.8 | 0.6828 | 0.7552 | 1.4381 | 650.0 |
| 700.0 | 503.08 | 0.02050 | 0.63505 | 0.65556 | 491.6 | 710.2 | 1201.8 | 0.6928 | 0.7377 | 1.4304 | 700.0 |
| 750.0 | 510.84 | 0.02069 | 0.58880 | 0.60949 | 500.9 | 699.8 | 1200.7 | 0.7022 | 0.7210 | 1.4232 | 750.0 |
| 800.0 | 518.21 | 0.02087 | 0.54809 | 0.56896 | 509.8 | 689.6 | 1199.4 | 0.7111 | 0.7051 | 1.4163 | 800.0 |
| 850.0 | 525.24 | 0.02105 | 0.51197 | 0.53302 | 518.4 | 679.5 | 1198.0 | 0.7197 | 0.6899 | 1.4096 | 850.0 |
| 900.0 | 531.95 | 0.02123 | 0.47968 | 0.50091 | 526.7 | 669.7 | 1196.4 | 0.7279 | 0.6753 | 1.4032 | 900.0 |
| 950.0 | 538.39 | 0.02141 | 0.45064 | 0.47205 | 534.7 | 660.0 | 1194.7 | 0.7358 | 0.6612 | 1.3970 | 950.0 |
| 1000.0 | 544.58 | 0.02159 | 0.42436 | 0.44596 | 542.6 | 650.4 | 1192.9 | 0.7434 | 0.6476 | 1.3910 | 1000.0 |
| 1050.0 | 550.53 | 0.02177 | 0.40047 | 0.42224 | 550.1 | 640.9 | 1191.0 | 0.7507 | 0.6344 | 1.3851 | 1050.0 |
| 1100.0 | 556.28 | 0.02195 | 0.37863 | 0.40058 | 557.5 | 631.5 | 1189.1 | 0.7578 | 0.6216 | 1.3794 | 1100.0 |
| 1150.0 | 561.82 | 0.02214 | 0.35859 | 0.38073 | 564.8 | 622.2 | 1187.0 | 0.7647 | 0.6091 | 1.3738 | 1150.0 |
| 1200.0 | 567.19 | 0.02232 | 0.34013 | 0.36245 | 571.9 | 613.0 | 1184.8 | 0.7714 | 0.5969 | 1.3683 | 1200.0 |
| 1250.0 | 572.38 | 0.02250 | 0.32306 | 0.34556 | 578.8 | 603.8 | 1182.6 | 0.7780 | 0.5850 | 1.3630 | 1250.0 |
| 1300.0 | 577.42 | 0.02269 | 0.30722 | 0.32991 | 585.6 | 594.6 | 1180.2 | 0.7843 | 0.5733 | 1.3577 | 1300.0 |
| 1350.0 | 582.32 | 0.02288 | 0.29250 | 0.31537 | 592.3 | 585.4 | 1177.8 | 0.7906 | 0.5620 | 1.3525 | 1350.0 |
| 1400.0 | 587.07 | 0.02307 | 0.27871 | 0.30178 | 598.8 | 576.5 | 1175.3 | 0.7966 | 0.5507 | 1.3474 | 1400.0 |
| 1450.0 | 591.70 | 0.02327 | 0.26584 | 0.28911 | 605.3 | 567.4 | 1172.8 | 0.8026 | 0.5397 | 1.3423 | 1450.0 |
| 1500.0 | 596.20 | 0.02346 | 0.25372 | 0.27719 | 611.7 | 558.4 | 1170.1 | 0.8085 | 0.5288 | 1.3373 | 1500.0 |
| 1550.0 | 600.59 | 0.02366 | 0.24235 | 0.26601 | 618.0 | 549.4 | 1167.4 | 0.8142 | 0.5182 | 1.3324 | 1550.0 |
| 1600.0 | 604.87 | 0.02387 | 0.23159 | 0.25545 | 624.2 | 540.3 | 1164.5 | 0.8199 | 0.5076 | 1.3274 | 1600.0 |
| 1650.0 | 609.05 | 0.02407 | 0.22143 | 0.24551 | 630.4 | 531.3 | 1161.6 | 0.8254 | 0.4971 | 1.3225 | 1650.0 |
| 1700.0 | 613.13 | 0.02428 | 0.21178 | 0.23607 | 636.5 | 522.2 | 1158.6 | 0.8309 | 0.4867 | 1.3176 | 1700.0 |
| 1750.0 | 617.12 | 0.02450 | 0.20263 | 0.22713 | 642.5 | 513.1 | 1155.6 | 0.8363 | 0.4765 | 1.3128 | 1750.0 |
| 1800.0 | 621.02 | 0.02472 | 0.19390 | 0.21861 | 648.5 | 503.8 | 1152.3 | 0.8417 | 0.4662 | 1.3079 | 1800.0 |
| 1850.0 | 624.83 | 0.02495 | 0.18558 | 0.21052 | 654.5 | 494.6 | 1149.0 | 0.8470 | 0.4561 | 1.3030 | 1850.0 |
| 1900.0 | 628.56 | 0.02517 | 0.17761 | 0.20278 | 660.4 | 485.2 | 1145.6 | 0.8522 | 0.4459 | 1.2981 | 1900.0 |
| 1950.0 | 632.22 | 0.02541 | 0.16999 | 0.19540 | 666.3 | 475.8 | 1142.0 | 0.8574 | 0.4358 | 1.2931 | 1950.0 |
| 2000.0 | 635.80 | 0.02565 | 0.16266 | 0.18831 | 672.1 | 466.2 | 1138.3 | 0.8625 | 0.4256 | 1.2881 | 2000.0 |
| 2100.0 | 642.76 | 0.02615 | 0.14885 | 0.17501 | 683.8 | 446.7 | 1130.5 | 0.8727 | 0.4053 | 1.2780 | 2100.0 |
| 2200.0 | 649.45 | 0.02669 | 0.13603 | 0.16272 | 695.5 | 426.7 | 1122.2 | 0.8828 | 0.3848 | 1.2676 | 2200.0 |
| 2300.0 | 655.89 | 0.02727 | 0.12406 | 0.15133 | 707.2 | 406.0 | 1113.2 | 0.8929 | 0.3640 | 1.2569 | 2300.0 |
| 2400.0 | 662.11 | 0.02790 | 0.11287 | 0.14076 | 719.0 | 384.8 | 1103.7 | 0.9031 | 0.3430 | 1.2460 | 2400.0 |
| 2500.0 | 668.11 | 0.02859 | 0.10209 | 0.13068 | 731.7 | 361.6 | 1093.3 | 0.9139 | 0.3206 | 1.2345 | 2500.0 |
| 2600.0 | 673.91 | 0.02938 | 0.09172 | 0.12110 | 744.5 | 337.6 | 1082.0 | 0.9247 | 0.2977 | 1.2225 | 2600.0 |
| 2700.0 | 679.53 | 0.03029 | 0.08165 | 0.11194 | 757.3 | 312.3 | 1069.7 | 0.9356 | 0.2741 | 1.2097 | 2700.0 |
| 2800.0 | 684.96 | 0.03134 | 0.07171 | 0.10305 | 770.7 | 285.1 | 1055.8 | 0.9468 | 0.2491 | 1.1958 | 2800.0 |
| 2900.0 | 690.22 | 0.03262 | 0.06158 | 0.09420 | 785.1 | 254.7 | 1039.8 | 0.9588 | 0.2215 | 1.1803 | 2900.0 |
| 3000.0 | 695.33 | 0.03428 | 0.05073 | 0.08500 | 801.8 | 218.4 | 1020.3 | 0.9728 | 0.1891 | 1.1619 | 3000.0 |
| 3100.0 | 700.28 | 0.03681 | 0.03771 | 0.07452 | 824.0 | 169.3 | 993.3 | 0.9914 | 0.1460 | 1.1373 | 3100.0 |
| 3200.0 | 705.08 | 0.04472 | 0.01191 | 0.05663 | 875.5 | 56.1 | 931.6 | 1.0351 | 0.0482 | 1.0832 | 3200.0 |
| 3208.2* | 705.47 | 0.05078 | 0.00000 | 0.05078 | 906.0 | 0.0 | 906.0 | 1.0612 | 0.0000 | 1.0612 | 3208.2* |

*Critical pressure

## Table 15-15. Superheated Steam

| Abs. Press. (psi) (Sat. Temp.) | | Sat. Water | Sat. Steam | Temperature (°F) | | | | | | | | | | | | | |
|---|---|---|---|---|---|---|---|---|---|---|---|---|---|---|---|---|---|
| | | | | 200 | 250 | 300 | 350 | 400 | 450 | 500 | 600 | 700 | 800 | 900 | 1000 | 1100 | 1200 |
| **1** (101.74) | Sh | | | 98.26 | 148.26 | 198.26 | 248.26 | 298.26 | 348.26 | 398.26 | 498.26 | 598.26 | 698.26 | 798.26 | 898.26 | 998.26 | 1098.26 |
| | v | 0.01614 | 333.6 | 392.5 | 422.4 | 452.3 | 482.1 | 511.9 | 541.7 | 571.5 | 631.1 | 690.7 | 750.3 | 809.8 | 869.4 | 929.0 | 988.6 |
| | h | 69.73 | 1105.8 | 1150.2 | 1172.9 | 1195.7 | 1218.7 | 1241.8 | 1265.1 | 1288.6 | 1336.1 | 1384.5 | 1433.7 | 1483.8 | 1534.9 | 1586.8 | 1639.7 |
| | s | 0.1326 | 1.9781 | 2.0509 | 2.0841 | 2.1152 | 2.1445 | 2.1722 | 2.1985 | 2.2237 | 2.2708 | 2.3144 | 2.3551 | 2.3934 | 2.4296 | 2.4640 | 2.4969 |
| **5** (162.24) | Sh | | | 37.76 | 87.76 | 137.76 | 187.76 | 237.76 | 287.76 | 337.76 | 437.76 | 537.76 | 637.76 | 737.76 | 837.76 | 937.76 | 1037.76 |
| | v | 0.01641 | 73.53 | 78.14 | 84.21 | 90.24 | 96.25 | 102.24 | 108.23 | 114.21 | 126.15 | 138.08 | 150.01 | 161.94 | 173.86 | 185.78 | 197.70 |
| | h | 130.20 | 1131.1 | 1148.6 | 1171.7 | 1194.8 | 1218.0 | 1241.3 | 1264.7 | 1288.2 | 1335.9 | 1384.3 | 1433.6 | 1483.7 | 1534.7 | 1586.7 | 1639.6 |
| | s | 0.2349 | 1.8443 | 1.8716 | 1.9054 | 1.9369 | 1.9664 | 1.9943 | 2.0208 | 2.0460 | 2.0932 | 2.1369 | 2.1776 | 2.2159 | 2.2521 | 2.2866 | 2.3194 |
| **10** (193.21) | Sh | | | 6.79 | 56.79 | 106.79 | 156.79 | 206.79 | 256.79 | 306.79 | 406.79 | 506.79 | 606.79 | 706.79 | 806.79 | 906.79 | 1006.79 |
| | v | 0.01659 | 38.42 | 38.84 | 41.93 | 44.98 | 48.02 | 51.03 | 54.04 | 57.04 | 63.03 | 69.00 | 74.98 | 80.94 | 86.91 | 92.87 | 98.84 |
| | h | 161.26 | 1143.3 | 1146.6 | 1170.2 | 1193.7 | 1217.1 | 1240.6 | 1264.1 | 1287.8 | 1335.5 | 1384.0 | 1433.4 | 1483.5 | 1534.6 | 1586.6 | 1639.5 |
| | s | 0.2836 | 1.7879 | 1.7928 | 1.8273 | 1.8593 | 1.8892 | 1.9173 | 1.9439 | 1.9692 | 2.0166 | 2.0603 | 2.1011 | 2.1394 | 2.1757 | 2.2101 | 2.2430 |
| **14.696** (212.00) | Sh | | | | 38.00 | 88.00 | 138.00 | 188.00 | 238.00 | 288.00 | 388.00 | 488.00 | 588.00 | 688.00 | 788.00 | 888.00 | 988.00 |
| | v | 0.0167 | 26.799 | | 28.42 | 30.52 | 32.60 | 34.67 | 36.72 | 38.77 | 42.86 | 46.93 | 51.00 | 55.06 | 59.13 | 63.19 | 67.25 |
| | h | 180.17 | 1150.5 | | 1168.8 | 1192.6 | 1216.3 | 1239.9 | 1263.6 | 1287.4 | 1335.2 | 1383.8 | 1433.2 | 1483.4 | 1534.5 | 1586.5 | 1639.4 |
| | s | 0.3121 | 1.7568 | | 1.7833 | 1.8158 | 1.8459 | 1.8743 | 1.9010 | 1.9265 | 1.9739 | 2.0177 | 2.0585 | 2.0969 | 2.1332 | 2.1676 | 2.2005 |
| **15** (213.03) | Sh | | | | 36.97 | 86.97 | 136.97 | 186.97 | 236.97 | 286.97 | 386.97 | 486.97 | 586.97 | 686.97 | 786.97 | 886.97 | 986.97 |
| | v | 0.01673 | 26.290 | | 27.837 | 29.899 | 31.939 | 33.963 | 35.977 | 37.985 | 41.986 | 45.978 | 49.964 | 53.946 | 57.926 | 61.905 | 65.882 |
| | h | 181.21 | 1150.9 | | 1168.7 | 1192.5 | 1216.2 | 1239.9 | 1263.6 | 1287.3 | 1335.2 | 1383.8 | 1433.2 | 1483.4 | 1534.5 | 1586.5 | 1639.4 |
| | s | 0.3137 | 1.7552 | | 1.7809 | 1.8134 | 1.8437 | 1.8720 | 1.8988 | 1.9242 | 1.9717 | 2.0155 | 2.0563 | 2.0946 | 2.1309 | 2.1653 | 2.1982 |
| **20** (227.96) | Sh | | | | 22.04 | 72.04 | 122.04 | 172.04 | 222.04 | 272.04 | 372.04 | 472.04 | 572.04 | 672.04 | 772.04 | 872.04 | 972.04 |
| | v | 0.01683 | 20.087 | | 20.788 | 22.356 | 23.900 | 25.428 | 26.946 | 28.457 | 31.466 | 34.465 | 37.458 | 40.447 | 43.435 | 46.420 | 49.405 |
| | h | 196.27 | 1156.3 | | 1167.1 | 1191.4 | 1215.4 | 1239.2 | 1263.0 | 1286.9 | 1334.9 | 1383.5 | 1432.9 | 1483.2 | 1534.3 | 1586.3 | 1639.3 |
| | s | 0.3358 | 1.7320 | | 1.7475 | 1.7805 | 1.8111 | 1.8397 | 1.8666 | 1.8921 | 1.9397 | 1.9836 | 2.0244 | 2.0628 | 2.0991 | 2.1336 | 2.1662 |
| **25** (240.07) | Sh | | | | 9.93 | 59.93 | 109.93 | 159.93 | 209.93 | 259.93 | 359.93 | 459.93 | 559.93 | 659.93 | 759.93 | 859.93 | 959.93 |
| | v | 0.01693 | 16.301 | | 16.558 | 17.829 | 19.076 | 20.307 | 21.527 | 22.740 | 25.153 | 27.557 | 29.954 | 32.348 | 34.740 | 37.130 | 39.518 |
| | h | 208.52 | 1160.6 | | 1165.6 | 1190.2 | 1214.5 | 1238.5 | 1262.5 | 1286.4 | 1334.6 | 1383.3 | 1432.7 | 1483.0 | 1534.2 | 1586.2 | 1639.2 |
| | s | 0.3535 | 1.7141 | | 1.7212 | 1.7547 | 1.7856 | 1.8145 | 1.8415 | 1.8672 | 1.9149 | 1.9588 | 1.9997 | 2.0381 | 2.0744 | 2.1089 | 2.1418 |
| **30** (250.34) | Sh | | | | | 49.66 | 99.66 | 149.66 | 199.66 | 249.66 | 349.66 | 449.66 | 549.66 | 649.66 | 749.66 | 849.66 | 949.66 |
| | v | 0.01701 | 13.744 | | | 14.810 | 15.859 | 16.892 | 17.914 | 18.929 | 20.945 | 22.951 | 24.952 | 26.949 | 28.943 | 30.936 | 32.927 |
| | h | 218.93 | 1164.1 | | | 1189.0 | 1213.6 | 1237.8 | 1261.9 | 1286.0 | 1334.2 | 1383.0 | 1432.5 | 1482.8 | 1534.0 | 1586.1 | 1639.0 |
| | s | 0.3682 | 1.6995 | | | 1.7334 | 1.7647 | 1.7937 | 1.8210 | 1.8467 | 1.8946 | 1.9386 | 1.9795 | 2.0179 | 2.0543 | 2.0888 | 2.1217 |

Abridged steam table (superheat degrees "Sh", specific volume $v$, enthalpy $h$, entropy $s$). Column headings are total temperature in °F.

| Abs Press (Sat Temp) | | Sat Liquid | Sat Vapor | 300 | 350 | 400 | 450 | 500 | 600 | 700 | 800 | 900 | 1000 | 1100 | 1200 |
|---|---|---|---|---|---|---|---|---|---|---|---|---|---|---|---|
| **35** (259.29) | Sh | | | 40.71 | 90.71 | 140.71 | 190.71 | 240.71 | 340.71 | 440.71 | 540.71 | 640.71 | 740.71 | 840.71 | 940.71 |
| | $v$ | 0.01708 | 11.896 | 12.654 | 13.562 | 14.453 | 15.334 | 16.207 | 17.939 | 19.662 | 21.379 | 23.092 | 24.803 | 26.512 | 28.220 |
| | $h$ | 228.03 | 1167.1 | 1187.8 | 1212.7 | 1237.1 | 1261.3 | 1285.5 | 1333.9 | 1382.8 | 1432.3 | 1482.7 | 1533.9 | 1586.0 | 1638.9 |
| | $s$ | 0.3809 | 1.6872 | 1.7152 | 1.7468 | 1.7761 | 1.8035 | 1.8294 | 1.8774 | 1.9214 | 1.9624 | 2.0009 | 2.0372 | 2.0717 | 2.1046 |
| **40** (267.25) | Sh | | | 32.75 | 82.75 | 132.75 | 182.75 | 232.75 | 332.75 | 432.75 | 532.75 | 632.75 | 732.75 | 832.75 | 932.75 |
| | $v$ | 0.01715 | 10.497 | 11.036 | 11.838 | 12.624 | 13.398 | 14.165 | 15.685 | 17.195 | 18.699 | 20.199 | 21.697 | 23.194 | 24.689 |
| | $h$ | 236.14 | 1169.8 | 1186.6 | 1211.7 | 1236.4 | 1260.8 | 1285.0 | 1333.6 | 1382.5 | 1432.1 | 1482.5 | 1533.7 | 1585.8 | 1638.8 |
| | $s$ | 0.3921 | 1.6765 | 1.6992 | 1.7312 | 1.7608 | 1.7883 | 1.8143 | 1.8624 | 1.9065 | 1.9476 | 1.9860 | 2.0224 | 2.0569 | 2.0899 |
| **45** (274.44) | Sh | | | 25.56 | 75.56 | 125.56 | 175.56 | 225.56 | 325.56 | 425.56 | 525.56 | 625.56 | 725.56 | 825.56 | 925.56 |
| | $v$ | 0.01721 | 9.399 | 9.777 | 10.497 | 11.201 | 11.892 | 12.577 | 13.932 | 15.276 | 16.614 | 17.950 | 19.282 | 20.613 | 21.943 |
| | $h$ | 243.49 | 1172.1 | 1185.4 | 1210.4 | 1235.7 | 1260.2 | 1284.6 | 1333.3 | 1382.3 | 1431.9 | 1482.3 | 1533.6 | 1585.7 | 1638.7 |
| | $s$ | 0.4021 | 1.6671 | 1.6849 | 1.7173 | 1.7471 | 1.7748 | 1.8010 | 1.8492 | 1.8934 | 1.9345 | 1.9730 | 2.0093 | 2.0439 | 2.0768 |
| **50** (281.02) | Sh | | | 18.98 | 68.98 | 118.98 | 168.98 | 218.98 | 318.98 | 418.98 | 518.98 | 618.98 | 718.98 | 818.98 | 918.98 |
| | $v$ | 0.01727 | 8.514 | 8.769 | 9.424 | 10.062 | 10.688 | 11.306 | 12.529 | 13.741 | 14.947 | 16.150 | 17.350 | 18.549 | 19.746 |
| | $h$ | 250.21 | 1174.1 | 1184.1 | 1209.9 | 1234.9 | 1259.6 | 1284.1 | 1332.9 | 1382.0 | 1431.7 | 1482.2 | 1533.4 | 1585.6 | 1638.6 |
| | $s$ | 0.4112 | 1.6586 | 1.6720 | 1.7048 | 1.7349 | 1.7628 | 1.7890 | 1.8374 | 1.8816 | 1.9227 | 1.9613 | 1.9977 | 2.0322 | 2.0652 |
| **55** (287.07) | Sh | | | 12.93 | 62.93 | 112.93 | 162.93 | 212.93 | 312.93 | 412.93 | 512.93 | 612.93 | 712.93 | 812.93 | 912.93 |
| | $v$ | 0.01733 | | 7.945 | 8.546 | 9.130 | 9.702 | 10.267 | 11.381 | 12.485 | 13.583 | 14.677 | 15.769 | 16.859 | 17.948 |
| | $h$ | 256.43 | | 1182.9 | 1208.9 | 1234.2 | 1259.1 | 1283.6 | 1332.6 | 1381.8 | 1431.5 | 1482.0 | 1533.3 | 1585.5 | 1638.5 |
| | $s$ | 0.4196 | | 1.6601 | 1.6933 | 1.7237 | 1.7518 | 1.7781 | 1.8266 | 1.8710 | 1.9121 | 1.9507 | 1.987 | 2.022 | 2.055 |
| **60** (292.71) | Sh | | | 7.29 | 57.29 | 107.29 | 157.29 | 207.29 | 307.29 | 407.29 | 507.29 | 607.29 | 707.29 | 807.29 | 907.29 |
| | $v$ | 0.01738 | 7.174 | 7.257 | 7.815 | 8.354 | 8.881 | 9.400 | 10.425 | 11.438 | 12.446 | 13.450 | 14.452 | 15.452 | 16.450 |
| | $h$ | 262.21 | 1177.6 | 1181.6 | 1208.0 | 1233.5 | 1258.5 | 1283.2 | 1332.3 | 1381.5 | 1431.3 | 1481.8 | 1533.2 | 1585.3 | 1638.4 |
| | $s$ | 0.4273 | 1.6440 | 1.6492 | 1.6934 | 1.7134 | 1.7417 | 1.7681 | 1.8168 | 1.8612 | 1.9024 | 1.9410 | 1.9774 | 2.0120 | 2.0450 |
| **65** (297.98) | Sh | | | 2.02 | 52.02 | 102.02 | 152.02 | 202.02 | 302.02 | 402.02 | 502.02 | 602.02 | 702.02 | 802.02 | 902.02 |
| | $v$ | 0.01743 | 6.653 | 6.675 | 7.195 | 7.697 | 8.186 | 8.667 | 9.615 | 10.552 | 11.484 | 12.412 | 13.337 | 14.261 | 15.183 |
| | $h$ | 267.63 | 1179.1 | 1180.3 | 1207.0 | 1232.7 | 1257.9 | 1282.7 | 1331.9 | 1381.3 | 1431.1 | 1481.6 | 1533.0 | 1585.2 | 1638.3 |
| | $s$ | 0.4344 | 1.6375 | 1.6390 | 1.6731 | 1.7040 | 1.7324 | 1.7590 | 1.8077 | 1.8522 | 1.8935 | 1.9321 | 1.9685 | 2.0031 | 2.0361 |
| **70** (302.93) | Sh | | | | 47.07 | 97.07 | 147.07 | 197.07 | 297.07 | 397.07 | 497.07 | 597.07 | 697.07 | 797.07 | 897.07 |
| | $v$ | 0.01748 | 6.205 | | 6.664 | 7.133 | 7.590 | 8.039 | 8.922 | 9.793 | 10.659 | 11.522 | 12.382 | 13.240 | 14.097 |
| | $h$ | 272.74 | 1180.6 | | 1206.0 | 1232.0 | 1257.3 | 1282.2 | 1331.6 | 1381.0 | 1430.9 | 1481.5 | 1532.9 | 1585.1 | 1638.2 |
| | $s$ | 0.4411 | 1.6316 | | 1.6640 | 1.6951 | 1.7237 | 1.7504 | 1.7993 | 1.8439 | 1.8852 | 1.9238 | 1.9603 | 1.9949 | 2.0279 |
| **75** (307.61) | Sh | | | | 42.39 | 92.39 | 142.39 | 192.39 | 292.39 | 392.39 | 492.39 | 592.39 | 692.39 | 792.39 | 892.39 |
| | $v$ | 0.01753 | 5.814 | | 6.204 | 6.645 | 7.074 | 7.494 | 8.320 | 9.135 | 9.945 | 10.750 | 11.553 | 12.355 | 13.155 |
| | $h$ | 277.56 | 1181.9 | | 1205.0 | 1231.2 | 1256.7 | 1281.7 | 1331.1 | 1380.7 | 1430.7 | 1481.3 | 1532.7 | 1585.0 | 1638.1 |
| | $s$ | 0.4474 | 1.6260 | | 1.6554 | 1.6868 | 1.7156 | 1.7424 | 1.7915 | 1.8361 | 1.8774 | 1.9161 | 1.9526 | 1.9872 | 2.0202 |

Table 15-15. Continued

| Abs. Press. (psi) (Sat. Temp.) | | Sat. Water | Sat. Steam | \multicolumn{14}{Temperature (°F)} | | | | | | | | | | | | | |
|---|---|---|---|---|---|---|---|---|---|---|---|---|---|---|---|---|---|
| | | | | 350 | 400 | 450 | 500 | 550 | 600 | 700 | 800 | 900 | 1000 | 1100 | 1200 | 1300 | 1400 |
| 80 (312.04) | Sh | | | 37.96 | 87.96 | 137.96 | 187.96 | 237.96 | 287.96 | 387.96 | 487.96 | 587.96 | 687.96 | 787.96 | 887.96 | 987.96 | 1087.96 |
| | v | 0.01757 | 5.471 | 5.801 | 6.218 | 6.622 | 7.018 | 7.408 | 7.794 | 8.560 | 9.319 | 10.075 | 10.829 | 11.581 | 12.331 | 13.081 | 13.829 |
| | h | 282.15 | 1183.1 | 1204.0 | 1230.5 | 1256.1 | 1281.3 | 1306.2 | 1330.9 | 1380.5 | 1430.5 | 1481.1 | 1532.6 | 1584.9 | 1638.0 | 1692.0 | 1746.8 |
| | s | 0.4534 | 1.6208 | 1.6473 | 1.6790 | 1.7080 | 1.7349 | 1.7602 | 1.7842 | 1.8289 | 1.8702 | 1.9089 | 1.9454 | 1.9800 | 2.0131 | 2.0446 | 2.0750 |
| 85 (316.26) | Sh | | | 33.74 | 83.74 | 133.74 | 183.74 | 233.74 | 283.74 | 383.74 | 483.74 | 583.74 | 683.74 | 783.74 | 883.74 | 983.74 | 1083.74 |
| | v | 0.01762 | 5.167 | 5.445 | 5.840 | 6.223 | 6.597 | 6.966 | 7.330 | 8.052 | 8.768 | 9.480 | 10.190 | 10.898 | 11.604 | 12.310 | 13.014 |
| | h | 286.52 | 1184.2 | 1203.0 | 1229.7 | 1255.5 | 1280.8 | 1305.8 | 1330.6 | 1380.2 | 1430.3 | 1481.0 | 1532.4 | 1584.7 | 1637.9 | 1691.9 | 1746.8 |
| | s | 0.4590 | 1.6159 | 1.6396 | 1.6716 | 1.7008 | 1.7279 | 1.7532 | 1.7772 | 1.8220 | 1.8634 | 1.9021 | 1.9386 | 1.9733 | 2.0063 | 2.0379 | 2.0682 |
| 90 (320.28) | Sh | | | 29.72 | 79.72 | 129.72 | 179.72 | 229.72 | 279.72 | 379.72 | 479.72 | 579.72 | 679.72 | 779.72 | 879.72 | 979.72 | 1079.72 |
| | v | 0.01766 | 4.895 | 5.128 | 5.505 | 5.869 | 6.223 | 6.572 | 6.917 | 7.600 | 8.277 | 8.950 | 9.621 | 10.290 | 10.958 | 11.625 | 12.290 |
| | h | 290.69 | 1185.3 | 1202.0 | 1228.9 | 1254.9 | 1280.3 | 1305.4 | 1330.2 | 1380.0 | 1430.1 | 1480.8 | 1532.3 | 1584.6 | 1637.8 | 1691.8 | 1746.7 |
| | s | 0.4643 | 1.6113 | 1.6323 | 1.6646 | 1.6940 | 1.7212 | 1.7467 | 1.7707 | 1.8156 | 1.8570 | 1.8957 | 1.9323 | 1.9669 | 2.0000 | 2.0316 | 2.0619 |
| 95 (324.13) | Sh | | | 25.87 | 75.87 | 125.87 | 175.87 | 225.87 | 275.87 | 375.87 | 475.87 | 575.87 | 675.87 | 775.87 | 875.87 | 975.87 | 1075.87 |
| | v | 0.01770 | 4.651 | 4.845 | 5.205 | 5.551 | 5.889 | 6.221 | 6.548 | 7.196 | 7.838 | 8.477 | 9.113 | 9.747 | 10.380 | 11.012 | 11.643 |
| | h | 294.70 | 1186.2 | 1200.9 | 1228.1 | 1254.3 | 1279.8 | 1305.0 | 1329.9 | 1379.7 | 1429.9 | 1480.6 | 1532.1 | 1584.5 | 1637.7 | 1691.7 | 1746.6 |
| | s | 0.4694 | 1.6069 | 1.6253 | 1.6580 | 1.6876 | 1.7149 | 1.7404 | 1.7645 | 1.8094 | 1.8509 | 1.8897 | 1.9262 | 1.9609 | 1.9940 | 2.0256 | 2.0559 |
| 100 (327.82) | Sh | | | 22.18 | 72.18 | 122.18 | 172.18 | 222.18 | 272.18 | 372.18 | 472.18 | 572.18 | 672.18 | 772.18 | 872.18 | 972.18 | 1072.18 |
| | v | 0.01774 | 4.431 | 4.590 | 4.935 | 5.266 | 5.588 | 5.904 | 6.216 | 6.833 | 7.443 | 8.050 | 8.655 | 9.258 | 9.860 | 10.460 | 11.060 |
| | h | 298.54 | 1187.2 | 1199.9 | 1227.4 | 1253.7 | 1279.3 | 1304.6 | 1329.6 | 1379.5 | 1429.7 | 1480.4 | 1532.0 | 1584.4 | 1637.6 | 1691.6 | 1746.5 |
| | s | 0.4743 | 1.6027 | 1.6187 | 1.6516 | 1.6814 | 1.7088 | 1.7344 | 1.7586 | 1.8036 | 1.8451 | 1.8839 | 1.9205 | 1.9552 | 1.9883 | 2.0199 | 2.0502 |
| 105 (331.37) | Sh | | | 18.63 | 68.63 | 118.63 | 168.63 | 218.63 | 268.63 | 368.63 | 468.63 | 568.63 | 668.63 | 768.63 | 868.63 | 968.63 | 1068.63 |
| | v | 0.01778 | 4.231 | 4.359 | 4.690 | 5.007 | 5.315 | 5.617 | 5.915 | 6.504 | 7.086 | 7.665 | 8.241 | 8.816 | 9.389 | 9.961 | 10.532 |
| | h | 302.24 | 1188.0 | 1198.8 | 1226.6 | 1253.1 | 1278.8 | 1304.2 | 1329.2 | 1379.2 | 1429.4 | 1480.3 | 1531.8 | 1584.2 | 1637.5 | 1691.5 | 1746.4 |
| | s | 0.4790 | 1.5988 | 1.6122 | 1.6455 | 1.6755 | 1.7031 | 1.7288 | 1.7530 | 1.7981 | 1.8396 | 1.8785 | 1.9151 | 1.9498 | 1.9823 | 2.0145 | 2.0448 |
| 110 (334.79) | Sh | | | 15.21 | 65.21 | 115.21 | 165.21 | 215.21 | 265.21 | 365.21 | 465.21 | 565.21 | 665.21 | 765.21 | 865.21 | 965.21 | 1065.21 |
| | v | 0.01782 | 4.048 | 4.149 | 4.468 | 4.772 | 5.068 | 5.357 | 5.642 | 6.205 | 6.761 | 7.314 | 7.865 | 8.413 | 8.961 | 9.507 | 10.053 |
| | h | 305.80 | 1188.9 | 1197.7 | 1225.8 | 1252.5 | 1278.3 | 1303.8 | 1328.9 | 1379.0 | 1429.2 | 1480.1 | 1531.7 | 1584.1 | 1637.4 | 1691.4 | 1746.4 |
| | s | 0.4834 | 1.5950 | 1.6061 | 1.6396 | 1.6698 | 1.6975 | 1.7233 | 1.7476 | 1.7928 | 1.8344 | 1.8732 | 1.9099 | 1.9446 | 1.9777 | 2.0093 | 2.0397 |
| 115 (338.08) | Sh | | | 11.92 | 61.92 | 111.92 | 161.92 | 211.92 | 261.92 | 361.92 | 461.92 | 561.92 | 661.92 | 761.92 | 861.92 | 961.92 | 1061.92 |
| | v | 0.01785 | 3.881 | 3.957 | 4.265 | 4.558 | 4.841 | 5.119 | 5.392 | 5.932 | 6.465 | 6.994 | 7.521 | 8.046 | 8.570 | 9.093 | 9.615 |
| | h | 309.25 | 1189.6 | 1196.7 | 1225.0 | 1251.8 | 1277.9 | 1303.3 | 1328.6 | 1378.7 | 1429.0 | 1479.9 | 1531.6 | 1584.0 | 1637.2 | 1691.4 | 1746.3 |
| | s | 0.4877 | 1.5913 | 1.6001 | 1.6340 | 1.6644 | 1.6922 | 1.7181 | 1.7425 | 1.7877 | 1.8294 | 1.8682 | 1.9049 | 1.9396 | 1.9727 | 2.0044 | 2.0347 |

**120 (341.27)**

| | Sat. Liq. | Sat. Vap. | 8.73 | 58.73 | 108.73 | 158.73 | 208.73 | 258.73 | 358.73 | 458.73 | 558.73 | 658.73 | 758.73 | 858.73 | 958.73 | 1058.73 |
|---|---|---|---|---|---|---|---|---|---|---|---|---|---|---|---|---|
| $v$ | 0.01789 | 3.7275 | 3.7815 | 4.0786 | 4.3610 | 4.6341 | 4.9009 | 5.1637 | 5.6813 | 6.1928 | 6.7006 | 7.2060 | 7.7096 | 8.2219 | 8.7130 | 9.2134 |
| $h$ | 312.58 | 1190.4 | 1195.6 | 1224.1 | 1251.2 | 1277.4 | 1302.9 | 1328.2 | 1378.4 | 1428.8 | 1479.8 | 1531.4 | 1583.9 | 1637.1 | 1691.3 | 1746.2 |
| $s$ | 0.4919 | 1.5879 | 1.5943 | 1.6286 | 1.6592 | 1.6872 | 1.7132 | 1.7376 | 1.7829 | 1.8246 | 1.8635 | 1.9001 | 1.9349 | 1.9680 | 1.9996 | 2.0300 |

**130 (347.33)**

| | Sat. Liq. | Sat. Vap. | 2.67 | 52.67 | 102.67 | 152.67 | 202.67 | 252.67 | 352.67 | 452.67 | 552.67 | 652.67 | 752.67 | 852.67 | 952.67 | 1052.67 |
|---|---|---|---|---|---|---|---|---|---|---|---|---|---|---|---|---|
| $v$ | 0.01796 | 3.4544 | 3.4699 | 3.7489 | 4.0129 | 4.2672 | 4.5151 | 4.7589 | 5.2384 | 5.7118 | 6.1814 | 6.6486 | 7.1140 | 7.5781 | 8.0411 | 8.5033 |
| $h$ | 318.95 | 1191.7 | 1193.4 | 1222.5 | 1249.9 | 1276.4 | 1302.1 | 1327.5 | 1377.9 | 1428.4 | 1479.4 | 1531.1 | 1583.6 | 1636.9 | 1691.1 | 1746.1 |
| $s$ | 0.4998 | 1.5813 | 1.5833 | 1.6182 | 1.6493 | 1.6775 | 1.7037 | 1.7283 | 1.7737 | 1.8155 | 1.8545 | 1.8911 | 1.9259 | 1.9591 | 1.9907 | 2.0211 |

**140 (353.04)**

| | Sat. Liq. | Sat. Vap. | 46.96 | 96.96 | 146.96 | 196.96 | 246.96 | 346.96 | 446.96 | 546.96 | 646.96 | 746.96 | 846.96 | 946.96 | 1046.96 |
|---|---|---|---|---|---|---|---|---|---|---|---|---|---|---|---|
| $v$ | 0.01803 | 3.2190 | 3.4661 | 3.7143 | 3.9526 | 4.1844 | 4.4119 | 4.8588 | 5.2995 | 5.7364 | 6.1709 | 6.6036 | 7.0349 | 7.4652 | 7.8946 |
| $h$ | 324.96 | 1193.0 | 1220.8 | 1248.7 | 1275.3 | 1301.3 | 1326.8 | 1377.4 | 1428.0 | 1479.1 | 1530.8 | 1583.4 | 1636.7 | 1690.9 | 1745.9 |
| $s$ | 0.5071 | 1.5752 | 1.6085 | 1.6400 | 1.6686 | 1.6949 | 1.7196 | 1.7652 | 1.8071 | 1.8461 | 1.8828 | 1.9176 | 1.9508 | 1.9825 | 2.0129 |

**150 (358.43)**

| | Sat. Liq. | Sat. Vap. | 41.57 | 91.57 | 141.57 | 191.57 | 241.57 | 341.57 | 441.57 | 541.57 | 641.57 | 741.57 | 841.57 | 941.57 | 1041.57 |
|---|---|---|---|---|---|---|---|---|---|---|---|---|---|---|---|
| $v$ | 0.01809 | 3.0139 | 3.2208 | 3.4555 | 3.6799 | 3.8978 | 4.1112 | 4.5298 | 4.9421 | 5.3507 | 5.7568 | 6.1612 | 6.5642 | 6.9661 | 7.3671 |
| $h$ | 330.65 | 1194.1 | 1219.1 | 1247.3 | 1274.3 | 1300.5 | 1326.1 | 1376.9 | 1427.6 | 1478.7 | 1530.5 | 1583.1 | 1636.5 | 1690.7 | 1745.7 |
| $s$ | 0.5141 | 1.5695 | 1.5993 | 1.6313 | 1.6602 | 1.6867 | 1.7115 | 1.7573 | 1.7992 | 1.8383 | 1.8751 | 1.9099 | 1.9431 | 1.9748 | 2.0052 |

**160 (363.55)**

| | Sat. Liq. | Sat. Vap. | 36.45 | 86.45 | 136.45 | 186.45 | 236.45 | 336.45 | 436.45 | 536.45 | 636.45 | 736.45 | 836.45 | 936.45 | 1036.45 |
|---|---|---|---|---|---|---|---|---|---|---|---|---|---|---|---|
| $v$ | 0.01815 | 2.8336 | 3.0060 | 3.2288 | 3.4413 | 3.6469 | 3.8480 | 4.2420 | 4.6295 | 5.0132 | 5.3945 | 5.7741 | 6.1522 | 6.5293 | 6.9055 |
| $h$ | 336.07 | 1195.1 | 1217.4 | 1246.0 | 1273.3 | 1299.6 | 1325.4 | 1376.4 | 1427.2 | 1478.4 | 1530.3 | 1582.9 | 1636.3 | 1690.5 | 1745.6 |
| $s$ | 0.5206 | 1.5641 | 1.5906 | 1.6231 | 1.6522 | 1.6790 | 1.7039 | 1.7499 | 1.7919 | 1.8310 | 1.8678 | 1.9027 | 1.9359 | 1.9676 | 1.9980 |

**170 (368.42)**

| | Sat. Liq. | Sat. Vap. | 31.58 | 81.58 | 131.58 | 181.58 | 231.58 | 331.58 | 431.58 | 531.58 | 631.58 | 731.58 | 831.58 | 931.58 | 1031.58 |
|---|---|---|---|---|---|---|---|---|---|---|---|---|---|---|---|
| $v$ | 0.01821 | 2.6738 | 2.8162 | 3.0288 | 3.2306 | 3.4255 | 3.6158 | 3.9879 | 4.3536 | 4.7155 | 5.0749 | 5.4325 | 5.7888 | 6.1440 | 6.4983 |
| $h$ | 341.24 | 1196.0 | 1215.6 | 1244.7 | 1272.2 | 1298.8 | 1324.7 | 1375.8 | 1426.8 | 1478.0 | 1530.0 | 1582.6 | 1636.1 | 1690.4 | 1745.4 |
| $s$ | 0.5269 | 1.5591 | 1.5823 | 1.6152 | 1.6447 | 1.6717 | 1.6968 | 1.7428 | 1.7850 | 1.8241 | 1.8610 | 1.8959 | 1.9291 | 1.9608 | 1.9913 |

**180 (373.08)**

| | Sat. Liq. | Sat. Vap. | 26.92 | 76.92 | 126.92 | 176.92 | 226.92 | 326.92 | 426.92 | 526.92 | 626.92 | 726.92 | 826.92 | 926.92 | 1026.92 |
|---|---|---|---|---|---|---|---|---|---|---|---|---|---|---|---|
| $v$ | 0.01827 | 2.5312 | 2.6474 | 2.8508 | 3.0433 | 3.2286 | 3.4093 | 3.7621 | 4.1084 | 4.4508 | 4.7907 | 5.1289 | 5.4657 | 5.8014 | 6.1363 |
| $h$ | 346.19 | 1196.9 | 1213.8 | 1243.4 | 1271.2 | 1297.9 | 1324.0 | 1375.3 | 1426.3 | 1477.7 | 1529.7 | 1582.4 | 1635.9 | 1690.2 | 1745.3 |
| $s$ | 0.5328 | 1.5543 | 1.5743 | 1.6078 | 1.6376 | 1.6647 | 1.6900 | 1.7362 | 1.7784 | 1.8176 | 1.8545 | 1.8894 | 1.9227 | 1.9545 | 1.9849 |

**190 (377.53)**

| | Sat. Liq. | Sat. Vap. | 22.47 | 72.47 | 122.47 | 172.47 | 222.47 | 322.47 | 422.47 | 522.47 | 622.47 | 722.47 | 822.47 | 922.47 | 1022.47 |
|---|---|---|---|---|---|---|---|---|---|---|---|---|---|---|---|
| $v$ | 0.01833 | 2.4030 | 2.4961 | 2.6915 | 2.8756 | 3.0525 | 3.2246 | 3.5601 | 3.8889 | 4.2140 | 4.5365 | 4.8572 | 5.1766 | 5.4949 | 5.8124 |
| $h$ | 350.94 | 1197.6 | 1212.0 | 1242.0 | 1270.1 | 1297.1 | 1323.3 | 1374.8 | 1425.9 | 1477.4 | 1529.4 | 1582.1 | 1635.7 | 1690.0 | 1745.1 |
| $s$ | 0.5384 | 1.5498 | 1.5667 | 1.6006 | 1.6307 | 1.6581 | 1.6835 | 1.7299 | 1.7722 | 1.8115 | 1.8484 | 1.8834 | 1.9166 | 1.9484 | 1.9789 |

**200 (381.80)**

| | Sat. Liq. | Sat. Vap. | 18.20 | 68.20 | 118.20 | 168.20 | 218.20 | 318.20 | 418.20 | 518.20 | 618.20 | 718.20 | 818.20 | 918.20 | 1018.20 |
|---|---|---|---|---|---|---|---|---|---|---|---|---|---|---|---|
| $v$ | 0.01839 | 2.2873 | 2.3598 | 2.5480 | 2.7247 | 2.8939 | 3.0583 | 3.3783 | 3.6915 | 4.0008 | 4.3077 | 4.6128 | 4.9165 | 5.2191 | 5.5209 |
| $h$ | 355.51 | 1198.3 | 1210.1 | 1240.6 | 1269.0 | 1296.2 | 1322.6 | 1374.3 | 1425.5 | 1477.0 | 1529.1 | 1581.9 | 1635.4 | 1689.8 | 1745.0 |
| $s$ | 0.5438 | 1.5454 | 1.5593 | 1.5938 | 1.6242 | 1.6518 | 1.6773 | 1.7239 | 1.7663 | 1.8057 | 1.8426 | 1.8776 | 1.9109 | 1.9427 | 1.9732 |

Table 15-15. Continued

| Abs. Press. (psi) (Sat. Temp.) | | Sat. Water | Sat. Steam | 400 | 450 | 500 | 550 | 600 | 700 | 800 | 900 | 1000 | 1100 | 1200 | 1300 | 1400 | 1500 |
|---|---|---|---|---|---|---|---|---|---|---|---|---|---|---|---|---|---|
| 210 (385.91) | Sh | | | 14.09 | 64.09 | 114.09 | 164.09 | 214.09 | 314.09 | 414.09 | 514.09 | 614.09 | 714.09 | 814.09 | 914.09 | 1014.09 | 1114.09 |
| | $v$ | 0.01844 | 2.1822 | 2.2364 | 2.4181 | 2.5880 | 2.7504 | 2.9078 | 3.2137 | 3.5128 | 3.8080 | 4.1007 | 4.3915 | 4.6811 | 4.9695 | 5.2571 | 5.5440 |
| | $h$ | 359.91 | 1199.0 | 1208.02 | 1239.2 | 1268.0 | 1295.3 | 1321.9 | 1373.7 | 1425.1 | 1476.7 | 1528.8 | 1581.6 | 1635.2 | 1689.6 | 1744.8 | 1800.8 |
| | $s$ | 0.5490 | 1.5413 | 1.5522 | 1.5872 | 1.6180 | 1.6458 | 1.6715 | 1.7182 | 1.7607 | 1.8001 | 1.8371 | 1.8721 | 1.9054 | 1.9372 | 1.9677 | 1.9970 |
| 220 (389.88) | Sh | | | 10.12 | 60.12 | 110.12 | 160.12 | 210.12 | 310.12 | 410.12 | 510.12 | 610.12 | 710.12 | 810.12 | 910.12 | 1010.12 | 1110.12 |
| | $v$ | 0.01850 | 2.0863 | 2.1240 | 2.2999 | 2.4638 | 2.6199 | 2.7710 | 3.0642 | 3.3504 | 3.6327 | 3.9125 | 4.1905 | 4.4671 | 4.7426 | 5.0173 | 5.2913 |
| | $h$ | 364.17 | 1199.6 | 1206.3 | 1237.8 | 1266.9 | 1294.5 | 1321.2 | 1373.2 | 1424.7 | 1476.3 | 1528.5 | 1581.4 | 1635.0 | 1689.4 | 1744.7 | 1800.6 |
| | $s$ | 0.5540 | 1.5374 | 1.5453 | 1.5808 | 1.6120 | 1.6400 | 1.6658 | 1.7128 | 1.7553 | 1.7948 | 1.8318 | 1.8668 | 1.9002 | 1.9320 | 1.9625 | 1.9919 |
| 230 (393.70) | Sh | | | 6.30 | 56.30 | 106.30 | 156.30 | 206.30 | 306.30 | 406.30 | 506.30 | 606.30 | 706.30 | 806.30 | 906.30 | 1006.30 | 1106.30 |
| | $v$ | 0.01855 | 1.9985 | 2.0212 | 2.1919 | 2.3503 | 2.5008 | 2.6461 | 2.9276 | 3.2020 | 3.4726 | 3.7406 | 4.0068 | 4.2717 | 4.5355 | 4.7984 | 5.0606 |
| | $h$ | 368.28 | 1200.1 | 1204.4 | 1236.3 | 1265.7 | 1293.6 | 1320.4 | 1372.7 | 1424.2 | 1476.0 | 1528.2 | 1581.1 | 1634.8 | 1689.3 | 1744.5 | 1800.5 |
| | $s$ | 0.5588 | 1.5336 | 1.5385 | 1.5747 | 1.6062 | 1.6344 | 1.6604 | 1.7075 | 1.7502 | 1.7897 | 1.8268 | 1.8618 | 1.8952 | 1.9270 | 1.9576 | 1.9869 |
| 240 (397.39) | Sh | | | 2.61 | 52.61 | 102.61 | 152.61 | 202.61 | 302.61 | 402.61 | 502.61 | 602.61 | 702.61 | 802.61 | 902.61 | 1002.61 | 1102.61 |
| | $v$ | 0.01860 | 1.9177 | 1.9268 | 2.0928 | 2.2462 | 2.3915 | 2.5316 | 2.8024 | 3.0661 | 3.3259 | 3.5831 | 3.8385 | 4.0926 | 4.3456 | 4.5977 | 4.8492 |
| | $h$ | 372.27 | 1200.6 | 1202.4 | 1234.9 | 1264.6 | 1292.7 | 1319.7 | 1372.1 | 1423.8 | 1475.6 | 1527.9 | 1580.9 | 1634.6 | 1689.1 | 1744.3 | 1800.4 |
| | $s$ | 0.5634 | 1.5299 | 1.5320 | 1.5687 | 1.6006 | 1.6291 | 1.6552 | 1.7025 | 1.7452 | 1.7848 | 1.8219 | 1.8570 | 1.8904 | 1.9223 | 1.9528 | 1.9822 |
| 250 (400.97) | Sh | | | | 49.03 | 99.03 | 149.03 | 199.03 | 299.03 | 399.03 | 499.03 | 599.03 | 699.03 | 799.03 | 899.03 | 999.03 | 1099.03 |
| | $v$ | 0.01865 | 1.8432 | | 2.0016 | 2.1504 | 2.2909 | 2.4262 | 2.6872 | 2.9410 | 3.1909 | 3.4382 | 3.6837 | 3.9278 | 4.1709 | 4.4131 | 4.6546 |
| | $h$ | 376.14 | 1201.1 | | 1233.5 | 1263.5 | 1291.8 | 1319.0 | 1371.6 | 1423.4 | 1475.3 | 1527.6 | 1580.6 | 1634.4 | 1688.9 | 1744.2 | 1800.2 |
| | $s$ | 0.5679 | 1.5264 | | 1.5629 | 1.5951 | 1.6239 | 1.6502 | 1.6976 | 1.7405 | 1.7801 | 1.8173 | 1.8524 | 1.8858 | 1.9177 | 1.9482 | 1.9776 |
| 260 (404.44) | Sh | | | | 45.56 | 95.56 | 145.56 | 195.56 | 295.56 | 395.56 | 495.56 | 595.56 | 695.56 | 795.56 | 895.56 | 995.56 | 1095.56 |
| | $v$ | 0.01870 | 1.7742 | | 1.9173 | 2.0619 | 2.1981 | 2.3289 | 2.5808 | 2.8256 | 3.0663 | 3.3044 | 3.5408 | 3.7758 | 4.0097 | 4.2427 | 4.4750 |
| | $h$ | 379.90 | 1201.5 | | 1231.9 | 1262.4 | 1290.9 | 1318.2 | 1371.1 | 1423.0 | 1474.9 | 1527.3 | 1580.4 | 1634.2 | 1688.7 | 1744.0 | 1800.1 |
| | $s$ | 0.5722 | 1.5230 | | 1.5573 | 1.5899 | 1.6189 | 1.6453 | 1.6930 | 1.7359 | 1.7756 | 1.8128 | 1.8480 | 1.8814 | 1.9133 | 1.9439 | 1.9732 |
| 270 (407.80) | Sh | | | | 42.20 | 92.20 | 142.20 | 192.20 | 292.20 | 392.20 | 492.20 | 592.20 | 692.20 | 792.20 | 892.20 | 992.20 | 1092.20 |
| | $v$ | 0.01875 | 1.7101 | | 1.8391 | 1.9799 | 2.1121 | 2.2388 | 2.4824 | 2.7186 | 2.9509 | 3.1806 | 3.4084 | 3.6349 | 3.8603 | 4.0849 | 4.3087 |
| | $h$ | 383.56 | 1201.9 | | 1230.4 | 1261.2 | 1290.0 | 1317.5 | 1370.5 | 1422.6 | 1474.6 | 1527.1 | 1580.1 | 1634.0 | 1688.5 | 1743.9 | 1800.0 |
| | $s$ | 0.5764 | 1.5197 | | 1.5518 | 1.5848 | 1.6140 | 1.6406 | 1.6885 | 1.7315 | 1.7713 | 1.8085 | 1.8437 | 1.8771 | 1.9090 | 1.9396 | 1.9690 |
| 280 (411.07) | Sh | | | | 38.93 | 88.93 | 138.93 | 188.93 | 288.93 | 388.93 | 488.93 | 588.93 | 688.93 | 788.93 | 888.93 | 988.93 | 1088.93 |
| | $v$ | 0.01880 | 1.6505 | | 1.7665 | 1.9037 | 2.0332 | 2.1551 | 2.3909 | 2.6194 | 2.8437 | 3.0655 | 3.2855 | 3.5042 | 3.7217 | 3.9384 | 4.1543 |
| | $h$ | 387.12 | 1202.3 | | 1228.8 | 1260.0 | 1289.1 | 1316.8 | 1370.0 | 1422.1 | 1474.2 | 1526.8 | 1579.9 | 1633.8 | 1688.4 | 1743.7 | 1799.8 |
| | $s$ | 0.5805 | 1.5166 | | 1.5464 | 1.5798 | 1.6093 | 1.6361 | 1.6841 | 1.7273 | 1.7671 | 1.8043 | 1.8395 | 1.8730 | 1.9050 | 1.9356 | 1.9649 |

Temperature (°F)

| P (Sat. Temp.) | Sh | Sat. Water | Sat. Steam | 35.75 | 85.75 | 135.75 | 185.75 | 285.75 | 385.75 | 485.75 | 585.75 | 685.75 | 785.75 | 885.75 | 985.75 | 1085.75 |
|---|---|---|---|---|---|---|---|---|---|---|---|---|---|---|---|---|
| **290** (414.25) | v | 0.01885 | 1.5948 | 1.6988 | 1.8327 | 1.9578 | 2.0772 | 2.3058 | 2.5269 | 2.7440 | 2.9585 | 3.1711 | 3.3824 | 3.5926 | 3.8019 | 4.0106 |
|  | h | 390.60 | 1202.6 | 1227.3 | 1258.9 | 1288.1 | 1316.0 | 1369.5 | 1421.7 | 1473.9 | 1526.5 | 1579.6 | 1633.5 | 1688.2 | 1743.6 | 1799.7 |
|  | s | 0.5844 | 1.5135 | 1.5412 | 1.5750 | 1.6048 | 1.6317 | 1.6799 | 1.7232 | 1.7630 | 1.8003 | 1.8356 | 1.8690 | 1.9010 | 1.9316 | 1.9610 |
| **300** (417.35) | v | 0.01889 | 1.5427 | 1.6356 | 1.7665 | 1.8883 | 2.0044 | 2.2263 | 2.4407 | 2.6509 | 2.8585 | 3.0643 | 3.2688 | 3.4721 | 3.6746 | 3.8764 |
|  | h | 393.99 | 1202.9 | 1225.7 | 1257.7 | 1287.2 | 1315.2 | 1368.9 | 1421.3 | 1473.6 | 1526.2 | 1579.4 | 1633.3 | 1688.0 | 1743.4 | 1799.6 |
|  | s | 0.5882 | 1.5105 | 1.5361 | 1.5703 | 1.6003 | 1.6274 | 1.6758 | 1.7192 | 1.7591 | 1.7964 | 1.8317 | 1.8652 | 1.8972 | 1.9278 | 1.9572 |
| **310** (420.36) | v | 0.01894 | 1.4939 | 1.5763 | 1.7044 | 1.8233 | 1.9363 | 2.1520 | 2.3600 | 2.5638 | 2.7650 | 2.9644 | 3.1625 | 3.3594 | 3.5555 | 3.7509 |
|  | h | 397.30 | 1203.2 | 1224.1 | 1256.5 | 1286.3 | 1314.5 | 1368.4 | 1420.9 | 1473.2 | 1525.9 | 1579.2 | 1633.1 | 1687.8 | 1743.3 | 1799.4 |
|  | s | 0.5920 | 1.5076 | 1.5311 | 1.5657 | 1.5960 | 1.6233 | 1.6719 | 1.7153 | 1.7553 | 1.7927 | 1.8280 | 1.8615 | 1.8935 | 1.9241 | 1.9536 |
| **320** (423.31) | v | 0.01899 | 1.4480 | 1.5207 | 1.6462 | 1.7623 | 1.8725 | 2.0823 | 2.2843 | 2.4821 | 2.6774 | 2.8708 | 3.0628 | 3.2538 | 3.4438 | 3.6332 |
|  | h | 400.53 | 1203.4 | 1222.5 | 1255.2 | 1285.3 | 1313.7 | 1367.8 | 1420.5 | 1472.9 | 1525.6 | 1578.9 | 1632.9 | 1687.6 | 1743.1 | 1799.3 |
|  | s | 0.5956 | 1.5048 | 1.5261 | 1.5612 | 1.5918 | 1.6192 | 1.6680 | 1.7116 | 1.7516 | 1.7890 | 1.8243 | 1.8579 | 1.8899 | 1.9206 | 1.9500 |
| **330** (426.18) | v | 0.01903 | 1.4048 | 1.4684 | 1.5915 | 1.7050 | 1.8125 | 2.0168 | 2.2132 | 2.4054 | 2.5950 | 2.7828 | 2.9692 | 3.1545 | 3.3389 | 3.5227 |
|  | h | 403.70 | 1203.6 | 1220.9 | 1254.0 | 1284.4 | 1313.0 | 1367.3 | 1420.0 | 1472.5 | 1525.3 | 1578.7 | 1632.7 | 1687.5 | 1742.9 | 1799.2 |
|  | s | 0.5991 | 1.5021 | 1.5213 | 1.5568 | 1.5876 | 1.6153 | 1.6643 | 1.7079 | 1.7480 | 1.7855 | 1.8208 | 1.8544 | 1.8864 | 1.9171 | 1.9466 |
| **340** (428.99) | v | 0.01908 | 1.3640 | 1.4191 | 1.5399 | 1.6511 | 1.7561 | 1.9552 | 2.1463 | 2.3333 | 2.5175 | 2.7000 | 2.8811 | 3.0611 | 3.2402 | 3.4186 |
|  | h | 406.80 | 1203.8 | 1219.2 | 1252.8 | 1283.4 | 1312.2 | 1366.7 | 1419.6 | 1472.2 | 1525.0 | 1578.4 | 1632.5 | 1687.3 | 1742.8 | 1799.0 |
|  | s | 0.6026 | 1.4994 | 1.5165 | 1.5525 | 1.5836 | 1.6114 | 1.6606 | 1.7044 | 1.7445 | 1.7820 | 1.8174 | 1.8510 | 1.8831 | 1.9138 | 1.9432 |
| **350** (431.73) | v | 0.01912 | 1.3255 | 1.3725 | 1.4913 | 1.6002 | 1.7028 | 1.8970 | 2.0832 | 2.2652 | 2.4445 | 2.6219 | 2.7980 | 2.9730 | 3.1471 | 3.3205 |
|  | h | 409.83 | 1204.0 | 1217.5 | 1251.5 | 1282.4 | 1311.4 | 1366.2 | 1419.2 | 1471.8 | 1524.7 | 1578.2 | 1632.3 | 1687.1 | 1742.6 | 1798.9 |
|  | s | 0.6059 | 1.4968 | 1.5119 | 1.5483 | 1.5797 | 1.6077 | 1.6571 | 1.7009 | 1.7411 | 1.7787 | 1.8141 | 1.8477 | 1.8798 | 1.9105 | 1.9400 |
| **360** (434.41) | v | 0.01917 | 1.2891 | 1.3285 | 1.4454 | 1.5521 | 1.6525 | 1.8421 | 2.0237 | 2.2009 | 2.3755 | 2.5482 | 2.7196 | 2.8898 | 3.0592 | 3.2279 |
|  | h | 412.81 | 1204.1 | 1215.8 | 1250.3 | 1281.5 | 1310.6 | 1365.6 | 1418.7 | 1471.5 | 1524.4 | 1577.9 | 1632.1 | 1686.9 | 1742.5 | 1798.8 |
|  | s | 0.6092 | 1.4943 | 1.5073 | 1.5441 | 1.5758 | 1.6040 | 1.6536 | 1.6976 | 1.7379 | 1.7754 | 1.8109 | 1.8445 | 1.8766 | 1.9073 | 1.9368 |
| **380** (439.61) | v | 0.01925 | 1.2218 | 1.2472 | 1.3606 | 1.4635 | 1.5598 | 1.7410 | 1.9139 | 2.0825 | 2.2484 | 2.4124 | 2.5750 | 2.7366 | 2.8973 | 3.0572 |
|  | h | 418.59 | 1204.4 | 1212.4 | 1247.7 | 1279.5 | 1309.0 | 1364.5 | 1417.9 | 1470.8 | 1523.8 | 1577.4 | 1631.6 | 1686.5 | 1742.2 | 1798.5 |
|  | s | 0.6156 | 1.4894 | 1.4982 | 1.5360 | 1.5683 | 1.5969 | 1.6470 | 1.6911 | 1.7315 | 1.7692 | 1.8047 | 1.8384 | 1.8705 | 1.9012 | 1.9307 |

Table 15-15. Continued

| Abs. Press. (psi) (Sat. Temp.) | | Sat. Water | Sat. Steam | 450 | 500 | 550 | 600 | 650 | 700 | 800 | 900 | 1000 | 1100 | 1200 | 1300 | 1400 | 1500 |
|---|---|---|---|---|---|---|---|---|---|---|---|---|---|---|---|---|---|
| 400 (444.60) | Sh | | | 5.40 | 55.40 | 105.40 | 155.40 | 205.40 | 255.40 | 355.40 | 455.40 | 555.40 | 655.40 | 755.40 | 855.40 | 955.40 | 1055.40 |
| | v | 0.01934 | 1.1610 | 1.1738 | 1.2841 | 1.3836 | 1.4763 | 1.5646 | 1.6499 | 1.8151 | 1.9759 | 2.1139 | 2.2901 | 2.4450 | 2.5987 | 2.7515 | 2.9037 |
| | h | 424.17 | 1204.6 | 1208.8 | 1245.1 | 1277.5 | 1307.4 | 1335.9 | 1363.4 | 1417.0 | 1470.1 | 1523.3 | 1576.9 | 1631.2 | 1686.2 | 1741.9 | 1798.2 |
| | s | 0.6217 | 1.4847 | 1.4894 | 1.5282 | 1.5611 | 1.5901 | 1.6163 | 1.6406 | 1.6850 | 1.7255 | 1.7632 | 1.7988 | 1.8325 | 1.8647 | 1.8955 | 1.9250 |
| 420 (449.40) | Sh | | | .60 | 50.60 | 100.60 | 150.60 | 200.60 | 250.60 | 350.60 | 450.60 | 550.60 | 650.60 | 750.60 | 850.60 | 950.60 | 1050.60 |
| | v | 0.01942 | 1.1057 | 1.1071 | 1.2148 | 1.3113 | 1.4007 | 1.4856 | 1.5676 | 1.7258 | 1.8795 | 2.0304 | 2.1795 | 2.3273 | 2.4739 | 2.6196 | 2.7647 |
| | h | 429.56 | 1204.7 | 1205.2 | 1242.4 | 1275.4 | 1305.8 | 1334.5 | 1362.3 | 1416.2 | 1469.4 | 1522.7 | 1576.4 | 1630.8 | 1685.8 | 1741.6 | 1798.0 |
| | s | 0.6276 | 1.4802 | 1.4808 | 1.5206 | 1.5542 | 1.5835 | 1.6100 | 1.6345 | 1.6791 | 1.7197 | 1.7575 | 1.7932 | 1.8269 | 1.8591 | 1.8899 | 1.9195 |
| 440 (454.03) | Sh | | | | 45.97 | 95.97 | 145.97 | 195.97 | 245.97 | 345.97 | 445.97 | 545.97 | 645.97 | 745.97 | 845.97 | 945.97 | 1045.97 |
| | v | 0.01950 | 1.0554 | | 1.1517 | 1.2454 | 1.3319 | 1.4138 | 1.4926 | 1.6445 | 1.7918 | 1.9363 | 2.0790 | 2.2203 | 2.3605 | 2.4998 | 2.6384 |
| | h | 434.77 | 1204.8 | | 1239.7 | 1273.4 | 1304.2 | 1333.2 | 1361.1 | 1415.3 | 1468.7 | 1522.1 | 1575.9 | 1630.4 | 1685.5 | 1741.2 | 1797.7 |
| | s | 0.6332 | 1.4759 | | 1.5132 | 1.5474 | 1.5772 | 1.6040 | 1.6286 | 1.6734 | 1.7142 | 1.7521 | 1.7878 | 1.8216 | 1.8538 | 1.8847 | 1.9143 |
| 460 (458.50) | Sh | | | | 41.50 | 91.50 | 141.50 | 191.50 | 241.50 | 341.50 | 441.50 | 541.50 | 641.50 | 741.50 | 841.50 | 941.50 | 1041.50 |
| | v | 0.01959 | 1.0092 | | 1.0939 | 1.1852 | 1.2691 | 1.3482 | 1.4242 | 1.5703 | 1.7117 | 1.8504 | 1.9872 | 2.1226 | 2.2569 | 2.3903 | 2.5230 |
| | h | 439.83 | 1204.8 | | 1236.9 | 1271.3 | 1302.5 | 1331.8 | 1360.0 | 1414.4 | 1468.0 | 1521.5 | 1575.4 | 1629.9 | 1685.1 | 1740.9 | 1797.4 |
| | s | 0.6387 | 1.4718 | | 1.5060 | 1.5409 | 1.5711 | 1.5982 | 1.6230 | 1.6680 | 1.7089 | 1.7469 | 1.7826 | 1.8165 | 1.8488 | 1.8797 | 1.9093 |
| 480 (462.82) | Sh | | | | 37.18 | 87.18 | 137.18 | 187.18 | 237.18 | 337.18 | 437.18 | 537.18 | 637.18 | 737.18 | 837.18 | 937.18 | 1037.18 |
| | v | 0.01967 | 0.9668 | | 1.0409 | 1.1300 | 1.2115 | 1.2881 | 1.3615 | 1.5023 | 1.6384 | 1.7716 | 1.9030 | 2.0330 | 2.1619 | 2.2900 | 2.4173 |
| | h | 444.75 | 1204.8 | | 1234.1 | 1269.1 | 1300.8 | 1330.5 | 1358.8 | 1413.6 | 1467.3 | 1520.9 | 1574.9 | 1629.5 | 1684.7 | 1740.6 | 1797.2 |
| | s | 0.6439 | 1.4677 | | 1.4990 | 1.5346 | 1.5652 | 1.5925 | 1.6176 | 1.6628 | 1.7038 | 1.7419 | 1.7777 | 1.8116 | 1.8439 | 1.8748 | 1.9045 |
| 500 (467.01) | Sh | | | | 32.99 | 82.99 | 132.99 | 182.99 | 232.99 | 332.99 | 432.99 | 532.99 | 632.99 | 732.99 | 832.99 | 932.99 | 1032.99 |
| | v | 0.01975 | 0.9276 | | 0.9919 | 1.0791 | 1.1584 | 1.2327 | 1.3037 | 1.4397 | 1.5708 | 1.6992 | 1.8256 | 1.9507 | 2.0746 | 2.1977 | 2.3200 |
| | h | 449.52 | 1204.7 | | 1231.2 | 1267.0 | 1299.1 | 1329.1 | 1357.7 | 1412.7 | 1466.6 | 1520.3 | 1574.4 | 1629.1 | 1684.4 | 1740.3 | 1796.9 |
| | s | 0.6490 | 1.4639 | | 1.4921 | 1.5284 | 1.5595 | 1.5871 | 1.6123 | 1.6578 | 1.6990 | 1.7371 | 1.7730 | 1.8069 | 1.8393 | 1.8702 | 1.8998 |
| 520 (471.07) | Sh | | | | 28.93 | 78.93 | 128.93 | 178.93 | 228.93 | 328.93 | 428.93 | 528.93 | 628.93 | 728.93 | 828.93 | 928.93 | 1028.93 |
| | v | 0.01982 | 0.8914 | | 0.9466 | 1.0321 | 1.1094 | 1.1816 | 1.2504 | 1.3819 | 1.5085 | 1.6323 | 1.7542 | 1.8746 | 1.9940 | 2.1125 | 2.2302 |
| | h | 454.18 | 1204.5 | | 1228.3 | 1264.8 | 1297.4 | 1327.7 | 1356.5 | 1411.8 | 1465.9 | 1519.7 | 1573.9 | 1628.7 | 1684.0 | 1740.0 | 1796.7 |
| | s | 0.6540 | 1.4601 | | 1.4853 | 1.5223 | 1.5539 | 1.5818 | 1.6072 | 1.6530 | 1.6943 | 1.7325 | 1.7684 | 1.8024 | 1.8348 | 1.8657 | 1.8954 |
| 540 (475.01) | Sh | | | | 24.99 | 74.99 | 124.99 | 174.99 | 224.99 | 324.99 | 424.99 | 524.99 | 624.99 | 724.99 | 824.99 | 924.99 | 1024.99 |
| | v | 0.01990 | 0.8577 | | 0.9045 | 0.9884 | 1.0640 | 1.1342 | 1.2010 | 1.3284 | 1.4508 | 1.5704 | 1.6880 | 1.8042 | 1.9193 | 2.0336 | 2.1471 |
| | h | 458.71 | 1204.4 | | 1225.3 | 1262.5 | 1295.7 | 1326.3 | 1355.3 | 1410.9 | 1465.1 | 1519.1 | 1573.4 | 1628.2 | 1683.6 | 1739.7 | 1796.4 |
| | s | 0.6587 | 1.4565 | | 1.4786 | 1.5164 | 1.5485 | 1.5767 | 1.6023 | 1.6483 | 1.6897 | 1.7280 | 1.7640 | 1.7981 | 1.8305 | 1.8615 | 1.8911 |

| Abs. Press. (Sat. Temp.) | | Sat. Water | Sat. Steam | 500 | 550 | 600 | 650 | 700 | 800 | 900 | 1000 | 1100 | 1200 | 1300 | 1400 | 1500 |
|---|---|---|---|---|---|---|---|---|---|---|---|---|---|---|---|---|
| **560** (478.84) | Sh | | | 21.16 | 71.16 | 121.16 | 171.16 | 221.16 | 321.16 | 421.16 | 521.16 | 621.16 | 721.16 | 821.16 | 921.16 | 1021.16 |
| | $v$ | 0.01998 | 0.8264 | 0.8653 | 0.9479 | 1.0217 | 1.0902 | 1.1552 | 1.2787 | 1.3972 | 1.5129 | 1.6266 | 1.7388 | 1.8500 | 1.9603 | 2.0699 |
| | $h$ | 463.14 | 1204.2 | 1222.2 | 1260.3 | 1293.9 | 1324.9 | 1354.2 | 1410.0 | 1464.4 | 1518.6 | 1572.9 | 1627.8 | 1683.3 | 1739.4 | 1796.1 |
| | $s$ | 0.6634 | 1.4529 | 1.4720 | 1.5106 | 1.5431 | 1.5717 | 1.5975 | 1.6438 | 1.6853 | 1.7237 | 1.7598 | 1.7939 | 1.8263 | 1.8573 | 1.8870 |
| **580** (482.57) | Sh | | | 17.43 | 67.43 | 117.43 | 167.43 | 217.43 | 317.43 | 417.43 | 517.43 | 617.43 | 717.43 | 817.43 | 917.43 | 1017.43 |
| | $v$ | 0.02006 | 0.7971 | 0.8287 | 0.9100 | 0.9824 | 1.0492 | 1.1125 | 1.2324 | 1.3473 | 1.4593 | 1.5693 | 1.6780 | 1.7855 | 1.8921 | 1.9980 |
| | $h$ | 467.47 | 1203.9 | 1219.1 | 1258.0 | 1292.1 | 1323.4 | 1353.0 | 1409.2 | 1463.7 | 1518.0 | 1572.4 | 1627.4 | 1682.9 | 1739.1 | 1795.9 |
| | $s$ | 0.6679 | 1.4495 | 1.4654 | 1.5049 | 1.5380 | 1.5668 | 1.5929 | 1.6394 | 1.6811 | 1.7196 | 1.7556 | 1.7898 | 1.8223 | 1.8533 | 1.8831 |
| **600** (486.20) | Sh | | | 13.80 | 63.80 | 113.80 | 163.80 | 213.80 | 313.80 | 413.80 | 513.80 | 613.80 | 713.80 | 813.80 | 913.80 | 1013.80 |
| | $v$ | 0.02013 | 0.7697 | 0.7944 | 0.8746 | 0.9456 | 1.0109 | 1.0726 | 1.1892 | 1.3008 | 1.4093 | 1.5160 | 1.6211 | 1.7252 | 1.8284 | 1.9309 |
| | $h$ | 471.70 | 1203.7 | 1215.9 | 1255.6 | 1290.3 | 1322.0 | 1351.8 | 1408.3 | 1463.0 | 1517.4 | 1571.9 | 1627.0 | 1682.6 | 1738.8 | 1795.6 |
| | $s$ | 0.6723 | 1.4461 | 1.4590 | 1.4993 | 1.5329 | 1.5621 | 1.5884 | 1.6351 | 1.6769 | 1.7155 | 1.7517 | 1.7859 | 1.8184 | 1.8494 | 1.8792 |
| **650** (494.89) | Sh | | | 5.11 | 55.11 | 105.11 | 155.11 | 205.11 | 305.11 | 405.11 | 505.11 | 605.11 | 705.11 | 805.11 | 905.11 | 1005.11 |
| | $v$ | 0.02032 | 0.7084 | 0.7173 | 0.7954 | 0.8634 | 0.9254 | 0.9835 | 1.0929 | 1.1969 | 1.2979 | 1.3969 | 1.4944 | 1.5909 | 1.6864 | 1.7813 |
| | $h$ | 481.89 | 1202.8 | 1207.6 | 1249.6 | 1285.7 | 1318.3 | 1348.7 | 1406.0 | 1461.2 | 1515.9 | 1570.7 | 1625.9 | 1681.6 | 1738.0 | 1794.9 |
| | $s$ | 1.6828 | 1.4381 | 1.4430 | 1.4858 | 1.5207 | 1.5507 | 1.5775 | 1.6249 | 1.6671 | 1.7059 | 1.7422 | 1.7765 | 1.8092 | 1.8403 | 1.8701 |
| **700** (503.08) | Sh | | | | 46.92 | 96.92 | 146.92 | 196.92 | 296.92 | 396.92 | 496.92 | 596.92 | 696.92 | 796.92 | 896.92 | 996.92 |
| | $v$ | 0.02050 | 0.6556 | | 0.7271 | 0.7928 | 0.8520 | 0.9072 | 1.0102 | 1.1078 | 1.2023 | 1.2948 | 1.3858 | 1.4757 | 1.5647 | 1.6530 |
| | $h$ | 491.60 | 1201.8 | | 1243.4 | 1281.0 | 1314.6 | 1345.6 | 1403.7 | 1459.4 | 1514.4 | 1569.4 | 1624.8 | 1680.7 | 1737.2 | 1794.3 |
| | $s$ | 1.6928 | 1.4304 | | 1.4726 | 1.5090 | 1.5399 | 1.5673 | 1.6154 | 1.6580 | 1.6970 | 1.7335 | 1.7679 | 1.8006 | 1.8318 | 1.8617 |
| **750** (510.84) | Sh | | | | 39.16 | 89.16 | 139.16 | 189.16 | 289.16 | 389.16 | 489.16 | 589.16 | 689.16 | 789.16 | 889.16 | 989.16 |
| | $v$ | 0.02069 | 0.6095 | | 0.6676 | 0.7313 | 0.7882 | 0.8409 | 0.9386 | 1.0306 | 1.1195 | 1.2063 | 1.2916 | 1.3759 | 1.4592 | 1.5419 |
| | $h$ | 500.89 | 1200.7 | | 1236.9 | 1276.1 | 1310.7 | 1342.5 | 1401.5 | 1457.6 | 1512.9 | 1568.2 | 1623.8 | 1679.8 | 1736.4 | 1793.6 |
| | $s$ | 0.7022 | 1.4232 | | 1.4598 | 1.4977 | 1.5296 | 1.5577 | 1.6065 | 1.6494 | 1.6886 | 1.7252 | 1.7598 | 1.7926 | 1.8239 | 1.8538 |
| **800** (518.21) | Sh | | | | 31.79 | 81.79 | 131.79 | 181.79 | 281.79 | 381.79 | 481.79 | 581.79 | 681.79 | 781.79 | 881.79 | 981.79 |
| | $v$ | 0.02087 | 0.5690 | | 0.6151 | 0.6774 | 0.7323 | 0.7828 | 0.8759 | 0.9631 | 1.0470 | 1.1289 | 1.2093 | 1.2885 | 1.3669 | 1.4446 |
| | $h$ | 509.81 | 1199.4 | | 1230.1 | 1271.1 | 1306.8 | 1339.3 | 1399.1 | 1455.8 | 1511.4 | 1566.9 | 1622.7 | 1678.9 | 1735.7 | 1792.9 |
| | $s$ | 0.7111 | 1.4163 | | 1.4472 | 1.4869 | 1.5198 | 1.5484 | 1.5980 | 1.6413 | 1.6807 | 1.7175 | 1.7522 | 1.7851 | 1.8164 | 1.8464 |
| **850** (525.24) | Sh | | | | 24.76 | 74.76 | 124.76 | 174.76 | 274.76 | 374.76 | 474.76 | 574.76 | 674.76 | 774.76 | 874.76 | 974.76 |
| | $v$ | 0.02105 | 0.5330 | | 0.5683 | 0.6296 | 0.6829 | 0.7315 | 0.8205 | 0.9034 | 0.9830 | 1.0606 | 1.1366 | 1.2115 | 1.2855 | 1.3588 |
| | $h$ | 518.40 | 1198.0 | | 1223.0 | 1265.9 | 1302.8 | 1336.0 | 1396.8 | 1454.0 | 1510.0 | 1565.7 | 1621.6 | 1678.0 | 1734.9 | 1792.3 |
| | $s$ | 0.7197 | 1.4096 | | 1.4347 | 1.4763 | 1.5102 | 1.5396 | 1.5899 | 1.6336 | 1.6733 | 1.7102 | 1.7450 | 1.7780 | 1.8094 | 1.8395 |
| **900** (531.95) | Sh | | | | 18.05 | 68.05 | 118.05 | 168.05 | 268.05 | 368.05 | 468.05 | 568.05 | 668.05 | 768.05 | 868.05 | 968.05 |
| | $v$ | 0.02123 | 0.5009 | | 0.5263 | 0.5869 | 0.6388 | 0.6858 | 0.7713 | 0.8504 | 0.9262 | 0.9998 | 1.0720 | 1.1430 | 1.2131 | 1.2825 |
| | $h$ | 526.70 | 1196.4 | | 1215.5 | 1260.6 | 1298.6 | 1332.7 | 1394.4 | 1452.2 | 1508.5 | 1564.4 | 1620.6 | 1677.1 | 1734.1 | 1791.6 |
| | $s$ | 0.7279 | 1.4032 | | 1.4223 | 1.4659 | 1.5010 | 1.5311 | 1.5822 | 1.6263 | 1.6662 | 1.7033 | 1.7382 | 1.7713 | 1.8028 | 1.8329 |

Table 15-15. Continued

| Abs. Press. (psi) (Sat. Temp.) | | Sat. Water | Sat. Steam | \| Temperature (°F) 550 | 600 | 650 | 700 | 750 | 800 | 850 | 900 | 1000 | 1100 | 1200 | 1300 | 1400 | 1500 |
|---|---|---|---|---|---|---|---|---|---|---|---|---|---|---|---|---|---|
| 950 (538.39) | Sh | | | 11.61 | 61.61 | 111.61 | 161.61 | 211.61 | 261.61 | 311.61 | 361.61 | 461.61 | 561.61 | 661.61 | 761.61 | 861.61 | 961.61 |
| | v | 0.02141 | 0.4721 | 0.4883 | 0.5485 | 0.5993 | 0.6449 | 0.6871 | 0.7272 | 0.7656 | 0.8030 | 0.8753 | 0.9455 | 1.0142 | 1.0817 | 1.1484 | 1.2143 |
| | h | 534.74 | 1194.7 | 1207.6 | 1255.1 | 1294.4 | 1329.3 | 1361.5 | 1392.0 | 1421.5 | 1450.3 | 1507.0 | 1563.2 | 1619.5 | 1676.2 | 1733.3 | 1791.0 |
| | s | 0.7358 | 1.3970 | 1.4098 | 1.4557 | 1.4921 | 1.5228 | 1.5500 | 1.5748 | 1.5977 | 1.6193 | 1.6595 | 1.6967 | 1.7317 | 1.7649 | 1.7965 | 1.8267 |
| 1000 (544.58) | Sh | | | 5.42 | 55.42 | 105.42 | 155.42 | 205.42 | 255.42 | 305.42 | 355.42 | 455.42 | 555.42 | 655.42 | 755.42 | 855.42 | 955.42 |
| | v | 0.02159 | 0.4460 | 0.4535 | 0.5137 | 0.5636 | 0.6080 | 0.6489 | 0.6875 | 0.7245 | 0.7603 | 0.8295 | 0.8966 | 0.9622 | 1.0266 | 1.0901 | 1.1529 |
| | h | 542.55 | 1192.9 | 1199.3 | 1249.3 | 1290.1 | 1325.9 | 1358.7 | 1389.6 | 1419.4 | 1448.5 | 1505.4 | 1561.9 | 1618.4 | 1675.3 | 1732.5 | 1790.3 |
| | s | 0.7434 | 1.3910 | 1.3973 | 1.4457 | 1.4833 | 1.5149 | 1.5426 | 1.5677 | 1.5908 | 1.6126 | 1.6530 | 1.6905 | 1.7256 | 1.7589 | 1.7905 | 1.8207 |
| 1050 (550.53) | Sh | | | | 49.47 | 99.47 | 149.47 | 199.47 | 249.47 | 299.47 | 349.47 | 449.47 | 549.47 | 649.47 | 749.47 | 849.47 | 949.47 |
| | v | 0.02177 | 0.4222 | | 0.4821 | 0.5312 | 0.5745 | 0.6142 | 0.6515 | 0.6872 | 0.7216 | 0.7881 | 0.8524 | 0.9151 | 0.9767 | 1.0373 | 1.0973 |
| | h | 550.15 | 1191.0 | | 1243.4 | 1285.7 | 1322.4 | 1355.8 | 1387.2 | 1417.3 | 1446.6 | 1503.9 | 1560.7 | 1617.4 | 1674.4 | 1731.8 | 1789.6 |
| | s | 0.7507 | 1.3851 | | 1.4358 | 1.4748 | 1.5072 | 1.5354 | 1.5608 | 1.5842 | 1.6062 | 1.6469 | 1.6845 | 1.7197 | 1.7531 | 1.7848 | 1.8151 |
| 1100 (556.28) | Sh | | | | 43.72 | 93.72 | 143.72 | 193.72 | 243.72 | 293.72 | 343.72 | 443.72 | 543.72 | 643.72 | 743.72 | 843.72 | 943.72 |
| | v | 0.02195 | 0.4006 | | 0.4531 | 0.5017 | 0.5440 | 0.5826 | 0.6188 | 0.6533 | 0.6865 | 0.7505 | 0.8121 | 0.8723 | 0.9313 | 0.9894 | 1.0468 |
| | h | 557.55 | 1189.1 | | 1237.3 | 1281.2 | 1318.8 | 1352.9 | 1384.7 | 1415.2 | 1444.7 | 1502.4 | 1559.4 | 1616.3 | 1673.5 | 1731.0 | 1789.0 |
| | s | 0.7578 | 1.3794 | | 1.4259 | 1.4664 | 1.4996 | 1.5284 | 1.5542 | 1.5779 | 1.6000 | 1.6410 | 1.6787 | 1.7141 | 1.7475 | 1.7793 | 1.8097 |
| 1150 (561.82) | Sh | | | | 39.18 | 89.18 | 139.18 | 189.18 | 239.18 | 289.18 | 339.18 | 439.18 | 539.18 | 639.18 | 739.18 | 839.18 | 939.18 |
| | v | 0.02214 | 0.3807 | | 0.4263 | 0.4746 | 0.5162 | 0.5538 | 0.5889 | 0.6223 | 0.6544 | 0.7161 | 0.7754 | 0.8332 | 0.8899 | 0.9456 | 1.0007 |
| | h | 564.78 | 1187.0 | | 1230.9 | 1276.6 | 1315.2 | 1349.9 | 1382.2 | 1413.0 | 1442.8 | 1500.9 | 1558.1 | 1615.2 | 1672.6 | 1730.2 | 1788.3 |
| | s | 0.7647 | 1.3738 | | 1.4160 | 1.4582 | 1.4923 | 1.5216 | 1.5478 | 1.5717 | 1.5941 | 1.6353 | 1.6732 | 1.7087 | 1.7422 | 1.7741 | 1.8045 |
| 1200 (567.19) | Sh | | | | 32.81 | 82.81 | 132.81 | 182.81 | 232.81 | 282.81 | 332.81 | 432.81 | 532.81 | 632.81 | 732.81 | 832.81 | 932.81 |
| | v | 0.02232 | 0.3624 | | 0.4016 | 0.4497 | 0.4905 | 0.5273 | 0.5615 | 0.5939 | 0.6250 | 0.6845 | 0.7418 | 0.7974 | 0.8519 | 0.9055 | 0.9584 |
| | h | 571.85 | 1184.8 | | 1224.2 | 1271.8 | 1311.5 | 1346.9 | 1379.7 | 1410.8 | 1440.9 | 1499.4 | 1556.9 | 1614.2 | 1671.6 | 1729.4 | 1787.6 |
| | s | 0.7714 | 1.3683 | | 1.4061 | 1.4501 | 1.4851 | 1.5150 | 1.5415 | 1.5658 | 1.5883 | 1.6298 | 1.6679 | 1.7035 | 1.7371 | 1.7691 | 1.7996 |
| 1300 (577.42) | Sh | | | | 22.58 | 72.58 | 122.58 | 172.58 | 222.58 | 272.58 | 322.58 | 422.58 | 522.58 | 622.58 | 722.58 | 822.58 | 922.58 |
| | v | 0.02269 | 0.3299 | | 0.3570 | 0.4052 | 0.4451 | 0.4804 | 0.5129 | 0.5436 | 0.5729 | 0.6287 | 0.6822 | 0.7341 | 0.7847 | 0.8345 | 0.8836 |
| | h | 585.58 | 1180.2 | | 1209.9 | 1261.9 | 1303.9 | 1340.8 | 1374.6 | 1406.4 | 1437.1 | 1496.3 | 1554.3 | 1612.0 | 1669.8 | 1727.9 | 1786.3 |
| | s | 0.7843 | 1.3577 | | 1.3860 | 1.4340 | 1.4711 | 1.5022 | 1.5296 | 1.5544 | 1.5773 | 1.6194 | 1.6578 | 1.6937 | 1.7275 | 1.7596 | 1.7902 |
| 1400 (587.07) | Sh | | | | 12.93 | 62.93 | 112.93 | 162.93 | 212.93 | 262.93 | 312.93 | 412.93 | 512.93 | 612.93 | 712.93 | 812.93 | 912.93 |
| | v | 0.02307 | 0.3018 | | 0.3176 | 0.3667 | 0.4059 | 0.4400 | 0.4712 | 0.5004 | 0.5282 | 0.5809 | 0.6311 | 0.6798 | 0.7272 | 0.7737 | 0.8195 |
| | h | 598.83 | 1175.3 | | 1194.1 | 1251.1 | 1296.1 | 1334.5 | 1369.3 | 1402.0 | 1433.2 | 1493.2 | 1551.8 | 1609.9 | 1668.0 | 1726.3 | 1785.0 |
| | s | 0.7966 | 1.3474 | | 1.3652 | 1.4181 | 1.4575 | 1.4900 | 1.5182 | 1.5436 | 1.5670 | 1.6096 | 1.6484 | 1.6845 | 1.7185 | 1.7508 | 1.7815 |

Properties of superheated steam (and saturation values). Column headings are degrees of superheat (Sh). For each pressure the rows give superheat (Sh), specific volume (v), enthalpy (h), and entropy (s). The first two value columns give saturated-liquid and saturated-vapor values.

| Press. (Sat. temp) | | Sat. liq | Sat. vap | 3.80 | 53.80 | 103.80 | 153.80 | 203.80 | 253.80 | 303.80 | 403.80 | 503.80 | 603.80 | 703.80 | 803.80 | 903.80 |
|---|---|---|---|---|---|---|---|---|---|---|---|---|---|---|---|---|
| **1500** (596.20) | Sh | | | 3.80 | 53.80 | 103.80 | 153.80 | 203.80 | 253.80 | 303.80 | 403.80 | 503.80 | 603.80 | 703.80 | 803.80 | 903.80 |
| | v | 0.02346 | 0.2772 | 0.2820 | 0.3333 | 0.3717 | 0.4049 | 0.4350 | 0.4629 | 0.4894 | 0.5394 | 0.5869 | 0.6327 | 0.6773 | 0.7210 | 0.7639 |
| | h | 611.68 | 1170.1 | 1176.3 | 1240.2 | 1287.9 | 1328.0 | 1364.0 | 1397.4 | 1429.2 | 1490.1 | 1549.2 | 1607.7 | 1666.2 | 1724.8 | 1783.7 |
| | s | 0.8085 | 1.3373 | 1.3431 | 1.4022 | 1.4443 | 1.4782 | 1.5073 | 1.5333 | 1.5572 | 1.6004 | 1.6395 | 1.6759 | 1.7101 | 1.7425 | 1.7734 |
| **1600** (604.87) | Sh | | | | 45.13 | 95.13 | 145.13 | 195.13 | 245.13 | 295.13 | 395.13 | 495.13 | 595.13 | 695.13 | 795.13 | 895.13 |
| | v | 0.02387 | 0.2555 | | 0.3026 | 0.3415 | 0.3741 | 0.4032 | 0.4301 | 0.4555 | 0.5031 | 0.5482 | 0.5915 | 0.6336 | 0.6748 | 0.7153 |
| | h | 624.20 | 1164.5 | | 1228.3 | 1279.4 | 1321.4 | 1358.5 | 1392.8 | 1425.2 | 1486.9 | 1546.6 | 1605.6 | 1664.3 | 1723.2 | 1782.3 |
| | s | 0.8199 | 1.3274 | | 1.3861 | 1.4312 | 1.4667 | 1.4968 | 1.5235 | 1.5478 | 1.5916 | 1.6312 | 1.6678 | 1.7022 | 1.7347 | 1.7657 |
| **1700** (613.13) | Sh | | | | 36.87 | 86.87 | 136.87 | 186.87 | 236.87 | 286.87 | 386.87 | 486.87 | 586.87 | 686.87 | 786.87 | 886.87 |
| | v | 0.02428 | 0.2361 | | 0.2754 | 0.3147 | 0.3468 | 0.3751 | 0.4011 | 0.4255 | 0.4711 | 0.5140 | 0.5552 | 0.5951 | 0.6341 | 0.6724 |
| | h | 636.45 | 1158.6 | | 1215.3 | 1270.5 | 1314.5 | 1352.9 | 1388.1 | 1421.2 | 1483.8 | 1544.0 | 1603.4 | 1662.5 | 1721.7 | 1781.0 |
| | s | 0.8309 | 1.3176 | | 1.3697 | 1.4183 | 1.4555 | 1.4867 | 1.5140 | 1.5388 | 1.5833 | 1.6232 | 1.6601 | 1.6947 | 1.7274 | 1.7585 |
| **1800** (621.02) | Sh | | | | 28.98 | 78.98 | 128.98 | 178.98 | 228.98 | 278.98 | 378.98 | 478.98 | 578.98 | 678.98 | 778.98 | 878.98 |
| | v | 0.02472 | 0.2186 | | 0.2505 | 0.2906 | 0.3223 | 0.3500 | 0.3752 | 0.3988 | 0.4426 | 0.4836 | 0.5229 | 0.5609 | 0.5980 | 0.6343 |
| | h | 648.49 | 1152.3 | | 1201.2 | 1261.1 | 1307.4 | 1347.2 | 1383.3 | 1417.1 | 1480.6 | 1541.4 | 1601.2 | 1660.7 | 1720.1 | 1779.7 |
| | s | 0.8417 | 1.3079 | | 1.3526 | 1.4054 | 1.4446 | 1.4768 | 1.5049 | 1.5302 | 1.5753 | 1.6156 | 1.6528 | 1.6876 | 1.7204 | 1.7516 |
| **1900** (628.56) | Sh | | | | 21.44 | 71.44 | 121.44 | 171.44 | 221.44 | 271.44 | 371.44 | 471.44 | 571.44 | 671.44 | 771.44 | 871.44 |
| | v | 0.02517 | 0.2028 | | 0.2274 | 0.2687 | 0.3004 | 0.3275 | 0.3521 | 0.3749 | 0.4171 | 0.4565 | 0.4940 | 0.5303 | 0.5656 | 0.6002 |
| | h | 660.36 | 1145.6 | | 1185.7 | 1251.3 | 1300.2 | 1341.4 | 1378.4 | 1412.9 | 1477.4 | 1538.8 | 1599.1 | 1658.8 | 1718.6 | 1778.4 |
| | s | 0.8522 | 1.2981 | | 1.3346 | 1.3925 | 1.4338 | 1.4672 | 1.4960 | 1.5219 | 1.5677 | 1.6084 | 1.6458 | 1.6808 | 1.7138 | 1.7451 |
| **2000** (635.80) | Sh | | | | 14.20 | 64.20 | 114.20 | 164.20 | 214.20 | 264.20 | 364.20 | 464.20 | 564.20 | 664.20 | 764.20 | 864.20 |
| | v | 0.02565 | 0.1883 | | 0.2056 | 0.2488 | 0.2805 | 0.3072 | 0.3312 | 0.3534 | 0.3942 | 0.4320 | 0.4680 | 0.5027 | 0.5365 | 0.5695 |
| | h | 672.11 | 1138.3 | | 1168.3 | 1240.9 | 1292.6 | 1335.4 | 1373.5 | 1408.7 | 1474.1 | 1536.2 | 1596.9 | 1657.0 | 1717.0 | 1777.1 |
| | s | 0.8625 | 1.2881 | | 1.3154 | 1.3794 | 1.4231 | 1.4578 | 1.4874 | 1.5138 | 1.5603 | 1.6014 | 1.6391 | 1.6743 | 1.7075 | 1.7389 |
| **2100** (642.76) | Sh | | | | 7.24 | 57.24 | 107.24 | 157.24 | 207.24 | 257.24 | 357.24 | 457.24 | 557.24 | 657.24 | 757.24 | 857.24 |
| | v | 0.02615 | 0.1750 | | 0.1847 | 0.2304 | 0.2624 | 0.2888 | 0.3123 | 0.3339 | 0.3734 | 0.4099 | 0.4445 | 0.4778 | 0.5101 | 0.5418 |
| | h | 683.79 | 1130.5 | | 1148.5 | 1229.8 | 1284.9 | 1329.3 | 1368.4 | 1404.4 | 1470.9 | 1533.6 | 1594.7 | 1655.2 | 1715.4 | 1775.7 |
| | s | 0.8727 | 1.2780 | | 1.2942 | 1.3661 | 1.4125 | 1.4486 | 1.4790 | 1.5060 | 1.5532 | 1.5948 | 1.6327 | 1.6681 | 1.7014 | 1.7330 |
| **2200** (649.45) | Sh | | | | 0.55 | 50.55 | 100.55 | 150.55 | 200.55 | 250.55 | 350.55 | 450.55 | 550.55 | 650.55 | 750.55 | 850.55 |
| | v | 0.02669 | 0.1627 | | 0.1636 | 0.2134 | 0.2458 | 0.2720 | 0.2950 | 0.3161 | 0.3545 | 0.3897 | 0.4231 | 0.4551 | 0.4862 | 0.5165 |
| | h | 695.46 | 1122.2 | | 1123.9 | 1218.0 | 1276.8 | 1323.1 | 1363.3 | 1400.0 | 1467.6 | 1530.9 | 1592.5 | 1653.3 | 1713.9 | 1774.4 |
| | s | 0.8828 | 1.2676 | | 1.2691 | 1.3523 | 1.4020 | 1.4395 | 1.4708 | 1.4984 | 1.5463 | 1.5883 | 1.6266 | 1.6622 | 1.6956 | 1.7273 |
| **2300** (655.89) | Sh | | | | | 44.11 | 94.11 | 144.11 | 194.11 | 244.11 | 344.11 | 444.11 | 544.11 | 644.11 | 744.11 | 844.11 |
| | v | 0.02727 | 0.1513 | | | 0.1975 | 0.2305 | 0.2566 | 0.2793 | 0.2999 | 0.3372 | 0.3714 | 0.4035 | 0.4344 | 0.4643 | 0.4935 |
| | h | 707.18 | 1113.2 | | | 1205.3 | 1268.4 | 1316.7 | 1358.1 | 1395.7 | 1464.2 | 1528.3 | 1590.3 | 1651.5 | 1712.3 | 1773.1 |
| | s | 0.8929 | 1.2569 | | | 1.3381 | 1.3914 | 1.4305 | 1.4628 | 1.4910 | 1.5397 | 1.5821 | 1.6207 | 1.6565 | 1.6901 | 1.7219 |

Table 15-15. Continued

| Abs. Press. (psi) (Sat. Temp.) | | Sat. Water | Sat. Steam | Temperature (°F) 700 | 750 | 800 | 850 | 900 | 950 | 1000 | 1050 | 1100 | 1150 | 1200 | 1300 | 1400 | 1500 |
|---|---|---|---|---|---|---|---|---|---|---|---|---|---|---|---|---|---|
| 2400 (662.11) | Sh | | | 37.89 | 87.89 | 137.89 | 187.89 | 237.89 | 287.89 | 337.89 | 387.89 | 437.89 | 487.89 | 537.89 | 637.89 | 737.89 | 837.89 |
| | v | 0.02790 | 0.1408 | 0.1824 | 0.2164 | 0.2424 | 0.2648 | 0.2850 | 0.3037 | 0.3214 | 0.3382 | 0.3545 | 0.3703 | 0.3856 | 0.4155 | 0.4443 | 0.4724 |
| | h | 718.95 | 1103.7 | 1191.6 | 1259.7 | 1310.1 | 1352.8 | 1391.2 | 1426.9 | 1460.9 | 1493.7 | 1525.6 | 1557.0 | 1588.1 | 1649.6 | 1710.8 | 1771.8 |
| | s | 0.9031 | 1.2460 | 1.3232 | 1.3808 | 1.4217 | 1.4549 | 1.4837 | 1.5095 | 1.5332 | 1.5553 | 1.5761 | 1.5959 | 1.6149 | 1.6509 | 1.6847 | 1.7167 |
| 2500 (668.11) | Sh | | | 31.89 | 81.89 | 131.89 | 181.89 | 231.89 | 281.89 | 331.89 | 381.89 | 431.89 | 481.89 | 531.89 | 631.89 | 731.89 | 831.89 |
| | v | 0.02859 | 0.1307 | 0.1681 | 0.2032 | 0.2293 | 0.2514 | 0.2712 | 0.2896 | 0.3068 | 0.3232 | 0.3390 | 0.3543 | 0.3692 | 0.3980 | 0.4259 | 0.4529 |
| | h | 731.71 | 1093.3 | 1176.7 | 1250.6 | 1303.4 | 1347.4 | 1386.7 | 1423.1 | 1457.5 | 1490.7 | 1522.9 | 1554.6 | 1585.9 | 1647.8 | 1709.2 | 1770.4 |
| | s | 0.9139 | 1.2345 | 1.3076 | 1.3701 | 1.4129 | 1.4472 | 1.4766 | 1.5029 | 1.5269 | 1.5492 | 1.5703 | 1.5903 | 1.6094 | 1.6456 | 1.6796 | 1.7116 |
| 2600 (673.91) | Sh | | | 26.09 | 76.09 | 126.09 | 176.09 | 226.09 | 276.09 | 326.09 | 376.09 | 426.09 | 476.09 | 526.09 | 626.09 | 726.09 | 826.09 |
| | v | 0.02938 | 0.1211 | 0.1544 | 0.1909 | 0.2171 | 0.2390 | 0.2585 | 0.2765 | 0.2933 | 0.3093 | 0.3247 | 0.3395 | 0.3540 | 0.3819 | 0.4088 | 0.4350 |
| | h | 744.47 | 1082.0 | 1160.2 | 1241.1 | 1296.5 | 1341.9 | 1382.1 | 1419.2 | 1454.1 | 1487.7 | 1520.2 | 1552.2 | 1583.7 | 1646.0 | 1707.7 | 1769.1 |
| | s | 0.9247 | 1.2225 | 1.2908 | 1.3592 | 1.4042 | 1.4395 | 1.4696 | 1.4964 | 1.5208 | 1.5434 | 1.5646 | 1.5848 | 1.6040 | 1.6405 | 1.6746 | 1.7068 |
| 2700 (679.53) | Sh | | | 20.47 | 70.47 | 120.47 | 170.47 | 220.47 | 270.47 | 320.47 | 370.47 | 420.47 | 470.47 | 520.47 | 620.47 | 720.47 | 820.47 |
| | v | 0.03029 | 0.1119 | 0.1411 | 0.1794 | 0.2058 | 0.2275 | 0.2468 | 0.2644 | 0.2809 | 0.2965 | 0.3114 | 0.3259 | 0.3399 | 0.3670 | 0.3931 | 0.4184 |
| | h | 757.34 | 1069.7 | 1142.0 | 1231.1 | 1289.5 | 1336.3 | 1377.5 | 1415.2 | 1450.7 | 1484.6 | 1517.5 | 1549.8 | 1581.5 | 1644.1 | 1706.1 | 1767.8 |
| | s | 0.9356 | 1.2097 | 1.2727 | 1.3481 | 1.3954 | 1.4319 | 1.4628 | 1.4900 | 1.5148 | 1.5376 | 1.5591 | 1.5794 | 1.5988 | 1.6355 | 1.6697 | 1.7021 |
| 2800 (684.96) | Sh | | | 15.04 | 65.04 | 115.04 | 165.04 | 215.04 | 265.04 | 315.04 | 365.04 | 415.04 | 465.04 | 515.04 | 615.04 | 715.04 | 815.04 |
| | v | 0.03134 | 0.1030 | 0.1278 | 0.1685 | 0.1952 | 0.2168 | 0.2358 | 0.2531 | 0.2693 | 0.2845 | 0.2991 | 0.3132 | 0.3268 | 0.3532 | 0.3785 | 0.4030 |
| | h | 770.69 | 1055.8 | 1121.2 | 1220.6 | 1282.2 | 1330.7 | 1372.8 | 1411.2 | 1447.2 | 1481.6 | 1514.8 | 1547.3 | 1579.3 | 1642.2 | 1704.5 | 1766.5 |
| | s | 0.9468 | 1.1958 | 1.2527 | 1.3368 | 1.3867 | 1.4245 | 1.4561 | 1.4838 | 1.5089 | 1.5321 | 1.5537 | 1.5742 | 1.5938 | 1.6306 | 1.6651 | 1.6975 |
| 2900 (690.22) | Sh | | | 9.78 | 59.78 | 109.78 | 159.78 | 209.78 | 259.78 | 309.78 | 359.78 | 409.78 | 459.78 | 509.78 | 609.78 | 709.78 | 809.78 |
| | v | 0.03262 | 0.0942 | 0.1138 | 0.1581 | 0.1853 | 0.2068 | 0.2256 | 0.2427 | 0.2585 | 0.2734 | 0.2877 | 0.3014 | 0.3147 | 0.3403 | 0.3649 | 0.3887 |
| | h | 785.13 | 1039.8 | 1095.3 | 1209.6 | 1274.7 | 1324.9 | 1368.0 | 1407.2 | 1443.7 | 1478.5 | 1512.1 | 1544.9 | 1577.0 | 1640.4 | 1703.0 | 1765.2 |
| | s | 0.9588 | 1.1803 | 1.2283 | 1.3251 | 1.3780 | 1.4171 | 1.4494 | 1.4777 | 1.5032 | 1.5266 | 1.5485 | 1.5692 | 1.5889 | 1.6259 | 1.6605 | 1.6931 |
| 3000 (695.33) | Sh | | | 4.67 | 54.67 | 104.67 | 154.67 | 204.67 | 254.67 | 304.67 | 354.67 | 404.67 | 454.67 | 504.67 | 604.67 | 704.67 | 804.67 |
| | v | 0.03428 | 0.0850 | 0.0982 | 0.1483 | 0.1759 | 0.1975 | 0.2161 | 0.2339 | 0.2484 | 0.2630 | 0.2770 | 0.2904 | 0.3033 | 0.3282 | 0.3522 | 0.3753 |
| | h | 801.84 | 1020.3 | 1060.5 | 1197.9 | 1267.0 | 1319.0 | 1363.2 | 1403.1 | 1440.2 | 1475.4 | 1509.4 | 1542.4 | 1574.8 | 1638.5 | 1701.4 | 1763.8 |
| | s | 0.9728 | 1.1619 | 1.1966 | 1.3131 | 1.3692 | 1.4097 | 1.4429 | 1.4717 | 1.4976 | 1.5213 | 1.5434 | 1.5642 | 1.5841 | 1.6214 | 1.6561 | 1.6888 |
| 3100 (700.28) | Sh | | | | 49.72 | 99.72 | 149.72 | 199.72 | 249.72 | 299.72 | 349.72 | 399.72 | 449.72 | 499.72 | 599.72 | 699.72 | 799.72 |
| | v | 0.03681 | 0.0745 | | 0.1389 | 0.1671 | 0.1887 | 0.2071 | 0.2237 | 0.2390 | 0.2533 | 0.2670 | 0.2800 | 0.2927 | 0.3170 | 0.3403 | 0.3628 |
| | h | 823.97 | 993.3 | | 1185.4 | 1259.1 | 1313.0 | 1358.4 | 1399.0 | 1436.7 | 1472.3 | 1506.6 | 1539.9 | 1572.6 | 1636.7 | 1699.8 | 1762.5 |
| | s | 0.9914 | 1.1373 | | 1.3007 | 1.3604 | 1.4024 | 1.4364 | 1.4658 | 1.4920 | 1.5161 | 1.5384 | 1.5594 | 1.5794 | 1.6169 | 1.6518 | 1.6847 |

| P (T_sat) | | Sh = 44.92 | 94.92 | 144.92 | 194.92 | 244.92 | 294.92 | 344.92 | 394.92 | 444.92 | 494.92 | 594.92 | 694.92 | 794.92 |
|---|---|---|---|---|---|---|---|---|---|---|---|---|---|---|
| 3200 (705.08) | Sh | | | | | | | | | | | | | |
| | v | 0.1300 | 0.1588 | 0.1804 | 0.1987 | 0.2151 | 0.2301 | 0.2442 | 0.2576 | 0.2704 | 0.2827 | 0.3065 | 0.3291 | 0.3510 |
| | h | 1172.3 | 1250.9 | 1306.9 | 1353.4 | 1394.9 | 1433.1 | 1469.2 | 1503.8 | 1537.4 | 1570.3 | 1634.8 | 1698.3 | 1761.2 |
| | s | 1.2877 | 1.3515 | 1.3951 | 1.4300 | 1.4600 | 1.4866 | 1.5110 | 1.5335 | 1.5547 | 1.5749 | 1.6126 | 1.6477 | 1.6806 |
| 3300 | Sh | | | | | | | | | | | | | |
| | v | 0.1213 | 0.1510 | 0.1727 | 0.1908 | 0.2070 | 0.2218 | 0.2357 | 0.2488 | 0.2613 | 0.2734 | 0.2966 | 0.3187 | 0.3400 |
| | h | 1158.2 | 1242.5 | 1300.7 | 1348.4 | 1390.7 | 1429.5 | 1466.1 | 1501.0 | 1534.9 | 1568.1 | 1632.9 | 1696.7 | 1759.9 |
| | s | 1.2742 | 1.3425 | 1.3879 | 1.4237 | 1.4542 | 1.4813 | 1.5059 | 1.5287 | 1.5501 | 1.5704 | 1.6084 | 1.6436 | 1.6767 |
| 3400 | Sh | | | | | | | | | | | | | |
| | v | 0.1129 | 0.1435 | 0.1653 | 0.1834 | 0.1994 | 0.2140 | 0.2276 | 0.2405 | 0.2528 | 0.2646 | 0.2872 | 0.3088 | 0.3296 |
| | h | 1143.2 | 1233.7 | 1294.3 | 1343.3 | 1386.4 | 1425.9 | 1462.9 | 1498.3 | 1532.4 | 1565.8 | 1631.1 | 1695.1 | 1758.2 |
| | s | 1.2600 | 1.3334 | 1.3807 | 1.4174 | 1.4486 | 1.4761 | 1.5010 | 1.5240 | 1.5456 | 1.5660 | 1.6042 | 1.6396 | 1.6728 |
| 3500 | Sh | | | | | | | | | | | | | |
| | v | 0.1048 | 0.1364 | 0.1583 | 0.1764 | 0.1922 | 0.2066 | 0.2200 | 0.2326 | 0.2447 | 0.2563 | 0.2784 | 0.2995 | 0.3198 |
| | h | 1127.1 | 1224.6 | 1287.8 | 1338.2 | 1382.2 | 1422.2 | 1459.7 | 1495.5 | 1529.9 | 1563.6 | 1629.2 | 1693.6 | 1757.2 |
| | s | 1.2450 | 1.3242 | 1.3734 | 1.4112 | 1.4430 | 1.4709 | 1.4962 | 1.5194 | 1.5412 | 1.5618 | 1.6002 | 1.6358 | 1.6691 |
| 3600 | Sh | | | | | | | | | | | | | |
| | v | 0.0966 | 0.1296 | 0.1517 | 0.1697 | 0.1854 | 0.1996 | 0.2128 | 0.2252 | 0.2371 | 0.2485 | 0.2702 | 0.2908 | 0.3106 |
| | h | 1108.6 | 1215.3 | 1281.2 | 1333.0 | 1377.9 | 1418.6 | 1456.5 | 1492.6 | 1527.4 | 1561.3 | 1627.3 | 1692.0 | 1755.9 |
| | s | 1.2281 | 1.3148 | 1.3662 | 1.4050 | 1.4374 | 1.4658 | 1.4914 | 1.5149 | 1.5369 | 1.5576 | 1.5962 | 1.6320 | 1.6654 |
| 3800 | Sh | | | | | | | | | | | | | |
| | v | 0.0799 | 0.1169 | 0.1395 | 0.1574 | 0.1729 | 0.1868 | 0.1996 | 0.2116 | 0.2231 | 0.2340 | 0.2549 | 0.2746 | 0.2936 |
| | h | 1064.2 | 1195.5 | 1267.6 | 1322.4 | 1369.1 | 1411.2 | 1450.1 | 1487.0 | 1522.4 | 1556.8 | 1623.6 | 1688.9 | 1753.2 |
| | s | 1.1888 | 1.2955 | 1.3517 | 1.3928 | 1.4265 | 1.4558 | 1.4821 | 1.5061 | 1.5284 | 1.5495 | 1.5886 | 1.6247 | 1.6584 |
| 4000 | Sh | | | | | | | | | | | | | |
| | v | 0.0631 | 0.1052 | 0.1284 | 0.1463 | 0.1616 | 0.1752 | 0.1877 | 0.1994 | 0.2105 | 0.2210 | 0.2411 | 0.2601 | 0.2783 |
| | h | 1007.4 | 1174.3 | 1253.4 | 1311.6 | 1360.2 | 1403.6 | 1443.6 | 1481.3 | 1517.3 | 1552.2 | 1619.8 | 1685.7 | 1750.6 |
| | s | 1.1396 | 1.2754 | 1.3371 | 1.3807 | 1.4158 | 1.4461 | 1.4730 | 1.4976 | 1.5203 | 1.5417 | 1.5812 | 1.6177 | 1.6516 |
| 4200 | Sh | | | | | | | | | | | | | |
| | v | 0.0498 | 0.0945 | 0.1183 | 0.1362 | 0.1513 | 0.1647 | 0.1769 | 0.1883 | 0.1991 | 0.2093 | 0.2287 | 0.2470 | 0.2645 |
| | h | 950.1 | 1151.6 | 1238.6 | 1300.4 | 1351.2 | 1396.0 | 1437.1 | 1475.5 | 1512.2 | 1547.6 | 1616.1 | 1682.6 | 1748.0 |
| | s | 1.0905 | 1.2544 | 1.3223 | 1.3686 | 1.4053 | 1.4366 | 1.4642 | 1.4893 | 1.5124 | 1.5341 | 1.5742 | 1.6109 | 1.6452 |
| 4400 | Sh | | | | | | | | | | | | | |
| | v | 0.0421 | 0.0846 | 0.1090 | 0.1270 | 0.1420 | 0.1552 | 0.1671 | 0.1782 | 0.1887 | 0.1986 | 0.2174 | 0.2351 | 0.2519 |
| | h | 909.5 | 1127.3 | 1223.3 | 1289.0 | 1342.0 | 1388.3 | 1430.4 | 1469.7 | 1507.1 | 1543.0 | 1612.3 | 1679.4 | 1745.3 |
| | s | 1.0556 | 1.2325 | 1.3073 | 1.3566 | 1.3949 | 1.4272 | 1.4556 | 1.4812 | 1.5048 | 1.5268 | 1.5673 | 1.6044 | 1.6389 |

Additional values for 3200 (705.08):

| | v | h | s |
|---|---|---|---|
| | 0.04472 | 875.54 | 1.0351 |
| | 0.0566 | 931.6 | 1.0832 |

Table 15-15. Continued

| Abs. Press. (psi) (Sat. Temp.) | | Sat. Water | Sat. Steam | 750 | 800 | 850 | 900 | 950 | 1000 | 1050 | 1100 | 1150 | 1200 | 1250 | 1300 | 1400 | 1500 |
|---|---|---|---|---|---|---|---|---|---|---|---|---|---|---|---|---|---|
| 4600 | Sh | | | | | | | | | | | | | | | | |
| | v | | | 0.0380 | 0.0751 | 0.1005 | 0.1186 | 0.1335 | 0.1465 | 0.1582 | 0.1691 | 0.1792 | 0.1889 | 0.1982 | 0.2071 | 0.2242 | 0.2404 |
| | h | | | 883.8 | 1100.0 | 1207.3 | 1277.2 | 1332.6 | 1380.5 | 1423.7 | 1463.9 | 1501.9 | 1538.4 | 1573.8 | 1608.5 | 1676.3 | 1742.7 |
| | s | | | 1.0331 | 1.2084 | 1.2922 | 1.3446 | 1.3847 | 1.4181 | 1.4472 | 1.4734 | 1.4974 | 1.5197 | 1.5407 | 1.5607 | 1.5982 | 1.6330 |
| 4800 | Sh | | | | | | | | | | | | | | | | |
| | v | | | 0.0355 | 0.0665 | 0.0927 | 0.1109 | 0.1257 | 0.1385 | 0.1500 | 0.1606 | 0.1706 | 0.1800 | 0.1890 | 0.1977 | 0.2142 | 0.2299 |
| | h | | | 866.9 | 1071.2 | 1190.7 | 1265.2 | 1323.1 | 1372.6 | 1417.0 | 1458.0 | 1496.7 | 1533.8 | 1569.7 | 1604.7 | 1673.1 | 1740.0 |
| | s | | | 1.0180 | 1.1835 | 1.2768 | 1.3327 | 1.3745 | 1.4090 | 1.4390 | 1.4657 | 1.4901 | 1.5128 | 1.5341 | 1.5543 | 1.5921 | 1.6272 |
| 5000 | Sh | | | | | | | | | | | | | | | | |
| | v | | | 0.0338 | 0.591 | 0.0855 | 0.1038 | 0.1185 | 0.1312 | 0.1425 | 0.1529 | 0.1626 | 0.1718 | 0.1806 | 0.1890 | 0.2050 | 0.2203 |
| | h | | | 854.9 | 1042.9 | 1173.6 | 1252.8 | 1313.5 | 1364.6 | 1410.2 | 1452.1 | 1491.5 | 1529.1 | 1565.5 | 1600.9 | 1670.0 | 1737.4 |
| | s | | | 1.0070 | 1.1593 | 1.2612 | 1.3207 | 1.3645 | 1.4001 | 1.4309 | 1.4582 | 1.4831 | 1.5061 | 1.5277 | 1.5481 | 1.5863 | 1.6216 |
| 5200 | Sh | | | | | | | | | | | | | | | | |
| | v | | | 0.0326 | 0.0531 | 0.0789 | 0.0973 | 0.1119 | 0.1244 | 0.1356 | 0.1458 | 0.1553 | 0.1642 | 0.1728 | 0.1810 | 0.1966 | 0.2114 |
| | h | | | 845.8 | 1016.9 | 1156.0 | 1240.4 | 1303.7 | 1356.6 | 1403.4 | 1446.2 | 1486.3 | 1524.5 | 1561.3 | 1597.2 | 1666.8 | 1734.7 |
| | s | | | 0.9985 | 1.1370 | 1.2455 | 1.3088 | 1.3545 | 1.3914 | 1.4229 | 1.4509 | 1.4762 | 1.4995 | 1.5214 | 1.5420 | 1.5806 | 1.6161 |
| 5400 | Sh | | | | | | | | | | | | | | | | |
| | v | | | 0.0317 | 0.0483 | 0.0728 | 0.0912 | 0.1058 | 0.1182 | 0.1292 | 0.1392 | 0.1485 | 0.1572 | 0.1656 | 0.1736 | 0.1888 | 0.2031 |
| | h | | | 838.5 | 994.3 | 1138.1 | 1227.7 | 1293.7 | 1348.4 | 1396.5 | 1440.3 | 1481.1 | 1519.8 | 1557.1 | 1593.4 | 1663.7 | 1732.1 |
| | s | | | 0.9915 | 1.1175 | 1.2296 | 1.2969 | 1.3446 | 1.3827 | 1.4151 | 1.4437 | 1.4694 | 1.4931 | 1.5153 | 1.5362 | 1.5750 | 1.6109 |
| 5600 | Sh | | | | | | | | | | | | | | | | |
| | v | | | 0.0309 | 0.0447 | 0.0672 | 0.0856 | 0.1001 | 0.1124 | 0.1232 | 0.1331 | 0.1422 | 0.1508 | 0.1589 | 0.1667 | 0.1815 | 0.1954 |
| | h | | | 832.4 | 975.0 | 1119.9 | 1214.8 | 1283.7 | 1340.2 | 1389.6 | 1434.3 | 1475.5 | 1515.2 | 1552.9 | 1589.6 | 1660.5 | 1729.5 |
| | s | | | 0.9855 | 1.1008 | 1.2137 | 1.2850 | 1.3348 | 1.3742 | 1.4075 | 1.4366 | 1.4628 | 1.4869 | 1.5093 | 1.5304 | 1.5697 | 1.6058 |
| 5800 | Sh | | | | | | | | | | | | | | | | |
| | v | | | 0.0303 | 0.0419 | 0.0622 | 0.0805 | 0.0949 | 0.1070 | 0.1177 | 0.1274 | 0.1363 | 0.1447 | 0.1527 | 0.1603 | 0.1747 | 0.1883 |
| | h | | | 827.3 | 958.8 | 1101.8 | 1201.8 | 1273.6 | 1332.0 | 1382.6 | 1428.3 | 1470.6 | 1510.5 | 1548.7 | 1585.8 | 1657.4 | 1726.8 |
| | s | | | 0.9803 | 1.0867 | 1.1981 | 1.2732 | 1.3250 | 1.3658 | 1.3999 | 1.4297 | 1.4564 | 1.4808 | 1.5035 | 1.5248 | 1.5644 | 1.6008 |
| 6000 | Sh | | | | | | | | | | | | | | | | |
| | v | | | 0.0298 | 0.0397 | 0.0579 | 0.0757 | 0.0900 | 0.1020 | 0.1126 | 0.1221 | 0.1309 | 0.1391 | 0.1469 | 0.1544 | 0.1684 | 0.1817 |
| | h | | | 822.9 | 945.1 | 1084.6 | 1188.8 | 1263.4 | 1323.6 | 1375.7 | 1422.3 | 1465.4 | 1505.9 | 1544.6 | 1582.0 | 1654.2 | 1724.2 |
| | s | | | 0.9758 | 1.0746 | 1.1833 | 1.2615 | 1.3154 | 1.3574 | 1.3925 | 1.4229 | 1.4500 | 1.4748 | 1.4978 | 1.5194 | 1.5593 | 1.5960 |

| | | | | | | | | | | | | | | | |
|---|---|---|---|---|---|---|---|---|---|---|---|---|---|---|---|
| 6500 | Sh | | | | | | | | | | | | | | |
| | v | 0.0287 | 0.0358 | 0.0495 | 0.0655 | 0.0793 | 0.0909 | 0.1012 | 0.1104 | 0.1188 | 0.1266 | 0.1340 | 0.1411 | 0.1544 | 0.1669 |
| | h | 813.9 | 919.5 | 1046.7 | 1156.3 | 1237.8 | 1302.7 | 1358.1 | 1407.3 | 1452.2 | 1494.2 | 1534.1 | 1572.5 | 1646.4 | 1717.6 |
| | s | 0.9661 | 1.0515 | 1.1506 | 1.2328 | 1.2917 | 1.3370 | 1.3743 | 1.4064 | 1.4347 | 1.4604 | 1.4841 | 1.5062 | 1.5471 | 1.5844 |
| 7000 | Sh | | | | | | | | | | | | | | |
| | v | 0.0279 | 0.0334 | 0.0438 | 0.0573 | 0.0704 | 0.0816 | 0.0915 | 0.1004 | 0.1085 | 0.1160 | 0.1231 | 0.1298 | 0.1424 | 0.1542 |
| | h | 806.9 | 901.8 | 1016.5 | 1124.9 | 1212.6 | 1281.7 | 1340.5 | 1392.2 | 1439.1 | 1482.6 | 1523.7 | 1563.1 | 1638.6 | 1711.1 |
| | s | 0.9582 | 1.0350 | 1.1243 | 1.2055 | 1.2689 | 1.3171 | 1.3567 | 1.3904 | 1.4200 | 1.4466 | 1.4710 | 1.4938 | 1.5355 | 1.5735 |
| 7500 | Sh | | | | | | | | | | | | | | |
| | v | 0.0272 | 0.0318 | 0.0399 | 0.0512 | 0.0631 | 0.0737 | 0.0833 | 0.0918 | 0.0996 | 0.1068 | 0.1136 | 0.1200 | 0.1321 | 0.1433 |
| | h | 801.3 | 889.0 | 992.9 | 1097.7 | 1188.3 | 1261.0 | 1322.9 | 1377.2 | 1426.0 | 1471.0 | 1513.3 | 1553.7 | 1630.8 | 1704.6 |
| | s | 0.9514 | 1.0224 | 1.1033 | 1.1818 | 1.2473 | 1.2980 | 1.3397 | 1.3751 | 1.4059 | 1.4335 | 1.4586 | 1.4819 | 1.5245 | 1.5632 |
| 8000 | Sh | | | | | | | | | | | | | | |
| | v | 0.0267 | 0.0306 | 0.0371 | 0.0465 | 0.0571 | 0.0671 | 0.0762 | 0.0845 | 0.0920 | 0.0989 | 0.1054 | 0.1115 | 0.1230 | 0.1338 |
| | h | 796.6 | 879.1 | 974.4 | 1074.3 | 1165.4 | 1241.0 | 1305.5 | 1362.2 | 1413.0 | 1459.6 | 1503.1 | 1544.5 | 1623.1 | 1698.1 |
| | s | 0.9455 | 1.0122 | 1.0864 | 1.1613 | 1.2271 | 1.2798 | 1.3233 | 1.3603 | 1.3924 | 1.4208 | 1.4467 | 1.4705 | 1.5140 | 1.5533 |
| 8500 | Sh | | | | | | | | | | | | | | |
| | v | 0.0262 | 0.0296 | 0.0350 | 0.0429 | 0.0522 | 0.0615 | 0.0701 | 0.0780 | 0.0853 | 0.0919 | 0.0982 | 0.1041 | 0.1151 | 0.1254 |
| | h | 792.7 | 871.2 | 959.8 | 1054.5 | 1144.0 | 1221.9 | 1288.5 | 1347.5 | 1400.2 | 1448.2 | 1492.9 | 1535.3 | 1615.4 | 1691.7 |
| | s | 0.9402 | 1.0037 | 1.0727 | 1.1437 | 1.2084 | 1.2627 | 1.3076 | 1.3460 | 1.3793 | 1.4087 | 1.4352 | 1.4597 | 1.5040 | 1.5439 |
| 9000 | Sh | | | | | | | | | | | | | | |
| | v | 0.0258 | 0.0288 | 0.0335 | 0.0402 | 0.0483 | 0.0568 | 0.0649 | 0.0724 | 0.0794 | 0.0858 | 0.0918 | 0.0975 | 0.1081 | 0.1179 |
| | h | 789.3 | 864.7 | 948.0 | 1037.6 | 1125.4 | 1204.1 | 1272.1 | 1333.0 | 1387.5 | 1437.1 | 1482.9 | 1526.3 | 1607.9 | 1685.3 |
| | s | 0.9354 | 0.9964 | 1.0613 | 1.1285 | 1.1918 | 1.2468 | 1.2926 | 1.3323 | 1.3667 | 1.3970 | 1.4243 | 1.4492 | 1.4944 | 1.5349 |
| 9500 | Sh | | | | | | | | | | | | | | |
| | v | 0.0254 | 0.0282 | 0.0322 | 0.0380 | 0.0451 | 0.0528 | 0.0603 | 0.0675 | 0.0742 | 0.0804 | 0.0862 | 0.0917 | 0.1019 | 0.1113 |
| | h | 786.4 | 859.2 | 938.3 | 1023.4 | 1108.9 | 1187.7 | 1256.6 | 1318.9 | 1375.1 | 1426.1 | 1473.1 | 1517.3 | 1600.4 | 1679.0 |
| | s | 0.9310 | 0.9900 | 1.0516 | 1.1153 | 1.1771 | 1.2320 | 1.2785 | 1.3191 | 1.3546 | 1.3858 | 1.4137 | 1.4392 | 1.4851 | 1.5263 |
| 10000 | Sh | | | | | | | | | | | | | | |
| | v | 0.0251 | 0.0276 | 0.0312 | 0.0362 | 0.0425 | 0.0495 | 0.0565 | 0.0633 | 0.0697 | 0.0757 | 0.0812 | 0.0865 | 0.0963 | 0.1054 |
| | h | 783.8 | 854.5 | 930.2 | 1011.3 | 1094.2 | 1172.6 | 1242.0 | 1305.3 | 1362.9 | 1415.3 | 1463.4 | 1508.6 | 1593.1 | 1672.8 |
| | s | 0.9270 | 0.9842 | 1.0432 | 1.1039 | 1.1638 | 1.2185 | 1.2652 | 1.3065 | 1.3429 | 1.3749 | 1.4035 | 1.4295 | 1.4763 | 1.5180 |
| 10500 | Sh | | | | | | | | | | | | | | |
| | v | 0.0248 | 0.0271 | 0.0303 | 0.0347 | 0.0404 | 0.0467 | 0.0532 | 0.0595 | 0.0656 | 0.0714 | 0.0768 | 0.0818 | 0.0913 | 0.1001 |
| | h | 781.5 | 850.5 | 923.4 | 1001.0 | 1081.3 | 1158.9 | 1228.4 | 1292.4 | 1351.1 | 1404.7 | 1453.9 | 1500.0 | 1585.8 | 1666.7 |
| | s | 0.9232 | 0.9790 | 1.0358 | 1.0939 | 1.1519 | 1.2060 | 1.2529 | 1.2946 | 1.3371 | 1.3644 | 1.3937 | 1.4202 | 1.4677 | 1.5100 |

Table 15-15. Continued

| Abs. Press. (psi) (Sat. Temp.) | | Sat. Water | Sat. Steam | Temperature (°F) | | | | | | | | | | | | | |
|---|---|---|---|---|---|---|---|---|---|---|---|---|---|---|---|---|---|
| | | | | 750 | 800 | 850 | 900 | 950 | 1000 | 1050 | 1100 | 1150 | 1200 | 1250 | 1300 | 1400 | 1500 |
| 11000 | v | | | 0.0245 | 0.0267 | 0.0296 | 0.0335 | 0.0386 | 0.0443 | 0.0503 | 0.0562 | 0.0620 | 0.0676 | 0.0727 | 0.0776 | 0.0868 | 0.0952 |
| | h | | | 779.5 | 846.9 | 917.5 | 992.1 | 1069.9 | 1146.3 | 1215.9 | 1280.2 | 1339.7 | 1394.4 | 1444.6 | 1491.5 | 1578.7 | 1660.6 |
| | s | | | 0.9196 | 0.9742 | 1.0292 | 1.0851 | 1.1412 | 1.1945 | 1.2414 | 1.2833 | 1.3209 | 1.3544 | 1.3842 | 1.4112 | 1.4595 | 1.5023 |
| 11500 | v | | | 0.0243 | 0.0263 | 0.0290 | 0.0325 | 0.0370 | 0.0423 | 0.0478 | 0.0534 | 0.0588 | 0.0641 | 0.0691 | 0.0739 | 0.0827 | 0.0909 |
| | h | | | 777.7 | 843.8 | 912.4 | 984.5 | 1059.8 | 1134.9 | 1204.3 | 1268.7 | 1328.8 | 1384.4 | 1435.5 | 1483.2 | 1571.8 | 1654.7 |
| | s | | | 0.9163 | 0.9698 | 1.0232 | 1.0772 | 1.1316 | 1.1840 | 1.2308 | 1.2727 | 1.3107 | 1.3446 | 1.3750 | 1.4025 | 1.4515 | 1.4949 |
| 12000 | v | | | 0.0241 | 0.0260 | 0.0284 | 0.0317 | 0.0357 | 0.0405 | 0.0456 | 0.0508 | 0.0560 | 0.0610 | 0.0659 | 0.0704 | 0.0790 | 0.0869 |
| | h | | | 776.1 | 841.0 | 907.9 | 977.8 | 1050.9 | 1124.5 | 1193.7 | 1258.0 | 1318.5 | 1374.7 | 1426.6 | 1475.1 | 1564.9 | 1648.8 |
| | s | | | 0.9131 | 0.9657 | 1.0177 | 1.0701 | 1.1229 | 1.1742 | 1.2209 | 1.2627 | 1.3010 | 1.3353 | 1.3662 | 1.3941 | 1.4438 | 1.4877 |
| 12500 | v | | | 0.0238 | 0.0256 | 0.0279 | 0.0309 | 0.0346 | 0.0390 | 0.0437 | 0.0486 | 0.0535 | 0.0583 | 0.0629 | 0.0673 | 0.0756 | 0.0832 |
| | h | | | 774.7 | 838.6 | 903.9 | 971.9 | 1043.1 | 1115.2 | 1184.1 | 1247.9 | 1308.8 | 1365.4 | 1418.0 | 1467.2 | 1558.2 | 1643.1 |
| | s | | | 0.9101 | 0.9618 | 1.0127 | 1.0637 | 1.1151 | 1.1653 | 1.2117 | 1.2534 | 1.2918 | 1.3264 | 1.3576 | 1.3860 | 1.4363 | 1.4808 |
| 13000 | v | | | 0.0236 | 0.0253 | 0.0275 | 0.0302 | 0.0336 | 0.0376 | 0.0420 | 0.0466 | 0.0512 | 0.0558 | 0.0602 | 0.0645 | 0.0725 | 0.0799 |
| | h | | | 773.5 | 836.3 | 900.4 | 966.8 | 1036.2 | 1106.7 | 1174.8 | 1238.5 | 1299.6 | 1356.5 | 1409.6 | 1459.4 | 1551.6 | 1637.4 |
| | s | | | 0.9073 | 0.9582 | 1.0080 | 1.0578 | 1.1079 | 1.1571 | 1.2030 | 1.2445 | 1.2831 | 1.3179 | 1.3494 | 1.3781 | 1.4291 | 1.4741 |
| 13500 | v | | | 0.0235 | 0.0251 | 0.0271 | 0.0297 | 0.0328 | 0.0364 | 0.0405 | 0.0448 | 0.0492 | 0.0535 | 0.0577 | 0.0619 | 0.0696 | 0.0768 |
| | h | | | 772.3 | 834.4 | 897.2 | 962.2 | 1030.0 | 1099.1 | 1166.3 | 1229.7 | 1291.0 | 1348.1 | 1401.5 | 1451.8 | 1545.2 | 1631.9 |
| | s | | | 0.9045 | 0.9548 | 1.0037 | 1.0524 | 1.1014 | 1.1495 | 1.1948 | 1.2361 | 1.2749 | 1.3098 | 1.3415 | 1.3705 | 1.4221 | 1.4675 |
| 14000 | v | | | 0.0233 | 0.0248 | 0.0267 | 0.0291 | 0.0320 | 0.0354 | 0.0392 | 0.0432 | 0.0474 | 0.0515 | 0.0555 | 0.0595 | 0.0670 | 0.0740 |
| | h | | | 771.1 | 832.6 | 894.3 | 958.0 | 1024.5 | 1092.3 | 1158.5 | 1221.4 | 1283.0 | 1340.2 | 1393.8 | 1444.4 | 1538.8 | 1626.5 |
| | s | | | 0.9019 | 0.9515 | 0.9996 | 1.0473 | 1.0953 | 1.1426 | 1.1872 | 1.2282 | 1.2671 | 1.3021 | 1.3339 | 1.3631 | 1.4153 | 1.4612 |
| 14500 | v | | | 0.0231 | 0.0246 | 0.0264 | 0.0287 | 0.0314 | 0.0345 | 0.0380 | 0.0418 | 0.0458 | 0.0496 | 0.0534 | 0.0573 | 0.0646 | 0.0714 |
| | h | | | 770.4 | 831.0 | 891.7 | 954.3 | 1019.6 | 1086.2 | 1151.4 | 1213.8 | 1275.4 | 1332.9 | 1386.4 | 1437.3 | 1532.6 | 1621.1 |
| | s | | | 0.8994 | 0.9484 | 0.9957 | 1.0426 | 1.0897 | 1.1362 | 1.1801 | 1.2208 | 1.2597 | 1.2949 | 1.3266 | 1.3560 | 1.4087 | 1.4551 |
| 15000 | v | | | 0.0230 | 0.0244 | 0.0261 | 0.0282 | 0.0308 | 0.0337 | 0.0369 | 0.0405 | 0.0443 | 0.0479 | 0.0516 | 0.0552 | 0.0624 | 0.0690 |
| | h | | | 769.6 | 829.5 | 889.3 | 950.9 | 1015.1 | 1080.6 | 1144.9 | 1206.8 | 1268.1 | 1326.0 | 1379.4 | 1430.3 | 1526.4 | 1615.9 |
| | s | | | 0.8970 | 0.9455 | 0.9920 | 1.0382 | 1.0846 | 1.1302 | 1.1735 | 1.2139 | 1.2525 | 1.2880 | 1.3197 | 1.3491 | 1.4022 | 1.4491 |
| 15500 | v | | | 0.0228 | 0.0242 | 0.0258 | 0.0278 | 0.0302 | 0.0329 | 0.0360 | 0.0393 | 0.0429 | 0.0464 | 0.0499 | 0.0534 | 0.0603 | 0.0668 |
| | h | | | 768.9 | 828.2 | 887.2 | 947.8 | 1011.1 | 1075.7 | 1139.0 | 1200.3 | 1261.1 | 1319.6 | 1372.8 | 1423.6 | 1520.4 | 1610.8 |
| | s | | | 0.8946 | 0.9427 | 0.9886 | 1.0340 | 1.0797 | 1.1247 | 1.1674 | 1.2073 | 1.2457 | 1.2815 | 1.3131 | 1.3424 | 1.3959 | 1.4433 |

*Sh=superheat. °F; *v* specific volume, ft³/lb; *h* enthalpy, Btu/lb; *s* entropy, Btu/°F · lb.

*Source:* Copyright 1967 ASME (Abridged); reprinted by permission.

# References

1. Verderber, R.R. and O. Morse. Lawrence Berkeley Laboratory, Applied Science Division. "Preformance of Electronic Ballasts and Other New Lighting Equipment (Phase II: The 34-Watt F40 Rapid Start T-12 Fluorescent Lamp)." University of California, February, 1988.
2. Morse, O. and R. Verderber. Lawrence Berkeley Laboratory, Applied Science Division. "Cost Effective Lighting." University of California, July, 1987.
3. Hollister, D.D. Lawrence Berkeley Laboratory, Applied Science Division. "Overview of Advances in Light Sources." University of California, June, 1986.
4. Verderber, R. and O. Morse. Lawrence Berkeley Laboratory, Applied Science Division. "Performance of Electronic Ballasts and Other New Lighting Equipment: Final Report." University of California, October, 1985.
5. Lovins, Amory B. "The State of the Art: Space Cooling." Rocky Mountain Institute, Old Snowmass, CO, November, 1986.
6. U.S. Department of Energy, Assistant Secretary, Conservation and Renewable Energy, Office of Buildings and Community Systems. "Energy Conservation Goals for Buildings." Washington, DC, May, 1988.
7. Rubinstein, F., T. Clark, M. Simonovitch, and R. Verderber. Lawrence Berkeley Laboratory, Applied Science Division. "The Effect of Lighting System Components on Lighting Quality, Energy Use, and Life-Cycle Cost." University of California, July, 1986.
8. Manufacturing Energy Consumption Survey: Consumption of Energy 1985 Energy Information Administration, November 18, 1988.
9. Heery Energy Consultants, Inc. "Georgia Cogeneration Handbook." The Governor's Office of Energy Resources, August, 1988.
10. Thermodynamics, New York, F. Sears, Reading, MA, Addison-Wesley Publishing Company, Inc., 1952.
11. Climatic Atlas of the U.S., reprinted under the title: Weather Atlas of the United States, Gale Research Co., 1975.
12. Thumann, Albert, "Plant Engineers and Managers Guide to Energy Conservation - 3rd Edition." Fairmont Press, 1987.
13. Thumann, Albert. "Handbook of Energy Audits - 2nd Edition." Fairmont Press, 1983.
14. Kreider and McNeil. "Waste Heat Management Guidebook." NBS Handbook 121.
15. U.S. Department of Energy, Office of the Assistant Secretary for Conservation and Solar, Office of Building and Community Systems, Oak Ridge National Laboratory, Oak Ridge, TN, University of Massachusetts Cooperative Extensive Service Energy Education Center, Amherst, MA, The Solar Energy Research Institute, Golden, CO. "Residential Energy Audit Manual." Fairmont Press, 1981.

16. American Society of Heating, Refrigerating and Air-Conditioning Engineers, Inc. "ASHRAE Handbook, 1982 Fundamentals." Chapter 2 - Heat Transfer; Chapter 11 - Applied Heat Pump Systems; Chapter 23 - Design Heat Transmission Coefficients; Chapter 25 - Heating Load; Chapter 26 - Air Conditioning Cooling Load. Atlanta, GA, 1982.
17. American Society of Heating, Refrigerating and Air-Conditioning Engineers, Inc. "ASHRAE Handbook, 1982 Applications." Chapter 57 - Solar Energy Utilization for Heating and Cooling. Atlanta, GA, 1982.
18. American Society of Heating, Refrigerating and Air-Conditioning Engineers, Inc. "ASHRAE Handbook & Product Directory, 1982 Systems." Chapter 7 - Heat Recovery Systems; Chapter 43 - Energy Estimating Methods. Atlanta, GA, 1982.
19. Trane Air Conditioning Manual. The Trane Company, LaCrosse WI.

# Index